# CRC Handbook
# of
# Laboratory Model
# Systems
# for
# Microbial Ecosystems

## Volume I

Editor

**Julian W. T. Wimpenny, Ph.D.**
Reader in Microbiology
Department of Microbiology
University College
Cardiff, Wales

**CRC Press**
Taylor & Francis Group
Boca Raton  London  New York

CRC Press is an imprint of the
Taylor & Francis Group, an **informa** business

First published 1988 by CRC Press
Taylor & Francis Group
6000 Broken Sound Parkway NW, Suite
300 Boca Raton, FL 33487-2742

Reissued 2018 by CRC Press

© 1988 by Taylor & Francis
CRC Press is an imprint of Taylor & Francis Group, an Informa business

No claim to original U.S. Government works

A Library of Congress record exists under LC control number: 88024089

Publisher's Note
The publisher has gone to great lengths to ensure the quality of this reprint but points out that some imperfections in the original copies may be apparent.

Disclaimer
The publisher has made every effort to trace copyright holders and welcomes correspondence from those they have been unable to contact.

ISBN 13: 978-1-138-10510-2 (hbk)
ISBN 13: 978-1-138-55836-6 (pbk)
ISBN 13: 978-1-315-15079-6 (ebk)

Visit the Taylor & Francis Web site at http://www.taylorandfrancis.com and the CRC Press Web site at http://www.crcpress.com

# PREFACE

Interactions between microbes and their environment define the study of microbial ecology. The small size of individual microbes and the great range and variety of structures found in the physical world mean that most microbial habitats are spatially and temporally heterogeneous. Such complex systems can be studied successfully *in situ* or they can be investigated in the laboratory. If the latter, they may be incorporated into microcosms or studied in model systems. Microcosms retain most of the complexity of the natural system while models are subsets (more analogues than homologues) of the ecosystem from which they derive. The model aims to investigate a few properties of the natural system while other aspects are either held constant or omitted from the system or simply ignored.

In this book I have tried to collect some of the experimental model systems that are used to investigate microbial behavior in nature. These include homogeneous culture systems like the chemostat which has itself proved such a powerful research tool over the last three or four decades. Linked homogeneous fermenters include the multistage chemostat, which can be used to investigate stratified systems like aquatic sediments, and the gradostat whose bidirectionality means that opposing solute gradients can be studied under steady-state conditions. Bidirectional systems are also available for studying interactions between pure cultures located in different compartments linked only by solute diffusion across a permeable membrane.

Five chapters relate to the attachment and growth of organisms to solid surfaces. These chapters cover attachment and growth itself, as well as the development and employment of microbial film fermenters to investigate such disparate habitats as wastewater treatment systems and dental plaque.

The use of agar as a gelling agent allows the development of model ecosystems where molecular diffusion is the only solute transfer mechanism. These systems are discussed, as is the use of gel-stabilized solute gradient plate methods to investigate the habitat domains of different microbial species.

Packed column reactors are useful model tools for studying systems where there is a unidirectional flow of solutes downward through the structure. These can be good models of soil ecosystems. The latter have been investigated using many other microcosms and models, and these are also discussed.

Other topics covered include specific devices such as the Perfil'ev convectional flow technique, the bacterial colony, and motility in bacteria.

I have deliberately included a discussion of the application of mathematical models to natural ecosystems, because there is a close link between experimental and numerical modeling. Both approaches aim to simplify natural systems by identifying key fundamental processes that dominate the operation of the natural process. The chapter on bacterial motility bridges the gap between experimental and numerical models, and the chapter by Tett and Droop clearly links numerical modeling with experimental investigations.

I hope that the approaches discussed here will be of value both to laboratory-based microbial physiologists and to ecologists who feel that there is some virtue in simplifying systems which can throw light on key processes that operate in nature.

I would like to express my gratitude to all the contributors to both volumes. Also to my wife Lee, and my children Ross, Joshua, and Joanna who might have seen more of me over the period when this book was gestating.

# THE EDITOR

**J. W. T. Wimpenny, Ph.D.,** is Reader in Microbiology in the Department of Microbiology, University College, Cardiff. He read Biochemistry at the University of Cambridge from which he graduated in 1958. His Ph.D. degree on the "Mode of Action of Anti-Tubercular Drugs" was awarded by the University of London in 1962 while he was at Guy's Hospital in London. After a spell as postdoctoral fellow at Dartmouth College, New Hampshire, Dr. Wimpenny returned to Oxford and then to Cardiff in 1965.

Over his research career Dr. Wimpenny has published more than 100 research papers and invited communications in national and international journals. His current research interests concern the development of spatially heterogeneous models systems and their application to relevant ecological problems. His interests in the applications of computing to microbiology led to the formation of the British Society for General Microbiology Computer Club. He was its first covener and editor of its publication *Binary*.

He lives in the country on the edge of the Wye valley in Wales with his wife Lee and three children.

# CONTRIBUTORS

## Volume I

**Douglas E. Caldwell, Ph.D.**
Professor
Department of Applied Microbiology
University of Saskatchewan
Saskatoon, Saskatchewan, Canada

**William G. Characklis, Ph.D.**
Director
Institute for Biological and Chemical
 Process Analysis
Montana State University
Bozeman, Montana

**Lubbert Dijkhuizen, Ph.D.**
Lecturer
Department of Microbiology
University of Groningen
Haren, Netherlands

**Harold W. Fowler, Ph.D.**
Industrial Research Fellow
Department of Biochemistry
School of Biological Sciences
University of Bath
Claverton Down, Bath, England

**Jan C. Gottschal, Ph.D.**
Lecturer
Department of Microbiology
Biological Center
Haren, Netherlands

**Rod A. Herbert, Ph.D.**
Senior Lecturer
Department of Biological Sciences
University of Dundee
Dundee, Angus, Scotland

**J. R. Lawrence, Ph.D.**
Research Associate
Department of Applied Microbiology and
 Food Science
University of Saskatchewan
Saskatoon, Saskatchewan, Canada

**R. John Parkes, Ph.D.**
Principal Scientist
Dunstaffnage Marine Research Laboratory
Oban, Argyll, England

**Adrian Peters**
Postdoctoral Fellow
Department of Microbiology
University College
Cardiff, Wales

**Eric Senior, Ph.D.**
Lecturer in Applied Microbiology
Department of Bioscience and
 Biotechnology
University of Strathclyde
Glasgow, Scotland

**Adolph Tatevossian, Ph.D.**
Senior Lecturer
Department of Physiology
University College
Cardiff, Wales

**Paul Waters, M.Sc.**
Research Technician
BIOTAL
Cardiff, Wales

# VOLUME OUTLINES

## Volume I

Introduction
The Place of the Continuous Culture in Ecological Research
Multistage Chemostats and Other Models for Studying Anoxic Systems
Bidirectionally Linked Continuous Culture: The Gradostat
Bidirectional Compound Chemostats: Applications of Compound Diffusion-Linked
 Chemostats in Microbial Ecology
Study of Attached Cells in Continuous-Flow Slide Culture
Microbial Adhesion to Surfaces
Model Biofilm Reactors
A Constant-Depth Laboratory Model Film Fermenter
Film Fermenters in Dental Research
Gel-Plate Methods in Microbiology

## Volume II

One-Dimensional Gel-Stabilized Model Systems
The Use of Packed Column Reactors to Study Microbial Transformations in the Soil
Experimental Models in the Study of Soil Microbiology
The Perfil'ev Conventional Flow Technique for Modeling Stratified Natural Aquatic
 Communities
The Bacterial Colony
Laboratory and Theoretical Models for the Effects of Bacterial Motility and Chemotaxis
 on Microbial Population Growth
Cell Quota Models and Planktonic Primary Production
The Role of Mathematical Models and Experimental Ecosystems in the Study of
 Microbial Ecology

# TABLE OF CONTENTS

## Volume I

Chapter 1

## INTRODUCTION

### Julian W. T. Wimpenny

This book is devoted to a description of some of the laboratory model systems which can usefully be applied to investigations of microbial growth and interactions where they are most at home, in their natural habitats.

It is accepted from the start that natural ecosystems are nearly always extraordinarily complex genotypically, structurally, and dynamically. It is for this reason that I and all the contributors to this volume advocate the judicious use of model systems, which, if chosen wisely and if their limitations are constantly born in mind, can throw light on the operation of microbes in the natural world.

## MODELS AND MICROCOSMS

First we must discuss terminology. The terms "microcosm" and "model system" are sometimes used incautiously to signify the same thing. It is clear however that the two are conceptually quite different. Parkes[1] suggests that the term microcosm defines " . . . a laboratory system which attempts to actually simulate as far as possible the conditions prevailing in the environment or part of the environment under study". As is pointed out by Burns (Volume II, Chapter 3), the term microcosm can be totally misinterpreted to mean "site of microbial activity". There may, however, be a consensus among other microbiologists, in which a microcosm possesses some or all of the following properties:

1. *Origin.* Microcosms derive from natural ecosystems.
2. *Isolation.* The microcosm, whatever its origin, is physically enclosed and no longer in contact with the natural ecosystems.
3. *Size.* Though variable in size, microcosms tend to be compact subsets of the natural system from which they came.
4. *Genotypic heterogeneity.* With some exceptions, most microcosm work uses natural mixed cultures of microorganisms.
5. *Spatial heterogeneity.* Although spatial heterogeneity is not in any way implied by the term microcosm, virtually all of the latter retain this property to some degree.
6. *Temporal heterogeneity.* The majority of microcosms are closed or partially closed systems, where time-dependent changes in the physical, chemical, and biological properties of the system are seen.

At first sight, a microcosm seems just to be a convenience, a piece of natural ecology tamed and brought into the laboratory without changing anything within it.

The main point in favor of the microcosm is that environmental factors (light, temperature, chemistry, etc.) can be manipulated, prolonging, for example, certain seasons or simplifying systems, perhaps eliminating diurnal variations.

It should be emphasized that however good the intention is to accurately reproduce the natural ecosystem, the microcosm is almost certain to suffer as a result of scale changes. As an "island" community isolated from the original system, there may be successional changes in species composition and concommitant changes in physical chemistry aggravated by the presence of the container boundary. For instance, if the latter is glass and the system is illuminated, photosynthetic species will proliferate. A good example of such a microcosm is the Winogradsky column[2] which is derived from a sediment ecosystem. Sediment and water are placed over a source of carbon (usually cellulose) and a mixture of calcium sulfate and calcium carbonate in a glass vessel. If the container is kept in normal daylight, colored photosynthetic species develop as a series of pigmented bands. Clearly they mark *vertical* heterogeneity. On the other hand, by their presence and by the input of light energy to these zones, they contribute another level of heterogeneity at right angles to the vertical stratification already present in the sediment. This secondary heterogeneity is of course absent from the natural sediment, since the latter is never illuminated. These wall effects plus the addition of cellulose, gypsum, and chalk to the system mean that the Winogradsky column is no longer homologous with the sediment from which it originated. So is it microcosm or model? It is apparent that the Parkes definition cannot help here. The Winogradsky column obviously does not aim to simulate as exactly as possible the natural sediment. It does start from the natural system but equally it manipulates it. Perhaps a better definition which embraces these feelings is

*A microcosm is a laboratory subset of the natural system from which it originates but from which it also evolves.*

Such a brief definition suggests that if we establish a sample of a natural microbial ecosystem in the laboratory, it should still be termed a microcosm even if it is altered spatially or in terms of its physical chemistry. Microcosms allow a systematic examination of the responses of natural communities to environmental manipulation.

The model, on the other hand, is always an abstraction. The model never aims to reproduce the entire system in the laboratory. It always seeks to examine the properties of a part of the system ignoring or holding constant all the remaining factors. There may be any number of models investigating particular properties of the system. Some of these may be extremely simple. To grow a pure culture of a sulfate-reducing bacterium in a chemostat is to create a restricted model of one component of a sedimentary ecosystem. Such a model is perfectly valid if its limitations (spatial and temporal homogeneity, the absence of other species, the absence of different phases, and so on) are accepted.

Another model of a sedimentary system might recognize the importance of molecular diffusion and species diversity. The sulfate reducer could be incorporated into a gel stabilized system together with a sulfide oxidizing species. The two operating together in counter gradients of oxygen and a carbon source could carry out the chemical reactions of the sulfur cycle found in the upper levels of natural sediment ecosystems.

The model could be made even more sophisticated. Specific mineral particles and a greater range of substrates and species could be incorporated into the gel. The model system becomes closer to the natural ecosystem with each level of sophistication. At some point of course one could conceivably say "the model is the system . . . !" This is probably not the main virtue of models, however. The modeler may well consider that to travel hopefully is better than to arrive. To understand completely a particular ecosystem is probably less important than to discern rules of behavior which apply at a fundamental level to many different ecosystems. The model, if designed well, can by virtue of abstraction and simplification offer this possibility.

The difference between model and microcosm can, for some people, epitomize the difference between an essentially reductionist approach to understanding ecology on the one hand, and a holistic approach on the other. The reductionist believes that understanding the simplified subunits allows one to predict the properties of system from which they derived. Thus, if the properties of pure cultures of bacteria are totally understood, then their behavior in a natural community and hence the behavior of the community itself can be confidently predicted. The holist, on the other hand, believes that breaking a system down into component parts loses part of its veracity. A holist philosophy considers that the whole is greater than the sum of its component parts.

It is likely that both approaches are valid, essential even, to the prosecution of experimental science. Clearly, higher levels of complexity bring the need for new rules, new properties, new types of order. After all, a complete understanding of the properties of bricks does not help to comprehend the rules of design or architecture needed to construct a Victorian railway station! On the other hand, systems can be too complicated to comprehend as they stand. It must then be right to take a reductionist stance, at least at first, if this gives a clear understanding of some part of the natural system. If the model is very simple, the information too will be simple. Perhaps if we understand some of the properties of bricks, for example their mechanical strength, we can at least predict how tall we can build our railway station before it collapses. In the same way, understanding the biochemical behavior of a single species can give some insight into the way in which it could behave in a complex community. Once the simple model has been deployed and all the relevant information collated, it is time to sophisticate the model. The next level of complexity can now be investigated and comprehended. This is not really a reductionist approach at all. Simplification merely establishes a platform upon which higher levels of sophistication can rest more securely.

Ought reductionism be distinguished from the normal scientific processes of analysis and synthesis? The reductionist studies elements of a system and is then content to predict the behavior of the system as a whole. The "normal" scientist carefully dissects the whole system before reassembling increasingly more complex subsets of it. Pursued to its logical conclusion, even this distinction fades and reductionism is seen as just one extreme of the normal processes of analysis and synthesis.

In the meantime, right at the other end of this scale, the holist is investigating the system just as it stands. Some of his findings must inevitably be descriptive and may not be easy to interpret at the most basic level; however the holist performs the vital function of searching for and documenting the emergent properties of the complete system. It seems clear that holism and the sort of "reductionism" described here move together and at some point lead to a unified interpretation of the complete system which will satisfy all sides.

Holism and reductionism are not in the end very useful terms, since they simply represent two extreme attitudes to science. The advocate of one believes in starting simple and only increasing the complexity of a problem when it is necessary to do this, while the advocate of the other can only comprehend the entire system and is motivated to study this from an aesthetic sense of unity and even beauty that the system as a whole displays. Finally, science as a process is flexible enough to embrace all sorts of approaches and can accomodate, and indeed gain from, the instincts of a wide range of people so long as they are honestly dedicated to discovering the truth about natural phenomena.

I hope that what I have said so far goes part of the way to justify the use of model systems. I would like next to discuss briefly some of the properties of natural microbial ecosystems which seem to be worth incorporating in simpler models.

## HOMOGENEITY VS. HETEROGENEITY

I have emphasized[3,4] that microbiology has pursued the "homogeneity paradigm" for too long. Thus, the dictum that a mono culture in a stirred fermenter, preferably operating as a chemostat, is really all that is necessary to understand all the properties of microorganisms, seems to be a gross simplification. This is not in any way to deny the enormous importance of homogeneous systems and the chemostat in particular to studies in microbial ecology. The excellent chapter by Gottschal and Dijkhuisen in this volume (Chapter 2) illustrates both the power of the technique and the areas in microbial ecology where continuous culture has played such an important part. If the chemostat could answer all the questions concerning microbial behavior in their natural environments, however, there really would be little need for a book of this sort.

To recognize the need for more elaborate model systems is to accept a conceptual shift away from homogeneity to the heterogeneous systems that seem to predominate in nature. This conceptual shift is not merely to accept the fact implicit in the term "heterogeneity" that physico-chemical conditions alter as a function of position at a given time and of time at a given position, but it is to recognize and to accept instinctively the extreme importance of the space *outside* a cell to the growth and physiological behavior of that cell. However subtle and intricate and intrinsically fascinating are the ordered chemical reactions that go on within the cell envelope, they are dominated and modulated by the chemistry and physics of the regions surrounding it, yet this area has received only the scantiest attention in recent decades compared to the volumes published on the physiology and molecular biology of microbes.

A few of the factors affecting cell behavior include the distribution and flux of substrates needed for growth, the prevailing physical conditions of temperature, pressure, pH value, water potential ionic strength, etc., the presence of different phases and their physical behavior, the presence of inhibitory solutes, the presence of other prokaryotic and eukaryotic organisms, and the range of interactions that each exhibits. In microbiology, especially that of prokaryotic organisms, the business of living revolves around the distribution of solute molecules in a narrow aqueous layer surrounding the cytoplasmic membrane of the organism.

Once the importance of spatial heterogeneity is recognized, one is forced to accept the need for simple laboratory systems which incorporate heterogeneity factors. In the next section, I will briefly discuss a few spatially heterogeneous ecosystems, hopefully to illustrate just where laboratory models may have a useful part to play. This list is not intended to be exhaustive.

## SOME SPATIALLY HETEROGENEOUS ECOSYSTEMS

### Aquatic Systems

In general, water bodies have little spatial heterogeneity and many therefore behave like homogeneous fermenter systems. The oceans, seas, and large inland water bodies fall into this category at least at a "local" level. There are large-scale differences in chemistry that should be mentioned: for example, deep ocean troughts, some littoral regions, etc., can have differences in temperature, salinity, organic load, dissolved oxygen tension, and so on, while light penetration is a function of both of depth and of the clarity of the water. Related to gradients in light energy and spectrum is photosynthetic activity, which will in turn modulate the oxygen content of the waters. It may be that the ecology of such water bodies is best modeled in the laboratory by single-stage continuous culture systems running at very low concentrations of substrate. This is because although physico-chemical gradients may be present, they are extremely shallow, and this implies that interactions which depend on such gradients will be negligible in importance.

Rapidly flowing water bodies, including streams and "clean" shallow rivers, are also well mixed and substantially homogeneous. However, their ecology as a whole is often heterogeneous, since water flow is adjacent to solid surfaces which can support a substantial sessile population or to sediments which may be rich in microbial flora.

It should also be stressed that freely suspended individual cells are a comparative rarity in even these substantially homogeneous environments. This is because the populations present are usually associated with particulate matter forming rafts containing groups of cells. The propensity of microbes to associate with surfaces is now well recognized and raises interesting questions regarding the distribution of solutes near phase boundaries.

Of considerable interest in terms of spatial order are stratified water bodies. Water may become stratified for a number of reasons. Commonest by far is the annual cycle of thermal stratification seen in many lake systems. Here, the upper layers of the water become heated in the summer months, and because they are now less dense than the lower layers, little vertical mixing takes place. The lower layers are now deprived of any significant atmospheric gas exchange and, depending on the organic load in the water, can often become anoxic. In such a lake system the most interesting region from the microbial ecologist's point of view is the interface between the two zones. This region shows the steepest change in temperature (thermocline) and often the steepest gradients in chemical species (chemocline). Such a region is normally the site of greatest microbial activities where anaerobic and aerobic species interact by the diffusive transfer of substrates and products.

During the warm summer months, the lower colder layers of these lakes slowly heat up. In the autumn, however, when the weather becomes chillier, the upper layers cool relatively quickly. At length, the system becomes unstable leading to convective "turnover" of the water masses and the anaerobic lower layers now rise to the surface. This mixing represents a catastrophic change in chemistry and hence in microbial populations in the lake.

Much more beautiful in their spatial differentiation are the permanently stratified water bodies. These meromictic lakes and inland seas are formed in general by dense salt water lying beneath a layer of lighter fresher water.

There are several good examples of intensively studied meromictic water bodies. They include the Black Sea which has been the focus of extensive investigations by Russian microbiologists (reviewed by Kriss[6]) and a Transcaucasian lake, Lake Gek Gel', both of which show beautiful patterns of stratification which are spatially very extensive. The Black Sea, for example, is more than 2000 m deep and the point at which sulfide is first detected is between 120 and 200 m below the surface. This point roughly marks the start of the anaerobic regions of the water body. Lake Gek Gel' was studied intensively by several workers. Sorokin[7] has reported patterns of growth and chemistry in this lake as a function of depth. The stability of the system allows the generation of organized solute gradients and the growth of a whole range of physiological types each at its own optimum position in gradients of light, oxygen, sulfur, carbon, nitrogen, and iron compounds. This lake, which is about 70 m deep, together with the Black Sea, pose interesting questions concerning solute transport mechanisms. These permanently stratified water bodies are assumed not to show much vertical exchange of material by mixing, however, molecular diffusion is too slow a process to account for the activities in the various growth regions. Possible mechanisms promoting exchange are the inflow of water from rivers, from surface and subterranean streams, and from shore line run-off; a "fall out" of dead cells and other organic matter from the surface to the bottom of the lake; the ascent of gas bubbles from the sediment at the bottom of the system to the surface, and the mixing activities of macroscopic organisms especially in the aerated surface regions. In the case of the Black Sea, some mixing may be due to the inflow of more-saline-dense water from the Sea of Marmara.

Such gradient systems may be established on a far smaller scale in laboratory microcosms. Kriss[6] cites the work of Egunov, who in 1895 established laboratory systems from estuarine, lake, and Black Sea sediments. In the overlying water, he noticed the development of a thin film of microorganisms which separated a sulfide-rich lower layer from an upper layer in which no sulfide could be detected. Egunov suggested that such a bacterial plate ought also to be present in the Black Sea, however according to Kriss, such a densely packed layer of organisms could not be found. Layering is seen in other lake systems. Thus a narrow band of anaerobic photosynthetic bacteria appears just below the aerobic-anaerobic interface in the waters of Lake Faro (Messina).[8] This phenomenon could also be seen in laboratory microcosm experiments. Suckow and Schwartz[9] incorporated Baltic seawater above a sulfide-generating mud in an aquarium. A bacterial plate which was, after several months, 10 mm thick formed between the aerobic and anaerobic layers and contained pigmented photosynthetic bacteria.

The use of the laboratory system in investigating such spatially extensive systems must be commended if only because it can speed up processes which, in stably stratified water bodies, are almost certainly very slow. It would be interesting to establish such a lake ecosystem in a gel-stabilized laboratory model system in which transport was by molecular diffusion alone. The resolution of such models can be extremely high, and it ought to be possible to map growth zones and chemistry with some precision, and from there to construct a numerical model of behavior of the natural system. The open models discussed in Chapters 3 and 4 (this volume) may also be useful in modeling the dynamic behavior of such systems.

Water, an essentially homogeneous medium, shows heterogeneity only when vertical stratification due to thermal or salinity gradients can occur. Heterogeneity is essentially one dimensional from the top to the bottom of the system, and because it is in this sense rather simple, it can be macroscopically obvious and at the same time beautiful. Ways are suggested to model such systems in the laboratory with a view to speeding up what are extremely slow processes.

**Liquid Plus Solid Phase Systems**

The addition of a solid phase to aquatic systems adds considerably to its complexity and to its potential for heterogeneity. One ecosystem which has been investigated extensively in the field is the sedimentary ecosystem. Such systems are found at the base of all water bodies including streams, rivers and estuaries, ponds, lakes, inland seas, and oceans. The solid phase of sediments consists of three separate types of component: detrital material derived directly or indirectly from geological erosion, biogenic material which is formed by biological activities, and authigenic components which are formed within the sediment itself. Sediments show several orders of heterogeneity. They are always vertically stratified with the surface, if it is aerobic at all, containing measurable oxygen over the first few millimeters to centimeters. A secondary spherical heterogeneity has been reported. Thus in the aerobic superficial layers, small foci of anaerobiosis due to the activities of sulfate-reducing bacteria on organic debris can be seen.[10] In the upper layer, the burrowing habits of oligochete and chironomid worms can lead to cylindrical burrows in the sediment whose walls are aerated. The additional surface area available for oxygen uptake has been estimated by Fry.[11] A medium population of benthic invertebrates, for example, in a Welsh reservoir, more than doubles oxygen uptake; however, a dense population in the River Thames can increase it by a staggering 116 times! The upper layers of many sediments, depending on their position, can be perturbed not only by the activities of animals but by hydraulic factors including water currents, tides, and breaking waves. Jones has stressed that horizontal heterogeneity (zonation) is also apparent in sediment systems usually around the shore line. This is therefore quantitatively more important the smaller the body of water. In fresh water lakes, the littoral regions can be overlayed with shallow warm aerated water, while the profundal zones during thermal stratification will be covered with cool oxygen-depleted water. Significant differences in metabolic activity result. The last level of heterogeneity of importance in sediments must be at the microscopic level and concerns the growth of organisms in the sediment itself. Here, the nature of microcolony development and the physical chemistry of available surfaces plays an important role.

The scale of the different classes of heterogeneity must now be mentioned. It has already been stated that molecular diffusion is an essentially slow process in liquid phases, slower still where solid particles increase the tortuosity of solute transport. Apart from bioturbation and hydraulic processes, molecular diffusion is the most significant force for solute transfer in sediments. The shorter the diffusion path from sources to sinks, the faster the reactions that can take place. Thus the upper aerobic layer of sediments is a zone of intense metabolic activity, while shallower gradients in the deeper levels mean slower rates of reaction.

Many of the reactions in sediments could profitably be examined using some of the spatially heterogeneous model systems described in this book. One line would be to emulate some of the possible interactions in a gel stabilized model, and it would be fruitful to investigate the effect of added mineral particles to such a system.

## Three Phases Systems: Soils

The addition of another phase to the two-phase system already discussed significantly increases the opportunities for spatial heterogeneity, and soil ecosystems are perhaps the most complex to understand and hence to investigate using model systems.

Soil consists of three phases: gas, liquid, and solid. The gas phase is generally very similar to air in composition; however, it may be depleted in oxygen and enriched in any or all of the following: carbon dioxide, nitrogen oxides, ethylene, and hydrogen sulfide, etc. The water phase is generally rainwater which has picked up solutes as it travels through the soil. This phase moves in one direction on the whole; that is downwards. Percolation is the major solute vector in soil, though local distribution is by capillarity and by molecular diffusion. When the surface layers become dry, water may move in the reverse direction by capillarity. The solid phase is highly heterogeneous and very complex . . . ! Soil contains two main classes of minerals: the rock-pebble-sand family and the clay family. The former provide a relatively small and chemically unimportant surface area; however, they do contribute significantly to the porosity and texture of the soil. Clay particles, on the other hand, are generally colloids having a vast surface area which is physically and chemically reactive, and these play an important part in the overal biology of the system. Organic constituents are divided into nonliving and living. The former consist of biological material in various stages of decomposition, especially the more recalcitrant compounds like lignin and cellulose, chitin, humic and fulvic acids. The living fraction ranges from multicellular eukaryotic animals and plants (especially roots of the latter), and a vast range of unicellular eukaryotic and prokaryotic species. Products from living cells and the cells themselves contribute structure to the system. Thus, fungal hyphae, roots and root hairs, and microbial polysaccharides all help to form and hold together soil crumbs.

Heterogeneity exists in a number of main areas each of which is characterized by differences in scale and hence in biological reactivity. Vertical stratification is usual but much more variable than seen in the solid-liquid system. Scale here is of the order of centimeters and meters. Stratification is dominated by water relationships. A dry porous soil will provide many channels for gaseous diffusion and exchange, so that deeper levels will remain aerobic. Heavy rain or flooding will fill soil pores, gradually blocking off diffusion pathways so that the lower levels will become anaerobic. Deliberate flooding of rice paddy fields has been followed by microbiologists who have noted the fairly predictable change from aerobic to predominantly anaerobic biology. Vertical stratification is not just a question of oxygen relationships. In reality, far more important are the gradients in organic matter and in the living soil flora and fauna which are at their richest near the surface and gradually fall off with depth.

The soil crumb represents another level of heterogeneity. Here the system has an irregular spherical geometry. The crumb consists of mineral components and organic matter held together by roots, mycelia, and by cell capsule components, usually polysaccharides. Bacterial growth is usually as isolated microcolonies associated with clay colloids or organic matter. Once more, depending on prevailing water constraints, the crumb may be saturated even though there may be paths between crumbs allowing gas diffusion deep into the soil profile. If the crumbs are water saturated, their centers may be anaerobic while outer layers may be well aerated. The juxtaposition of aerobic and anaerobic ''spaces'' allows interesting chemical interactions to take place. These will be discussed later. A third level of heterogeneity is at the boundary between the microcolony and water in contact with it, and there is a final zone of interest at clay mineral interfaces where cation exchange activities associated with the clay lattices lead to steep pH gradients over a few nanometers.

Besides these "structural" levels of heterogeneity, the activities of plant and animals in the soil play a vitally important part in the distribution of substrates and products. Plant roots are so important here that we recognize an entire ecosystem, the rhizosphere, adjacent to them. Because of transfer of nutrients from the root, the microbial population in the rhizosphere is usually much higher than elsewhere. In flooded systems like paddy fields, plants can translocate oxygen into the root associated microflora. Root commensals contribute to the fertility of the soil by the fixation of gaseous nitrogen. The soil fauna, on the other hand, are responsible for physically turning soil over and aerating especially the heavier denser soils. They also play their own part in the energy flow of soil systems through consumption and turnover of other organisms.

For a full understanding of the soil ecosystem, each level of heterogeneity needs to be carefully investigated. Once more it seems likely that carefully selected experimental models ought to play an important part in this work. A model with many of the properties of soil is the percolating column discussed in detail in Chapter 2, Volume II, by Prosser and Bazin. It is perhaps interesting that systems discussed by these authors range from microcosm to model in relatively small steps. Thus some of the earlier work has been carried out on soil columns which must be defined as microcosm, while later work by Prosser and his colleagues, among others, has reduced the complexity in the system by using glass beads and restricted pure cultures of bacteria. It is easy to see that a large number of possible models could bridge the gap between microcosm and model in such systems. Both Caldwell and Fowler (Chapters 6 and 7, this volume) have considered surface attachment and growth, and of course such work is directly relevant to the growth of microcolonies on mineral constituents in soil. Diffusion-linked model systems ought to be employed to look at interactions possible in soil crumbs whose centers have become anaerobic since aerobic-anaerobic interfaces allow important biochemical pathways and cycles to take place. Finally, Burns discusses the soil ecosystem in some detail and illustrates some applications of models and microcosms to such systems (Volume II, Chapter 3). It is clear that there is scope for the application of relevant laboratory models in this area too.

**Microbial Film**

A family of microbial ecosystems are found attached to solid surfaces which are exposed to aqueous solutions containing nutrients. These microbial films are ubiquitous and have an important economic role besides their obvious interest ecologically. Characklis[12] has emphasized the problems that microbial film can cause. Examples range from the colonization of boat hulls, leading to fouling by larger organisms; films causing corrosion of marine steel or concrete installations; growth in water pipes leading to reduced flow, blockages, infection with pathogenic bacteria like *Legionella* species; or a reduction in heat conduction properties. Microbial films can be the bane of the fermentation technologist's life if they develop in a fermentation system. They can form enormously thick growths on agitator turbines and on baffles and so on in the fermenter, obviously interfering with the dynamics of the whole process.

Other films are associated with animals, including human surfaces. Thus dental caries is entirely correlated with the presence of a bacterial film of dental plaque which will always grow on teeth given the chance. Microbial film is not always bad, however. Most effluent treatment plants encourage the growth of microbial films, for example, on aerobic or anaerobic filter systems and on rotating disc aerators which are all used to recycle organic pollutants. Films of *Acetobacter* covering beech twigs or chips in the "quick" vinegar process or as a film at the air-liquid interface in the traditional Orleans wine vinegar process.

Besides films of economic importance, however, microbial films and slimes associate with numerous other surfaces. Most solid-liquid interfaces can become coated with microbes which tend to attach to a thin layer of adsorbed macromolecules, which quickly bind to any "clean" surface immersed in natural aquatic systems. Films are found on the gastric mucosa and internal epithelial linings of many animals. While many oral organisms attach to dental enamel, others "prefer" the cheek and tongue epithelial cells, for example. Films are formed on most surfaces immersed in any of the natural water systems. These surfaces are not simply mineral or wooden structures, but the surfaces of plant stems and leaves and of aquatic animals. Often, a thin film of biological origin can be found in the "neuston" at air-water interfaces. It can be shown that this film can grow and thicken if adequate nutrients are present. Microbes develop on terrestrrial surfaces, too. Thus the phylloplane is a habitat on the surface of leaves, which shows a succession of organisms throughout the growing season. Sometimes these proliferate enough to form a coherent film. A corresponding region around plant roots, the rhizoplane, leads to a cylindrical film-like proliferation of microbes using root exudates as nutrients. There is even a region around germinating seeds called the spermosphere which has some of the characteristics of a biofilm.

Research into biofilm has predictably followed two main routes: first, how to eliminate it from sensitive equipment/areas, and second, how to control its formation and activity where it performs a useful function. Microbiologists have become increasingly interested not so much in the film itself but in mechanisms for the attachment of microbes to surfaces. Really very little is known about the structure and physiological functioning of microbial films, and it is at this point that model film fermenters will play an increasingly important part. The ubiquity of microbial films makes any research into this area of fundamental importance. Each film is normally composed not just of a single species, but of a group of different genotypes each having some part to play in the overall behavior of the structure. It is of interest to find out what comprises a minimal structure that is capable of quasi steady-state growth. Biofilms usually become thick enough that certain solutes, in particular oxygen, become exhausted before the base of the film is reached. This region is therefore a suitable "space" for the proliferation of anaerobic species, so long as sufficient nutrents are available to them. The close juxtaposition of aerobic and anaerobic spaces allows all sorts of interactions to take place. For example, in films where anaerobic corrosion due to sulfate-reducing bacteria takes place, sulfide formed in the anaerobic regions can be reoxidized in the aerobic surface layers by sulfide-oxidizing bacteria (Hamilton[13]). A laboratory model film fermenter is badly needed that can allow the growth and maintenance of a highly reproducible microbial film. Such fermenters are described by Characklis (Chapter 8, this volume) and by Peters and Wimpenny (Chapter 9, this volume) and for dental plaque (a specific film important to human well being) by Tatevossian (Chapter 10) in this book.

## TEMPORAL HETEROGENEITY

Succession is a direct result of environmental physical chemistry changing at a point in real space. Given a wide range of genotypes at the location in question, the habitat will be appropriate for maximal growth of a particular species or group of interacting species. This will be the point from which successional changes can start. The physical chemistry of the point in question now alters. If external environmental changes occur the succession is said to be *allogenic;* if conditions change due to the activities of the community itself, then the succession is *autogenic.*

There are numerous examples of successional changes. Indeed, the latter are more the rule than the exception in natural ecosystems. The "classical" examples include the changes in population in haystacks or in compost heaps which result from a rise in temperature. The populations can change from predominantly mesophilic to thermophilic types such as *Bacillus stearothermophilus* and the thermoactinomyces. Successional changes may be caused by pH fluctuations. The silage pit and fermentations producing sauerkraut and the Korean "kimchee" are of this type. Successions in the gut population of newly born mammals have also been studied intensively. Breast-fed infants often support large populations of *Bifidobacterium* species at first. These are later replaced by the more typical gut populations of the adult.

A particularly interesting example is surface colonization leading to microbial film formation. The topic is reviewed by Fletcher and Marshall[14] whose account leads to the following conclusions: a clean surface is initially coated with a conditioning film of organic molecules. In natural aquatic systems, small copiotrophic species then attach. The latter are almost always small, Gram-negative, rod-shaped bacteria including pseudomonads, flavobacteria, and achromobacteria. Secondary colonizers then appear such as *Caulobacter, Hyphomicrobium,* and *Saprospira.* The diversity of attached species increases with time, and electron microscopy may reveal the presence of numerus eukaryotic organisms including fungi, diatoms, and protozoa. Successions in human dental plaque film are also documented (Tatevossian, Chapter 10, this volume).

Seasonal changes can lead to allogenic successions. The latter have been discussed for the phyto- and zooplanktonic populations of lakes and ponds, for example, by Cairns in 1982.[15]

If the spatial and temporal heterogenity of the physicochemical environment represent the stage, its sets and effects, then microbes and their interrelationships are the players.

## GENOTYPIC HETEROGENEITY AND MICROBIAL INTERACTIONS

There is little point in citing examples of genotypic heterogeneity in nature since almost every conceivable habitat is populated by a wide range of different classes of organism. More interesting are the possible interactions between species. Interactions can be discussed and subdivided on a number of different criteria. Thus the best known classifications based mainly on nutrition can be found in almost any textbook of microbial ecology and will not be summarized here. Another way to consider microbial interactions is based on their space relationships. Thus we can distinguish same-space from different-space interactions. Same-space interactions can proceed in a homogeneous environment and may be investigated in stirred fermentation systems, more particularly continuous flow devices. Such interactions give rise to the term "consortium" being applied to a steady state culture of organisms interacting to form a stable community. Consortia may be simple two-membered communities engaging perhaps in cross feeding, or they may be more complex communities such as the Dalapon consortium which involved seven different species. Such communities are discussed in more detail by Gotschall and Dijkhuisen (Chapter 2) in this volume.

Another important group of same-space interactions are the syntrophic associations in which the metabolism of two organisms are tightly coupled by the transfer of nutrients between them. Hydrogen scavenging, leading to an alteration in the available free energy relationships of certain organisms, is perhaps the classic syntrophic system, most clearly seen in the *Methanobacillus omelianskii* association.[16] The latter originally isolated as a single species was later shown to consist of two species, a methanogen and the so called "S" organism. The "S" organism oxidized ethanol to acetic acid and hydrogen, a reaction which only became thermodynamically probable once the hydrogen concentration was reduced to a very low level by the methanogen. Other syntrophic associations between photosynthetic bacteria and sulfate-reducing species are just as tightly coupled. Looser associations are also common, especially in anaerobic environments where hydrogen scavenging is an important function for the community as a whole.

Different-space systems offer other opportunities for profitable interactions. The most common interface between physico-chemically different spaces are at the boundary between aerobiosis and anaerobiosis. This is because oxygen is both a rapidly metabolized substrate and a very sparingly soluble molecule in aqueous media. Aerobic species rapidly remove oxygen leading to boundaries beyond which anaerobic bacteria can proliferate. The depth of aerobic zones in stratified ecosystems can vary in ecological systems over at least seven orders of magnitude: from tens to hundreds of meters in meromictic lakes and seas down to tens of micrometers in rapidly respiring bacterial colonies or microbial films. The juxtaposition of aerobic and anaerobic spaces can lead to many fundamentally important metabolic chains and cycles that would not otherwise operate. The simplest of these, for example, is the sulfur cycle which could operate with just two species, an aerobic, sulfur-oxidizing autotroph and an anaerobic, sulfate-reducing species. More complex is a nitrogen cycle consisting of nitrifying aerobes, the products of whose metabolism diffuse into an anaerobic region where denitrifying species reverse the direction of electron flow. This is not a perfect metabolic cycle since denitrification generates nitrogen gas which then leaves the system. Metabolic chains operate in one direction across such interfaces. Anaerobic bacteria hydrolyzing polymers to simple sugars, can ferment the latter to end products such as acetic or other fatty acids. These may diffuse across the oxygen interface to be further metabolized by aerobic bacteria.

It is easy to make a case for oxygen interfaces to illustrate different-space interactions. This is because oxygen is such a dominant solute in natural ecosystems and its gradients can be steep enough to provide sharp discontinuities. Less obvious but just as important are gradients of other factors. Light, for example, generates a vitally important family of physical gradients. These can also be steep where light impinges on and penetrates the surfaces of intrinsically opaque materials like sediments, soils, and even certain solid mineral conglomerates, where they support the growth of an endolithic community. Light gradients have the unique property of undergoing physical changes as they pass through liquid or solid media. That is, there are selective changes in wavelength depending on what absorbing materials are present. It is well known that anaerobic photosynthetic species absorb at longer wavelengths than do green and blue-green algae. The latter are nearly always located above the former and hence have first "go" at the light impinging on the system. Interactions based on an input of light energy will then take place. Thus, the anaerobic sulfur bacteria, for example, *Chromatium* species, will photooxidize sulfur generated by sulfate-reducing species back to sulfate.

## COMPARTMENTS AND DOMAINS

Much of this chapter has emphasized the occupation of spatially, temporally, and genotypically heterogeneous space as a good description of the majority of microbial habitats. The sheer complexity of such systems necessitates some form of intellectual abstraction. The terms "compartment" and "domain" have been used to provide a rather simple interpretation of the importance of extracellular space. These concepts appear most useful when applied to spatially heterogeneous systems dominanted by molecular diffusion as the major solute transport process. The discussion that follows gives a word picture of the aspect of microbial ecology which has really led to the perceived importance of model systems and hence to the need for a book of this type.

### Compartments

Compartments can be defined as discrete units of activity. In ecological terms, the most common compartment is a single cell. It is characterized by having an "inside" which is the site of chemical reactivity, and an "outside" which is an area immediately surrounding the compartment with which the latter interacts. While the cell is the most suitable paradigm of a compartment, groups of cells (colonies, microcolonies, floc, microbial film, layers of cells at interfaces, etc.) can all be regarded as compartments. In the same way, from a molecular biologist's viewpoint, organelles, the periplasm, multienzyme complexes, the active centers of enzymes and so on, are all also compartments. Compartments are all associated with specific chemical activities. They use substrates and manufacture products. The former are imported from outside to inside, while the latter flow in the opposite direction. In terms of solute distribution, the compartment is a black-box or chemical-transducer, converting substrates to products and usually synthesizing more of itself.

**Habitat Domains**

Every compartment lies within a region of "real" space. If the physico-chemical composition of this space allows it to proliferate, the compartment is within its "habitat domain". Habitat domains can be actual or potential. An actual habitat domain characterizes the space occupied by a specific compartment at a given instant in time. The potential habitat domain for an organism embraces the set of all possible spaces in which an organism can grow. It is clear that any description of habitat must be multifactorial. Thus there exists a range of pH values over which an organism can grow. In addition, it responds to a range of water activities, temperatures, ionic strengths, concentrations of many different chemical agents, and so on. To use the terminology of Whittaker, organisms exist in an "m-dimensional hyperspace" of physico-chemical effectors, one dimension for each factor. The experimental determination of habitat domains for particular species has only rarely been tackled systematically. Perhaps the best-known observations are those of Baas-Becking and his colleagues[17] who mapped a wide range of estuarine microorganisms in the two dimensions of redox potential and pH. Chapter 11 (this volume) outlines a simple approach to mapping multi-dimensional habitat domains using two-dimensional gradient plates.

Two subsets of the habitat domain definition can be discerned. They deal with factors essential to the life of the organism, on the one hand, or to agents that influence the organism adversely only when they are present. Among the first group are pH, temperature, water potential, ionic strength, and essential substrates. The latter must be present for growth to occur though they could be inhibitory at high concentrations. Similarly, there are optima for pH, temperature, water activity, and ionic strength, as well as values at which growth is impossible. Responses to inhibitory agents are different. Here, the habitat domain reflects concentrations from zero to a specified threshold value. Naturally, in the absence of inhibitor, growth rates are unaffected, so that a definition of habitat domains for these agents is redundant in their absence.

The habitat domain of a species has a number of other interpretations which deserve a brief discussion. First, growth rates of an organism in its habitat domain will be position dependent. The actual boundary of a habitat domain for a given character is that point at which growth rate falls to an arbitrarily low value compared to its maximum. Traversing across a parameter field will yield a curve of growth rate vs. position if the habitat domain for a single organism is encountered. If habitat domains for more than one species are present (and these may or may not overlap one another), such a traverse may cross them all. If the changes are space dependent, this will lead to separation of species in space. At any one point, changes in habitat parameters with time will lead to a succession in microbial growth rates. Clearly, different organisms may span a parameter field at any one *time* in just the same way that parameter fields may "slide" across organisms at any one *position*.

All that has been said here also applies to multi-dimensional parameter fields. This is easy to see for two- and three-dimensional systems. Where two factors are varied, the habitat domain consists of an area within a line separating growth from no-growth. If a measure of population or growth rate were plotted for each of the two environmental variables, a three-dimensional contour map (Gaussian response curve) defines the habitat domain.

It is possible to construct a cube for three separate habitat parameters. For a single organism, the habitat domain is now represented by a solid surface separating growth from no-growth. Where more than one organism is mapped, drawing a line in any direction

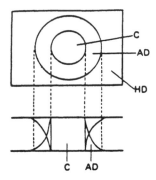

FIGURE 1. Compartment (C) and activity and habitat domains (AD and HD). The compartment may have source or sink domains for particular solutes as shown in the lower part of the figure.

through this cube may cross a number of habitat domains in its path. Spatial separation or temporal successions may be predicted from the cube depending on which is held constant, time or position. Consideration of a single cube containing habitat domain data for a dozen or so different species will indicate the huge range of distributions possible without invoking additional dimensions or stressing that the path through the cube will almost never be a straight line in a natural ecosystem. The three-dimensional cube just described could be made more complex if actual population data or growth rates were known: now a fourth dimension is necessary to define the *actual* growth rate of organisms at each point within it.

Obviously, as more environmental variables are added, the data become more complex and can no longer be represented in "real" space coordinates. More sophisticated mathematical analysis is needed to compare such habitat domain data. Whittaker imagines an m-dimensional "cloud" as defining the habitat characteristics of an organism. The term has a misleadingly friendly feeling to it, concealing as it does shapes that cannot easily be imagined by our three-dimensionally oriented brains: it may be the best we can do in the circumstances!

Finally, the habitat domain does not imply that a cell will *necessarily* be able to grow within its periphery: only that conditions are permissive for the single environmental factor. Only when a cell is within the permissive region for *all* the necessary environmental factors can growth actually take place.

**Activity Domains**

Habitat domains express only the set of global parameters which permit or restrict growth. The habitat domain says nothing about the effect of the cell itself on the space outside it. I chose the term "activity domain" to denote the property of interacting with components in this outer space. Activity in this sense means the ability of the cell to influence the composition of this region and as such is a local function. The cell contributes chemicals to or removes chemicals from the activity domain. The compartment is either a *source* or a *sink* for solute molecules and the space over which the compartment has some influence is therefore a source activity domain or a sink activity domain. These notions are represented by the diagrams in Figure 1. The definitions indicate no absolute boundary to these domains, however, a boundary could be drawn to denote a position in space around the compartment where the activity of the latter falls to some arbitrary and insignificant level. For example, if the cell is using glucose from the environment, the boundary of its sink domain is that point where there is no measurable depletion in glucose concentration in the bulk solution. Similarly, if it ferments glucose to lactic acid, the source domain boundary for lactic acid is that point where lactic acid concentration is no longer measurable.

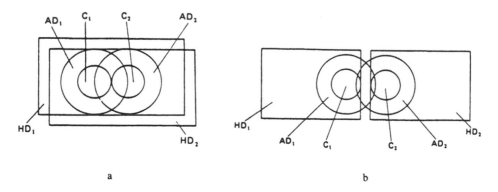

FIGURE 2. (a) Interacting compartments with overlapping habitat and activity domains. (b) Interacting compartments with exclusive habitat domains. Terminology as in Figure 1.

Activity domains are multidimensional, one for each source or sink activity of the cell. There are therefore *actual* activity domains and there are *potential* activity domains, just as there are actual and potential habitat domains.

Finally, the activity domain is a local function in that it describes what the cell is doing at specific positions in space at any one instant. The habitat domain, on the other hand, is a global function indicating ability to grow, and is free to move over the cell as environmental physical chemistry changes. The two are of course related to one another. While changes in activity domain may cause the habitat domain to move across the cell, altering the latter will almost certainly influence the activity domain for a given character.

### Domains and Interactions

The value of the concepts expounded so far are that they may be used to describe interactions between different cells or groups of cells. Many possibilities exist, however, two in particular are worth discussing, that is, the same-space and different-space interactions that were introduced earlier in this chapter. Same-space indicates that two interacting species occupy the same actual habitat domain. Because of this, they are able to grow very closely together. They interact because their activity domains overlap in the first place; they interact better because their habitat domains also overlap. It should be stressed here that there need be no complete correspondence of habitat domains, only that the latter overlap at some point where the two species are actually growing.

Different-space interactions occur when two compartments interact, even though they are unable to occupy the same space because their habitat domains do not overlap. Provided that their activity domains overlap, then interaction is possible. Depending on precisely which domains do overlap, interactions may be unidirectional or bidirectional. These situations are also illustrated in Figure 2.

### Vectorial Solute Flow

In homogeneous well-stirred reactors, the exchange of solutes between cells and their surrounding medium can only be defined in scalar terms. However, in heterogeneous systems where we assume that molecular diffusion is the sole solute transport process, another term must be included. Solute transfer occurs through *vectorial* processes. That is, solutes will flow down concentration gradients from sources to sinks. Thus two isolated compartments will interact by a vectored flow of solutes from one to the other. Of course, if the two compartments are located in an infinite diffusion field, the sinks will include not only an acceptor compartment but the whole diffusion field as well. The transfer process will be "lossy", the loss depending on the geometry of the system. There is naturally a bonus to be had by being as close as possible to a cooperating compartment. Not for nothing do interacting bacteria like the syntrophic *Pelochromatium roseo-viride*[18] consist of a central sulfate reducing bacterium entirely surrounded by photosynthetic species. Transfer of sulfide from the former to the latter must be virtually 100% efficient here.

Geometrical conditions can also make interactions between different-space systems quite efficient. The classical example is the sulfur cycle, which at its simplest, can be a single sulfide oxidizing aerobic autotroph whose metabolism is coupled by diffusion to a sulfate reducer. Such interactions are probably the rule in marine or estuarine sediments. Each partner can be thought of here as a sheet of cells occupying aerobic and anaerobic spaces one above the other. The whole two-dimensional array becomes a chemical transducer where sulfur compounds couple the oxidation of organic substrates from below to molecular oxygen from above.

## Spatial Determinants in Microbial Systems

The vectored flow of solutes have other consequences of course: they, together with habitat domain identities, are spatial determinants for the proliferation of the genotypes that are present. To pursue the word picture, if one imagines a cell growing in a diffusion field of a single essential solute, it will rather uninterestingly "attempt" to grow as close to the source as it can. It might simply be that organisms close to the source grow, while others further away grow more slowly or not at all. On the other hand, motile organisms attracted by the substrate will move up the gradient and grow at the same time. This subject is dealt with in detail by Lauffenburger in Chapter 6, Volume II.

Perhaps more interesting is the situation where an organism depends on two solutes for growth. If these solutes approach the cell from opposite directions, the only position where growth can take place is at the junction of the two gradients. Here we see another property of these heterogeneous systems. We cannot expect growth to occur at any position in the system. In the end a diffusion field consists of two main zones. These are zones of reaction where cells can grow (habitat domains), and transfer zones where they cannot grow but through which solutes may diffuse. In strongly diffusion-limited systems, the reaction zones may be spatially tiny compared to transfer zones. The effects of inhibitory components in the system are to shift growth, if possible, away from the source of inhibitor. The inhibitor, as has been stressed already, is one element of the habitat domain of the cell. The latter is constricted in the direction of the inhibitor source and the cell will grow up to the boundary of this new habitat domain.

The patterns discussed so far are clearly rather simple involving a single organism or two interacting species. The situation becomes far more complex in a natural ecosystem. Only rarely will systems fall into neat predictable categories, two clear examples where they do include the stratification seen in sediment systems and the beautiful organization on a huge scale of the different physiological types in a stably stratified meromictic lake. In these systems it is possible to see metabolic sequences aligned on redox potential or free-energy gradients. One can discern the effects, too, of gradients in light energy, in oxygen availability, and so on.

It is much more likely that an ecosystem is heterogeneous in three dimensions. Here, interacting species form a mosaic of domains and the range of interactions possible may be enormous.

## Solute Transfer Other Than by Diffusion

In most cases, molecular diffusion is not the sole solute vector. Natural ecosystems show a wide range of solute transport processes. All these share the common property that they involve motion of the solution relative to the cellular compartment. In soil, flow is due to downward percolation after rain or upward transport due to capillarity and surface evaporation. In streams, rivers, estuaries, and oceans, it is due to water currents. On the shorelines and beaches of water bodies it can be due to wave action and to bioturbation due to fauna in the environment.

Relative flow distorts the activity domains of the cellular compartments in a system. These domains are compressed upstream and extended in the direction of flow. It is clear that interactions may now become unidirectional, i.e., a compartment upstream can affect one lower down, but not vice versa. Such systems can be modeled in the laboratory using percolating columns. These are discussed in detail by Prosser & Basin (Chapter 2, Volume II).

Motility is a special case of relative motion. Motility is a process that leads to movement towards or away from specific solute sources which may be growth substrates or inhibitors. In addition to this obvious effect of motility, it is clear that motion distorts the sink domain for a nutrient around the compartment. The domain shrinks in the direction of motion, raising the effective concentration that the compartment sees, to something approaching the bulk concentration. If the latter is low anyway, movement will reduce the effects of nutrient depletion around the cell.

To add further to the complexity of natural ecosystems, the influence of the other two phases, gas and solid, must be included. Some of the properties of solids in sedimentary and in soil ecosystems have been described briefly already. They contribute to solute distribution in three main ways. First, their presence act as barriers to transfer, and solute molecules must follow a longer path to reach particular cellular compartments; second, their numbers, size, geometry, and distribution affect water availability due to matric potential; third, they impart a physico-chemical heterogeneity to the system leading to steep cation gradients or to secondary and even tertiary pools of certain solutes with different degrees of availability to the biota of the system.

Effects of the gas phase are easier to assess. Gas transfer to and from the atmosphere is comparatively straightforward. Mass transfer rates vary up to a factor of about ten between calm water and highly turbulent water surfaces. Otherwise it is only in soil that drastic changes in atmospheric components, in particular oxygen, are found. These changes are really due to changes in the aqueous phase where wet conditions block gas transfer and dry conditions promote it.

The general tenor of all that has been said here is that microbial ecology, in sharp contrast to animal and plant ecology, is to do with behavior of aqueous solutions in one-, two-, and three-phase systems. I have tried to emphasize the vital importance of the composition of the small space just outside microbial compartments to the fate of the cell itself. If there is a general theme in the way chapters for this book have been selected, it is this belief that laboratory models can help to understand many of the processes that contribute microbial ecology at the very basic level of the behavior of solutes and the solvents in which they are dissolved. Included also are chapters on numerical modeling, since it is clear that the latter complements the experimental approach, in particular, the predicted behavior of an experimental model can help in deciding what the critical experiments are that need to be done.

# REFERENCES

1. **Parkes, R. J.,** Methods for enriching, isolating and analysing microbial communities in laboratory systems, in *Microbial Interactions and Communities*, Vol. 1, Bull, A. T. and Slater, J. H., Eds., Academic Press, London, 1982, 45.
2. **Winogradsky, S.,** *Microbiologie du Sol*, Masson, Paris, 1949.
3. **Wimpenny, J. W. T.,** Spatial order in microbial ecosystems, *Biol. Rev.*, 56, 295, 1981.
4. **Wimpenny, J. W. T.,** Responses of microorganisms to physical and chemical gradients, *Philos. Trans. R. Soc. London Ser. B*, 297, 497, 1982.

5. **Wimpenny, J. W. T., Lovitt, R. W., and Coombs, J. P.,** Laboratory model systems for the investigation of spatially and temporally organized microbial ecosystems, *Symp. Soc. Gen. Microbiol.,* 34, 67, 1984.
6. **Kriss, A. E.,** *Marine Microbiology,* Shewan, J. M. and Kabata, Z., (Trans.), Oliver and Boyd, Edinburgh, 1962.
7. **Sorokin, Y. I.,** Primary production and microbiological processes in Lake Gek' Gel (transl.), *Microbiol.,* 37, 289, 1968.
8. **Genovese, S.,** The distribution of $H_2S$ in the Lake of Faro (Messina) with particular regard to the presence of 'red water', in *Marine Microbiology,* Oppenheimer, C. H., Ed., Charles C Thomas, Springfield, Ill., 1963, 194.
9. **Suckow, R. and Schwartz, W.,** Redox conditions and precipitations of iron and copper in sulphureta, in *Marine Microbiology,* Oppenheimer, C. H., Ed., Charles C Thomas, Springfield, Ill., 1963, 187.
10. **Jorgensen, B. B.,** Bacterial sulphate reduction within reduced microniches of oxidized marine sediments, *Marine Biol.,* 41, 7, 1977.
11. **Fry, J. C.,** Interactions between bacteria and benthic invertebrates, in *Sediment Microbiology,* Nedwell, D. B. and Brown, C. M., Eds., Academic Press, London, 1982, 171.
12. **Characklis, W. G.,** Fouling biofilm development: a process analysis, *Biotechnol. Bioeng.,* 23, 1923, 1980.
13. **Hamilton, W. A.,** Sulphate reducing bacteria and anaerobic corrosion, *Ann. Rev. Microbiol.,* 39, 195, 1985.
14. **Fletcher, M. and Marshall, K. C.,** Are solid surfaces of ecological significance to aquatic bacteria, *Adv. Microbial Ecol.,* 6, 199, 1982.
15. **Cairns, J.,** Freshwater protozoan communities, in *Microbial Interactions and Communities,* Bull, A. T. and Slater, J. H., Eds., Vol. 1, Academic Press, London, 1982, 249.
16. **Bryant, M. P., Wolin, E. A., Wolin, M. J., and Wolfe, R. S.,** *Methanobacillus omelianskii,* a symbiotic association of two species of bacteria, *Arch. Microbiol.,* 59, 20, 1967.
17. **Baas-Becking, L. G. M. and Ferguson Wood, E. J.,** Biological processes in the estuarine environment. I. Ecology of the sulphur cycle, *K. Ned. Akad. Wet.* B, 59, 160, 1955.
18. **Kuznetsov, S. I.,** Trends in the development of ecological microbiology, *Adv. Aquat. Microbiol.,* 1, 1, 1975.

Chapter 2

# THE PLACE OF THE CONTINUOUS CULTURE IN ECOLOGICAL RESEARCH

**Jan C. Gottschal and Lubbert Dijkhuizen**

## INTRODUCTION

In nature, a multiplicity of different microorganisms is involved in cycling the various elements. In order to come to some understanding of their place and function in the natural environment, it has become established practice to isolate microorganisms in pure cultures and to study their properties under well-defined conditions in the laboratory. Over the years this has led to the accumulation of a wealth of information on the metabolic activities of microbes. In these studies, the application of continuous cultivation techniques has, in particular, received considerable attention since this approach allows the study of microbial growth under conditions similar to those found in nature.

It is evident, however, that in nature a pure culture of microorganisms, though possible in specific micro-environments, will be very rare, as will growth limitation by a single nutrient. In fact, natural ecosystems are often spatially heterogeneous and growth conditions may change continuously in time with mixtures of various species competing among themselves for multiple substrates.[1-3] Heterogeneous systems will be dealt with elsewhere in this volume, and our objective is to discuss the experimental possibilities offered by homogeneous continuous culture techniques, to study the various ways in which microbes respond to ecologically relevant conditions. In doing so, the status of our present knowledge of growth in continuous culture, employing the various combinations of single or mixed substrates with pure or mixed cultures, will be reviewed and presented in order of increasing complexity. This not only reflects developments which over the years have taken place in the application of that technique, but also illustrates our firm conviction that a profound understanding of microbial life in nature can only be achieved when a clear picture of microbial behavior under each of the separate conditions already mentioned is available.

## BASIC THEORY OF THE CONTINUOUS CULTURE

The basic continuous culture system is characterized by a culture vessel, the contents of which is well mixed, to which fresh medium is added at a constant rate while the working volume is kept constant via an overflow device. The system is thus open. All the components required for growth are present in excess in the medium, except one: the growth-limiting substrate. Ideally, this cultivation system allows continuous exponential growth of microorganisms under constant environmental conditions, i.e., in a steady state, with no fluctuations in either the substrate concentration or the microbial population with respect to time (see below). This single stage, flow-controlled system was first described by Monod[4] and Novick and Szilard,[5,6] and in the past 30 years, it has gained recognition as one of the cultivation techniques best suited to the study of a wide spectrum of microbiological problems.[7-13] Continuous culture theory has been discussed extensively in several papers,[14-17] and only the basic principles will be briefly reiterated here. For detailed practical information on the use of this technique, the reader is referred to the papers by Evans et al.,[18] Calcott,[19] and Drew.[20]

Upon inoculation of the vessel, an organism will start growing at a rate determined by the substrate concentrations provided. This is described empirically by the Monod equation:

$$\mu = \mu_{max} \left( \frac{S}{K_s + S} \right) \tag{1}$$

where $\mu$ is the specific growth rate (per hour), S is the concentration of the growth rate-limiting substrate, and $K_s$ is a constant, numerically equal to the substrate concentration at which $\mu = \frac{1}{2} \mu_{max}$. Not only the growth rate but also the extent of growth in the culture per unit of time, $\mu x$ where x = dry weight of organisms per liter, is determined by the concentration of the growth-limiting substrate and, as is the case in batch cultures, growth would cease if this substrate were exhausted. In the continuous flow system, however, the supply of fresh medium ensures prolonged multiplication, which in turn leads to a constant dilution of the cell suspension. Dilution rate D is in units per hour and is defined as flow rate divided by volume of the culture vessel. Dilution of the cell suspension is simply Dx. The combination of growth with dilution therefore governs the change in concentration of microorganisms present in the culture vessel in time:

change in biomass with time = growth − output

$$^{dx}/_{dt} = \mu x - Dx = (\mu - D)x \tag{2}$$

It is clear from Equation 2 that only if $\mu$ = D will the level of organisms remain constant with time ($^{dx}/_{dt}$ = 0, a steady state). If $\mu > D$, the substrate comsumption rate will exceed its supply rate, leading to a drop in the residual substrate concentration of the culture. According to the Monod equation, Equation 1, this will cause a fall in growth rate $\mu$, a process that continues until $\mu$ = D. The same reasoning can be applied to the situation in which initially $\mu < D$. In this case, an increasing concentration of growth-limiting substrate allows an increase in the growth rate. In time, this will result in the establishment of a steady state, provided D does not exceed the critical dilution rate ($D_c$):

$$D_c = \mu_{max} \left( \frac{S_R}{K_s + S_R} \right) \tag{3}$$

where $S_R$ is the concentration of the growth-limiting substrate in the medium reservoir. In general, $S_R \gg K_s$ and it follows from Equation 3 that $D_c$ values will be approximately equal to $\mu_{max}$. Operation of continuous culture at dilution rates above $D_c$ will lead to a complete washout of the microbial population.

As already mentioned, the extent of growth is dependent on concentration of the growth-limiting substrate. The relation between these parameters is considered to be a constant[21] and is defined as the growth yield coefficient Y:

$$^{dx}/_{ds} = \frac{\text{weight of biomass formed}}{\text{weight of substrate consumed}} = Y \tag{4}$$

The overall change in the concentration of the growth-limiting substrate in the culture in time is then described by:

change in substrate concentration with time = substrate input

− substrate output − substrate consumed

$$^{ds}/_{dt} = DS_R - Ds - \mu x/_y \tag{5}$$

Now substitution of Equation 1 in 2 gives:

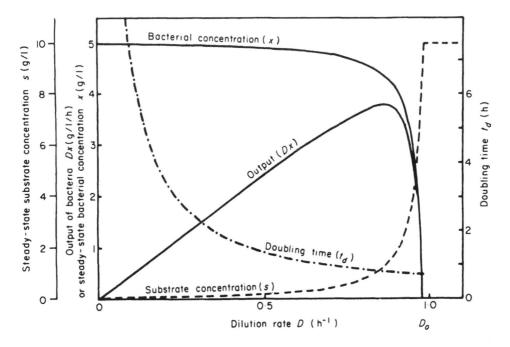

FIGURE 1. Theoretical relationship of the dilution rate (D) and the steady state values of bacterial concentration (x), substrate concentration (s), doubling time (t_d) and output (Dx). Data were calculated from Equations 7 and 8, for an organism with the following growth constants: $\mu_{max}$ = 1.0 per hour, Y = 0.5, and $K_s$ = 0.2 g/$\ell$, and a substrate concentration in the inflowing medium of $S_R$ = 10 g/$\ell$. (From Herbert, D., Elsworth, R., and Telling, R. C., *J. Gen. Microbiol.*, 14, 601, 1956. With permission.)

$$\frac{dx}{dt} = x\left[\left(\frac{\mu_{max} \cdot S}{K_S + S}\right) - D\right] \qquad (6)$$

Under steady state conditions, with $\mu = D$, not only $\frac{dx}{dt} = 0$ but also $\frac{ds}{dt} = 0$, and it follows from Equations 5 and 6, respectively:

$$\bar{x} = Y(S_R - \bar{s}) \qquad (7)$$

$$\bar{s} = \frac{K_s \cdot D}{\mu_{max} - D} \qquad (8)$$

where $\bar{x}$ and $\bar{s}$ are steady state cell and substrate concentrations, respectively. From equation 7 it is clear that $\bar{x}$ is primarily determined by the concentration of the growth-limiting substrate in the medium reservoir ($S_R$), whereas Equation 8 indicates that $\bar{s}$ is directly related to the dilution rate employed, all other parameters being considered constant. This is further illustrated in Figure 1 which shows typical D-$\bar{x}$ and D-$\bar{s}$ curves for some experimental values of the growth constants Y, $\mu_{max}$, and $K_s$. Once these values are known, the steady state values $\bar{x}$ and $\bar{s}$ at the various dilution rates can be predicted for any value of $S_R$. The shape of the D-$\bar{s}$ curve is largely determined by the $K_s$ value of the organism for its growth-limiting substrate. Since $K_s$ values for most substrates are in the $\mu$-molar range or even lower, values for $\bar{s}$ will generally be very low over quite a range of dilution rates. Steady state conditions in this system will therefore be most stable under conditions in which small changes in the dilution rate cause only minor changes in $\bar{x}$ and $\bar{s}$, i.e., clearly below $D_c$ (Figure 1). Thus, it is possible to grow cells in a continuous culture, at submaximal rates and with growth-limiting substrate concentrations, but up to high cell densities.

Over the years it has become apparent that deviations from Monod growth kinetics may occur in continuous culture systems. Most important, it is generally accepted now that the overall growth yield Y is not a constant. This is in contrast with the original notion of Monod[21] who concluded on the basis of batch culture studies that for a certain organism the overall growth yield remains the same depending only on the nature of the limiting nutrient. Ample evidence is now available in the literature that yields may vary with dilution rate due to accumulation of products, changes in cell composition or in viability, and maintenance energy requirements which will cause a drop in biomass yield especially at lower growth rates.[16] A detailed discussion of the latter phenomenon and of the overall bioenergetics of microbial growth in continuous culture is outside the scope of this review. For further information the reader should see papers by Pirt,[16] Stouthamer,[22,23] and Hempfling and Rice.[24]

## PURE CULTURE STUDIES

### Single Nutrient Limitation

Among the environmental parameters that commonly influence the properties of microbial cells in nature, the availability of essential nutrients, in particular carbon, nitrogen, and phosphorus, is of paramount importance. Natural ecosystems are frequently virtually depleted of one or more of these nutrients as a consequence of the potentially vigorous metabolic activities of the indigenous microbial populations.[25-27] Consequently, if one wants to understand the behavior of microorganisms in nature, it is necessary to study the mechanisms adopted to cope with nutrient-limiting conditions. Experimentally this can be approached in a straightforward fashion in continuous culture by controlling the composition of the inflowing medium in such a way that one of the nutrients is growth-limiting. Such a procedure can be executed at any of a whole range of growth rates and concentrations of the limiting substrate, followed by a detailed analysis of the properties of the microbial population once a steady state has been established. Since any substrate that a microorganism needs can be made growth-limiting at will, this has enabled the identification of a number of the physiological problems posed by nutrient limitation and some of the strategies that have evolved in the microbial world to deal with these situations.

Over the years, a wealth of information has accumulated on the responses of microbes to single nutrient limitation and a number of reviews on this aspect of microbial ecophysiology have been published.[27-33] In this section, a brief account of the most important points will be presented.

Microorganisms possess a wide variety of mechanisms for structural and functional phenotypic adaptation to their environment.[33] Not surprisingly, the main objectives of these mechanisms appear to aim at prolonged survival of the organism yet enable it to grow as fast as possible at low environmental concentrations of the limiting nutrient.[34,35] It is evident that physiologically this could be brought about if the organism were able to take up and metabolize that nutrient at the highest possible rate under conditions where its concentration outside the cell was very low, and to produce cell material with an optimal efficiency with respect to the limiting nutrient. Ample evidence available in the literature indicates that mechanisms to this effect have indeed evolved in the microbial world. They include the ability to increase the rate of transport of a nutrient when its concentration in the environment decreases, the facility to increase the rate of initial metabolism of a nutrient once it has accumulated inside the cell even when its intracellular concentration is low, and the ability to rearrange the chemical composition of cellular structures by redirecting fluxes of metabolites containing the limiting chemical element (carbon, nitrogen, or phosphorus). The rationale behind these strategies of phenotypic adaptation have been discussed extensively elsewhere[28,31,32] and will not be reiterated here. It should be stressed, though, that the potential of microorganisms to adapt themselves to changing environments ultimately resides in their genome and will be different for different genotypes.

Uptake and initial steps in the metabolism of the growth-limiting nutrient may be enhanced simply by the synthesis of more of the existing enzyme systems, i.e., an increase in $V_{max}$. This is generally observed under carbon-limiting conditions where catabolic enzymes usually increase in activity with decreasing dilution rates and thus with falling concentrations of the limiting carbon source.[31,36-39] Those enzymes involved in biosynthetic reactions, on the other hand, remain tuned to the anabolic requirements of the cells and generally increase with increasing dilution rates under carbon-limitation. The latter is also reflected in the macro-molecular composition of those cell components involved in biosynthetic reactions, for instance the RNA content, which increases almost linearly with growth rate.[40] Under these conditions, carbon conversion efficiency is high and diversion of substrate carbon to extra-cellular products is minimal.[41] Structurally, many bacteria respond to a decrease in the supply of nutrients by increasing their surface to volume ratio, thus enhancing cellular solute uptake capacity.[9,42,43] A very clear example of this behavior is the sphere to rod transition observed with *Arthrobacter* species.[44-46] Under nutrient limitation, especially at low growth rates, the coccoid morphology dominates, and this is the shape which is thought to be most abundant in nature as well. Enhanced metabolic capacity can be achieved not only by the synthesis of higher levels of existing enzymes but also by the production of a different, high-affinity enzyme system for uptake and metabolism of the growth-limiting nutrient. Well-known examples of the presence of dual pathways with a similar function but different substrate affinities in a single organism include glucose metabolism (via periplasmic and intracellular pathways) in *Pseudomonas aeruginosa*[47,48] and glycerol catabolism[49] and ammonia assimilation[50,51] in *Klebsiella aerogenes*.

The most pronounced response of microorganisms to the presence of growth-limiting concentrations of an essential nutrient other than the carbon and energy source is that of a redirection of fluxes of metabolites containing the limiting element. Under these conditions the carbon conversion efficiency is generally low and this may lead to a significant accu-mulation of carbon in intracellular reserve materials,[48,52] extracellular polymers,[53] or a variety of low molecular weight metabolites.[54] Other changes are possible, for example phosphorus-limitation effects the chemical composition of the bacterial cell envelope.[55] This may involve both a replacement of phosphorus-containing glycerol teichoic acids by teichuronic acids in the walls of Gram-positive bacteria[56] and of phospholipids by ornithine-containing lipids and acidic glycolipids as observed in a marine strain of *Pseudomonas fluorescens*.[57]

**Mixed Substrate Limitation**

In natural environments, microorganisms often grow in the presence of low concentrations of a diversity of compounds, several of which may make up combinations of homologous substrates, i.e., substrates that serve a similar metabolic function, for instance, carbon sources, nitrogen sources, etc. In order to gain more insight in the behavior of microbes in such environments, their metabolic responses to these more complex situations must be investigated. When a microbial species is placed under substrate-sufficient conditions in batch culture in a medium with, for instance, two utilizable carbon sources, the cells often, although not invariably, adapt themselves in such a way that the substrate that supports the highest growth rate is used preferentially while synthesis of enzymes involved in the utili-zation of the second substrate remains repressed. This leads to a sequential utilization of the two substrates and diauxic growth of the organism.[21] It should be stressed that this may be the most pronounced response in batch culture but it is certainly not the only one. Sequential utilization with no intervening lag and simultaneous utilization have been observed as well.[58] These batch culture studies have been extremely useful for the identification and detailed investigation of a number of cellular control mechanisms,[33,39] such as carbon catabolite repression[59] and, in case of mixtures of nitrogen sources, nitrogen metabolite repression.[60] It is clear, however, that such a distinct preference for only one compound to the exclusion of all others serving a similar metabolic function would be of little value in a nutritionally poor environment. One might expect instead that under these latter conditions selective pressure would favor those microbes capable of using a multiplicity of homologous substrates simultaneously. One must therefore ask whether the regulatory mechanisms observed at high substrate concentrations also play an important part under substrate-limiting conditions. For this purpose, continuous culture is a useful tool since it enables the investigation of microbial metabolism under steady state conditions at low concentrations of the growth-limiting substrate. Using this approach, experimental evidence has accumulated in recent years that many microorganisms indeed use a multiplicity of nutrients serving a similar physiological function at the same time, under conditions where their supply limits growth.[11,39] In a number of these experiments, true multiple substrate limitation was demonstrated, since an increase in reservoir concentration of either one of the supposedly limiting nutrients caused a proportional increase in cell density. From this, it can be concluded that for a mathematical description of growth under these conditions,[17,61,62] summation of the specific growth rates sustained by the single nutrients may be most appropriate, i.e.,

$$\mu = \mu(s_1) = \mu(s_2) \tag{9}$$

Generalizations on the responses of organisms when faced with mixtures of carbon sources, however, cannot be made. First, as is the case in batch culture, preferential or simultaneous utilization patterns may be observed. Second, when both substrates are apparently used together, the situation actually may be far more complex, in that the utilization of one of the limiting substrates is influenced strongly by the presence and utilization of the second substrate.[39] There is evidence emerging that these kinds of interaction even may affect the growth yields of the organisms involved (see below) and the steady state concentrations of the substrates.[63,64] Specific growth rate may therefore be expressed interactively, for example, as the product of individual growth rates.

$$\mu = \mu(s_1) \cdot \mu(s_2) \tag{10}$$

In this case, Equation 10 appears justified, while for preferential utilization, where only one of the substrates is used to completion, a noninteractive form, such as

$$\mu = \mu(s_1) \qquad \text{or} \qquad \mu = \mu(s_2) \tag{11}$$

may be more suitable. In general, the type of responses observed in studies with mixed substrates in continuous cultures depends on a number of factors: (1) specific properties of the microorganisms themselves (i.e., whether they are specialists or versatile organisms); (2) the combination of carbon sources chosen, since this determines the metabolic pathways involved and their possible interaction; (3) the maximum specific growth rates attainable on each substrate; (4) the medium reservoir substrate concentrations, and (5) the dilution rates employed.

A few examples may serve to illustrate some of these points. During growth of *Escherichia coli* on mixtures of glucose and fructose in batch culture, glucose is used before fructose.[65] Growth of the same organism on this mixture in carbon-limited continuous cultures showed that both sugars were used simultaneously and completely at low dilution rates, whereas fructose utilization became progressively impaired at the higher dilution rates.[66] Similar observations have been made with a variety of sugars in a number of microorganisms.[66-68] This response, however, is not restricted to mixtures of sugars but has also been reported

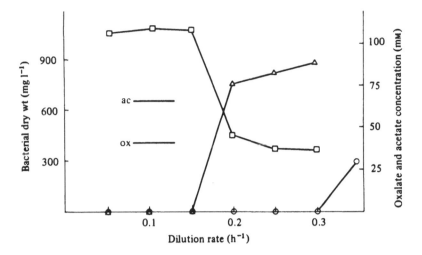

FIGURE 2. Growth of *Pseudomonas oxalaticus* OX1 on a mixture of acetate ($S_R$ = 30 m*M* and oxalate ($S_R$ = 100 m*M* in a continuous culture limited by carbon and energy. □, bacterial dry weight on the mixture; △, residual oxalate concentration; ○, residual acetate concentration. The bacterial dry weight values observed during growth of the organism on 30 m*M* acetate (ac) or 100 m*M* oxalate (ox) separately, are also given. (From Harder, W. and Dijkhuizen, L., *Philos. Trans. R. Soc. London Ser. B*, 297, 459, 1982. With permission.)

with other combinations of organic compounds.[39,58] *Pseudomonas oxalaticus* grown on a mixture of acetate and oxalate in batch culture showed diauxic growth, with acetate used first.[69] In carbon-limited continuous cultures, however, both acids were used simultaneously and were undetectable in the culture, provided the dilution rates were low (Figure 2). Only at dilution rates close to the $\mu_{max}$ on oxalate alone (0.20 per hour) did increasing amounts of oxalate accumulate, while no residual acetate was detectable in the culture supernatant up to a dilution rate of 0.30 per hour ($\mu_{max}$ acetate = 0.35 per hour). At first sight one could conclude from these data that the inability of the organism to use oxalate at dilution rates of approximately 0.20 per hour is a direct reflection of the maximum specific growth rate attainable on this substrate alone. Although this is one possibility, it is unlikely that this is the only factor involved, since studies of growth in batch cultures on the same mixture, with cells pregrown on oxalate, showed that for unknown reasons the presence of acetate blocked oxalate utilization almost completely within two hours.[69] Thus, *P. oxalaticus* can use these two substrates simultaneously, provided their concentrations are low as is the case at the lower dilution rates. Concurrent utilization of mixtures of carbon sources under carbon limitation, and especially at the lower dilution rates, is a widespread phenomenon. The available evidence suggests a relief from catabolite repression under these conditions. One example, already referred to in the previous section, is that the levels of catabolic enzyme systems generally increase with decreasing dilution rates. This effect is probably caused by a fall in intracellular concentrations of the relevant repressor molecules at the lower dilution rates, which parallels a fall in concentrations of the growth-limiting carbon sources in the culture (Figure 1). The most detailed information on this phenomenon has come from recent studies in which the regulation of autotrophic and methylotrophic metabolism in the so-called facultative autotrophs and methylotrophs was studied.[39,70,71] From this work it was clear that even though two substrates are used simultaneously, their metabolism may be influenced considerably by the presence of the other substrate. The following example illustrates this point:

Synthesis of the enzyme ribulose-1,5-bisphosphate carboxylase (RuBPCase), which catalyzes carbon dioxide fixation in the Calvin cycle in autotrophic bacteria[72] appears to be regulated by a repression/derepression mechanism in response to the intracellular concentrations of one or more intermediary metabolites. In carbon-limited continuous cultures, repression of the synthesis of this enzyme is lowest at the lower dilution rates. During growth of the facultative autotroph *P. oxalaticus* on formate alone in formate-limited continuous cultures,[73] the specific activities of RuBPCase increased with increasing dilution rates (D = 0.02 to 0.15 per hour), as is generally the case with biosynthetic enzymes (see above). Based on the dry weight values at the various dilution rates, the protein content and the elementary composition of cell material of *P. oxalaticus,* it could be calculated, however, that at the highest dilution rate tested, the specific activity of RuBPCase is in excess of that required to explain the growth rate. At decreasing dilution rates, even though specific activity fell, the overcapacity of the enzyme increased even further, and at D = 0.02 per hour there was about five times more enzyme available than that needed to maintain a specific growth rate of 0.02 per hour. When grown in batch culture on acetate (a "heterotrophic" substrate), synthesis of RuBPCase remained completely repressed, even when formate was present in the medium as well.[74] Interestingly, formate and acetate were used simultaneously under these conditions, although formate was now used only as an ancillary energy source, via formate dehydrogenase enzymes. Addition of up to 40 m$M$ formate to the medium reservoir of an acetate-limited ($S_R$ = 30 m$M$) continuous culture of *P. oxalaticus* (D = 0.10 per hour) showed a similar pattern, since formate utilization led to a clear increase in bacterial dey weight in the culture at steady state, while RuBPCase remained repressed.[75] Synthesis of the enzyme only occurred at formate concentrations ($S_R$) of 50 m$M$ and higher. During growth of the organism on a constant mixture of acetate ($S_R$ = 30 m$M$) plus formate ($S_R$ = 100 m$M$), but at dilution rates from 0.02 to 0.30 per hour, repression of RuBPCase synthesis increased with increasing dilution rates. It was also noted that, although the maximum specific growth rate on formate alone is close to 0.20 per hour, not only acetate but also formate was still being used up to the highest dilution rate tested (0.30 per hour). It is clear from batch and continuous culture studies that synthesis of RuBPCase is very sensitive to repression exerted by acetate metabolism. They also reaffirm that under substrate-limiting conditions the degree of repression diminished with decreasing dilution rates. Similar observations have been made in a number of studies on mixed substrate utilization by various facultative autotrophs.[39]

These experiments indicate that these metabolically flexible bacteria are able to adapt their enzyme profiles to ambient environmental conditions. Depending on the ratio of concentrations of "autotrophic" and "heterotrophic" substrates present in the medium reservoir, and on the dilution rate employed, such adaptation may lead to optimization of biomass yield. In other words, the biomass synthesized during growth on the mixtures may reach higher levels than expected when the two substrates are supplied separately. Apart from experiments with *P. oxalaticus,*[75] such phenomena have been observed with *Paracoccus denitrificans*[76] growing on mixtures of methanol and mannitol and *Thiobacillus versutus*[77] (formerly named *Thiobacillus* A2),[78] growing on thiosulphate/acetate mixtures. A main conclusion from the above work is that energy is saved by fixing less $CO_2$ via the Calvin cycle as more carbon is assimilated from the "heterotrophic" substrate supplied.

Although much remains to be discovered about both the regulatory mechanisms controlling the utilization of mixtures of homologous substrates, and the effects of this on the overall metabolism in microbes, the information currently available has already disclosed aspects of mixed substrate utilization relevant to natural ecosystems. First, it is clear that under carbon- (and/or energy-) limitation, the simultaneous utilization of a mixture of carbon sources may be a widespread phenomenon, provided their concentrations are low. Second, as outlined above, the ability of microorganisms to use various homologous substrates simultaneously may have a positive effect on cell yields. Although cell yield may not play a significant role in competition under single nutrient limitations, it definitely does affect competitiveness under mixed substrate testing conditions[64,79,80] (see section on defined mixed cultures)

So far, we have dealt only with the responses of microorganisms to mixed homologous substrate limitations. Growth conditions in which two nonhomologous substrates, e.g., a carbon source and a nitrogen source, are both present in concentrations that simultaneously limit the growth rate of the organism, are, however, feasible as well. As already predicted theoretically, the actual realization of growth under what is generally termed double or dual substrate limitation, appears to be hard since true simultaneous limitation seems only to occur over a very narrow range of experimental conditions.[61,62,81,82] It is clear that double substrate limitation, dependent on precisely defined concentrations of the two substrates, is unlikely to be maintained for a prolonged period of time in natural environments. This is particularly so since fluctuations in the availability of nutrients appear to be the rule rather than the exception here. Thus, although a detailed exploration of the responses of microorganisms to double substrate limitation may be interesting from a physiologist's point of view, they may not be relevant to growth in nature.

## Transient States

As pointed out by van Es et al.,[3] one objection to the application of continuous culture techniques to study microbial ecology is that microorganisms rarely grow under truly steady state conditions in nature. Rather, continual changes in the composition of natural microenvironments mean that the metabolic machinery of bacteria in such sites is generally in a transient state. However, continuous culture really *does* allow a study of the microbial responses to chemical and physical changes and has the additional experimental advantage of starting out from a steady-state situation, with a well-analyzed population of cells growing at submaximal rates in media with nutrient concentrations as low as those found in natural environments.[8] In recent years numerous studies on perturbing environmental parameters governing growth in chemostat cultures have been made with a variety of microorganisms. This approach, in which various perturbations were deliberately introduced, e.g., a shift-up or shift-down in nutrient,[83] oxygen,[84-86] or light[87,88] availability, or physical changes, e.g., in temperature or in pH,[89,90] has given important new insights in the reactivity and flexibility of different microbial species to conditions that may be found in nature. Moreover, it has added a new dimension to our understanding of how different cellular control mechanisms may allow optimal adaptation of different organisms to their specific ecological niches. In this section, we will restrict ourselves to discussing a few examples of the effects of a shift-down or a shift-up in growth-limiting substrate concentrations on pure cultures.

Many microorganisms can adapt phenotypically to a considerable range of variations in their environment, and this has led to the recognition that a microorganism of a given genotype is very much a product of its environment,[28,33,40] The ability of an organism to survive and to multiply over a range of environmental conditions is ultimately restricted by its genotype. Thus, at one extreme organisms have been found able to grow on many different substrates over a wide range of temperatures, oxygen partial pressures, and pH values, while at the other side there are organisms dependent on the availability of one specific carbon and energy source under strictly specified physical conditions. In some organisms, special survival mechanisms, i.e., endospore formation[91] or the swarmer cells of the budding and prosthecate bacteria,[92] become apparent when growth conditions become strenuous. In a continuous culture study of sporulation of *Bacillus subtilis* 168, Dawes and Mandelstam[93] observed that under glucose- or ammonia-limiting conditions, sporulation increased as dilution rates were reduced. In transient state experiments, the fraction of refractile spores in the culture, following a shift-down in dilution rates, increased markedly after 3 hr for both types of nutrient limitation. This was explained as an increase in the fraction of cells initiating sporulation, expressed as onset of refractility after 3 hr. In shift-up experiments, it was apparent that partial relief of glucose limitation, due to an increased dilution rate, stopped further initiation of sporulation but not further development, whereas partial relief of nitrogen limitation blocked both.

Organisms unable to alter structurally in this way may, following a shift-down in energy supply, have to face a situation with a rapid dwindling of their energy stores. Under these conditions, they may rely at first on the availability and metabolism of endogenous reserve materials, but ultimately this will be followed by a breakdown of essential cellular structures (RNA, protein, DNA, etc.) as well as the dissipation of important intracellular metabolite pools, resulting in loss of viability.[29,52,94]

Synthesis of reserve material is commonly seen when carbon is in excess and nitrogen, phosphorus, or oxygen, etc. limits growth. In a few cases, however, it has also been observed under carbon-limiting conditions. An example of this is synthesis of polyhydroxybutyrate by a *Hyphomicrobium* species, grown under methylamine limitation in continuous culture at D = 0.02 per hour.[95] A second example was reported by Matin et al.[96] where a *Spirillum* sp. accumulated polyhydroxybutyrate during lactate-limited growth especially at the lower dilution rates tested. Following growth of the organism at various dilution rates, the medium flow was switched off and its resistance to starvation was monitored. The results showed that survival was proportional to the polyhydroxybutyrate content of the culture and increased with decreasing dilution rates.[96] Although accumulation of reserve materials under carbon limitation is rather unexpected, it may reflect some special adaptation, i.e., a high rate of nutrient uptake relative to intracellular metabolism, in these organisms, both of which possess a number of other physiological characteristics considered desirable for existence in oligo-trophic environments.[42,43,97]

Endogenous metabolism in energy-starved organisms can thus allow a prolonged period of survival, but in a dynamically changing environment its main function may be to sustain a low level of metabolic activity. The latter enables the cells to accumulate energy sources rapidly, once they become available again, and, by their dissimilation, to generate a sufficiently large electrochemical proton gradient ($\Delta\bar{\mu}_{H^+}$) to act as a driving force for the uptake of many compounds from the environment.[98] Regeneration of the proton gradient will further facilitate the uptake of essential nutrients and thus determine the ability of the cells to start growing again. The effect of starvation on the cellular energy status of *Streptococcus cremoris* (Otto et al.[99,100,100a]) serves as an illustration of this thesis. Following growth of *S. cremoris* in a lactose-limited continuous culture (pH 6.8, D = 0.10 per hour), starvation was imposed by switching off the lactose supply (Figure 3). In the first hour of the experiment, the intracellular ATP pool and the phosphate potential decreased rapidly. In contrast, the intra-cellular pool of phosphoenolpyruvate (PEP) initially increased considerably, from 6.9 m$M$ at zero time to 11 m$M$ after 5 min, and subsequently dropped to approximately 0.25 m$M$ where it remained stable for many more hours. In another experiment (D = 0.15 per hour; pH 7.6) it was shown that the membrane potential ($\Delta\psi$ component of the $\Delta\bar{\mu}_{H^+}$ collapsed within 30 min following lactose exhaustion. Since the transmembrane pH gradient ($\Delta$pH) was zero under these growth conditions, absence of lactose caused a complete dissipation of the $\Delta\bar{\mu}_{H^+}$. Similar results were obtained in an experiment performed at pH 5.7, except that in this case a $\Delta$pH was present and remained constant for at least 1 hr.[99] Following readdition of lactose to cultures of *S. cremoris*, starved for periods up to 24 hr, a rapid synthesis of ATP (Figure 4) and regeneration of the electrochemical proton gradient was observed. In *S. cremoris*, lactose uptake is mediated by a PEP-dependent sugar phospho-transferase system. It thus appeared that in this organism the intracellular PEP pool was still sufficiently high, not only after 5 min but even after 24 hr of starvation, to allow the uptake of at least some lactose molecules, their subsequent intracellular oxidation leading to a reestablishment of the $\Delta\bar{\mu}_{H^+}$. On the basis of the observations of Otto et al.,[99] Konings and

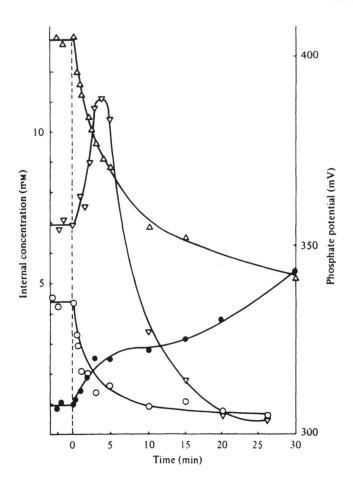

FIGURE 3. Changes in phosphate potential ($\Delta G'_p$, $\triangle$) and concentrations of ATP ($\circ$), phosphoenolpyruvate ($\triangledown$) and AMP ($\bullet$) in *Streptococcus cremoris* following a switch off of the medium supply at zero time. The organism was grown in a lactose-limited continuous culture at a dilution rate of 0.10 per hour, at pH 6.8 and 30°C, and a steady state was established before switching off the supply of fresh medium. (From Harder, W., Dijkhuizen, L., and Veldkamp, H., *The Microbe 1984: Part II Prokaryotes and Eukaryotes,* Kelly, D. P. and Carr, N. G., Eds., Cambridge University Press, Cambridge, 1984, 51. With permission.)

Veldkamp[27] postulated that in those bacteria for which sugars are important energy sources, transport systems have developed which do not depend on the $\Delta\bar{\mu}_{H^+}$, since energy starvation may lead to the dissipation of $\Delta\bar{\mu}_{H^+}$, but on other sources of energy such as PEP.

Over the years, many studies have dealt with the responses of bacterial populations, growing in continuous cultures under steady state conditions, to a nutritional shift-up, either by increasing the concentration of the growth-limiting substrate or by increasing dilution rate. The results of these studies show that microbial cells, when growing at submaximal rates under nutrient limitation, can often increase their growth rate considerably and instantaneously upon relief of the limitation.[83,101] The transient response observed, i.e., the duration and the extent of physiological adaptation, however, is strongly dependent on the prechange conditions and the magnitude and nature of the change imposed. Attempts to develop mathematical models of this transient behavior in microorganisms are discussed by Singh,[102,103] Barford et al.,[104] Pickett,[105] and Daigger and Grady.[106]

Following a shift-up in nutrient supply allowing considerably faster growth, large changes

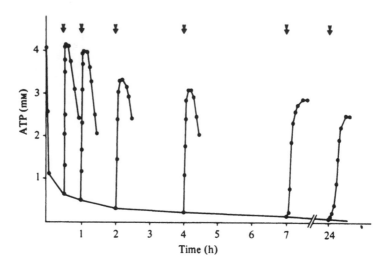

FIGURE 4.   Generation of ATP in energy-starved *Streptococcus cremoris* upon addition of lactose (indicated by the arrows; 0.9 m$M$ per pulse) at different times after the start of starvation (t = 0). Culture conditions are as in Figure 3, except that the dilution rate was 0.14 per hour. (From Harder, W., Dijkhuizen, L., and Veldkamp, H., *The Microbe 1984: Part II Prokaryotes and Eukaryotes*, Kelly, D. P. and Carr, N. G., Eds., Cambridge University Press, Cambridge, 1984, 51. With permission.)

in the macromolecular composition of the cells are generally observed in the ensuing period of adaptation. This applies in particular to the RNA content which is an almost linear function of the growth rate.[40] Since 80 to 85% of the RNA is ribosomal, it was concluded that ribosome content increases approximately in proportion to growth rate.[107] Later studies, however, showed that ribosomes do not function with a constant efficiency, but rather that the fraction of ribosomes actively involved in protein synthesis in exponentially growing bacteria increases progressively with the growth rate.[108-110] Small changes in growth rate may therefore be due to changes in the efficiency of ribosome function, but for larger increases, additional synthesis of ribosomal RNA will be needed as well. In recent years, it has become clear that 5'-diphosphoguanosine 3'-diphosphate (ppGpp) plays an important part in regulating the expression of the genes coding both for ribosomal RNA and for ribosomal proteins, and possibly for various other proteins involved in the protein-synthesizing machinery.[111] A survey of the literature shows that transient responses of different species to a shift-up to nutritionally excess conditions varies considerably. The experimental evidence currently available strongly suggests that this phenomenon of what is generally termed reactivity[28] is primarily based on the catabolic potential of the organism, i.e., the potential to rapidly generate extra energy by the oxidation of the extra carbon and energy supplied. Thus, depending on levels and activities of the enzyme systems involved in the uptake and initial metabolism of the growth-limiting substrate, a shift-up in its availability may lead directly to a rapidly increasing growth rate or may require *de novo* synthesis of these enzymes in order to increase the rate of energy generation sufficiently to sustain an enhanced rate of anabolism. A few examples illustrate this. Harvey[112] observed that only at dilution rates below 0.3 per hour (60% $\mu_{max}$) did cultures of *E. coli* react immediately to relief from glucose-limitation by increasing the specific rate of RNA, protein, and cell mass synthesis. Recently, Leegwater[113] reported that only at these lower dilution rates did potential rates of glucose uptake strongly exceed actual glucose uptake rates at steady state in glucose-limited cultures of *E. coli*. A somewhat different situation was shown in glucose-limited cultures of *Cytophaga johnsonae*[114] growing at D = 0.15 per hour (75% $\mu_{max}$). When this organism was transferred to a medium with excess glucose, growth started with no lag at the maximum specific growth rate. However, cells grown at a low dilution rate (D = 0.03 per hour) showed a transient lag before slowly starting to increase their growth rate. It transpires that *C. johnsonae* is able to synthesize two different glucose uptake systems.[115] At high glucose concentrations, i.e., in batch culture cells or in glucose-limited cultures adapted for 5 to 20 generations to higher D values, a low affinity but high $V_{max}$ glucose transport system operates. At low glucose concentrations, on the other hand, i.e., in glucose-limited cultures at the lower D values, a different glucose transport system with a high affinity for glucose but a low $V_{max}$ operates. Höfle[114] concluded that cells using the low $V_{max}$ system, can only increase their growth rate significantly following a shift-up in glucose availability by switching to the high $V_{max}$ uptake system. Comparable results were obtained in experiments with lysine-limited cultures of *Pseudomonas putida*.[106] This organism responded most rapidly to a nutritional shift-up when grown first at a moderate growth rate (D = 0.194 per hour; 33% of $\mu_{max}$). The shift-up led to an increase in the specific rate of RNA synthesis, which reached a value close to the maximum specific growth rate, while synthesis of other macromolecules and the cell mass as a whole lagged somewhat behind. Cultures grown at D = 0.067 per hour, on the other hand, showed a transient lag before gradually increasing their rate of macromolecule synthesis. No other physiological data were reported. In experiments with glucose-limited continuous cultures of *Klebsiella aerogenes*, Leegwater[113] observed that following addition of excess glucose, the rate of RNA synthesis instantly almost equalled the maximum specific growth rate, while the rate of protein synthesis and the growth rate itself varied from 35 to 70% of $\mu_{max}$. With this organism, the observed response was virtually independent of the dilution rate proceeding addition of the glucose. Under glucose-limitation, *K. aerogenes* is catabolically a highly reactive organism. Following relief from glucose-limitation, this organism has at any submaximal growth rate the capacity to consume glucose at a very high rate, close to the rate for glucose uptake at $\mu_{max}$.[113,116] While this organism can rapidly increase its rate of energy generation from glucose oxidation, excess glucose can clearly uncouple catabolism and anabolism, especially at lower growth rates.[113] This leads to transient extracellular accumulation of a variety of products of glucose metabolism, including acetate with or without pyruvate. Leegwater[113] also studied growth of *K. aerogenes* in a discontinuous glucose-limited culture, i.e., in a continuous culture to which limiting amounts of glucose were added pulsewise at regular time intervals. Addition of excess glucose to such a culture gave an even more pronounced response, since the rate of RNA synthesis and the actual specific growth rate immediately equalled the $\mu_{max}$ attainable on glucose. These results indicate that *K. aerogenes*, when confronted with fluctuations in carbon- and energy-source availability, maximizes its metabolic reactivity. Such conditions are precisely those that are thought to predominate in natural environments.

## MIXED CULTURE STUDIES

Only exceptionally will microbial populations exist in pure isolation, and most natural habitats harbor a wealth of different bacterial species. If we can accept Gause's principle of competitive exclusion,[117] it follows that all of these different organisms must occupy slightly different niches. Although pure culture studies are definitely needed to explain the complete spectrum of physiological behavior of pure cultures, this cannot give complete insight into their true niche. This must be so, since defining the niche of a given bacterium implies information of its relationships towards other (micro)organisms in the same habitat. It is necessary therefore to study the behavior of the bacterium in question in coculture with its presumed natural "neighbors", not only under controlled laboratory conditions but in its natural environment as well.

In the following sections we have presented a representative choice of examples illustrating the use of continuous culture as a tool for the study of mixed microbial populations.

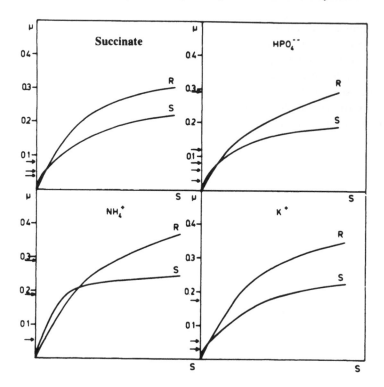

FIGURE 5.    Specific growth rate of *Spirillum* sp. (S) and a rod-shaped bacterium (R) as a function of the growth-limiting substrate concentration. Lactate was the carbon- and energy source in experiments with inorganic ion-limitations ($PO_4^{3-}$, $NH_4^+$, and $K^+$). Arrows indicate the dilution rates at which competition experiments were carried out. (From Kuenen, J. G., Boonstra, J., Schroder, H. G. J., and Veldkamp, H., *Microb. Ecol.*, 3, 119, 1977. With permission.)

## Defined Mixed Cultures

In 1958 Powell[118] published an important paper in which he demonstrated that the competition between microbial species for a single growth-limiting substrate could be described mathematically using Monod-type[4] growth kinetics. Thus, it was shown theoretically that *in the absence of any other interactions* the outcome of competition at a given dilution rate depended solely on the relationship between specific growth rate and substrate concentration of each member of the mixed culture. Jannasch[119] was the first to illustrate this principle practically, by showing that in enrichments with lactate as the growth-limiting substrate, the dilution rate selected determined which microbial species predominated. Further study of a *Pseudomonas* sp. and a *Spirillum* sp. isolated at high and low dilution rates, respectively, showed that the growth rate vs. substrate concentration curves (μ-s relationships) of the two species crossed. At dilution rates above the crossing point, the *Pseudomonas* sp. displaced the *Spirillum* sp., whereas at lower dilution rates the opposite was observed. Similar results have since been reported for many different species and substrates.[9,35,42,120-125] For example, Kuenen et al.[123] noted that two freshwater chemoorganotrophs showed only minor differences with respect to substrate specificity (more than 20 different carbon and energy sources were shown to support growth). One of the two, a *Spirillum* sp., outcompeted the other species (a rod-shaped bacterium) at relatively low dilution rates when grown with aspartate, lactate, succinate, potassium, ammonium, or phosphate as the limiting substrate in a chemostat. At relatively high dilution rates, the rod shaped bacterium was always dominant. The exact position of the crossover point, however, depended on the nature of the growth limiting substrate (see Figure 5). The relative abundance of examples of such competitive behavior in substrate-limited chemostats suggests that this may reflect an important principle in natural environments too. In almost any natural habitat, growth rate will be nutrient-limited (see for example Reference 27) and one is therefore tempted to believe that in habitats with relatively high or low rates of nutrient supply, appropriate different species will dominate. This fundamental physiological distinction may actually be comparable to the difference between autochthonous and zymogenous bacteria as originally proposed by Winogradsky.[126]

At this point we must stress that although growth conditions in chemostats and in the natural environment may both be nutrient limited, direct extrapolation of the laboratory situation to the natural milieu is not usually warranted. A major difficulty here is our poor knowledge of the nature, number, and relative importance of the substrates available to the equally unknown, but usually large, number of different bacteria present in one and the same habitat.

The chemostat offers a choice of two fundamentally different approaches for studying such complex mixed culture situations. Well-defined two-, three- or perhaps four-membered systems may be constructed and studied during growth on well-characterized substrate mixtures. The second approach is to endeavor to maintain entire communities in the chemostat, with the expectation that the population structure resembles that of the natural community as closely as possible. It is hoped, taking the first approach, that key principles concerning microbial interactions are disclosed, while using the second route, direct information on how entire communities grow and respond to environmental factors will be gained. A few examples using both approaches follow.

Perhaps the first well-documented example of a continuous culture experiment in which two species competed for two substrates was reported by Chian and Mateles.[127] Chemostat enrichment with glucose and butyrate as limiting nutrients yielded a two-membered mixed culture consisting of a pseudomonad and a coliform. The coliform was dominant (>90%) at dilution rates below 0.8 per hour. At higher rates, the pseudomonad became numerically dominant (up to 75%). Further conclusions were not possible as other noncompetitive interactions interfered. For example, at high dilution rates considerable quantities of butyrate and glucose remained unused, while acetate was excreted. Furthermore, the pseudomonad was shown to require an unidentified growth factor provided by the coliform.

Pirt[16] referred to some experiments with a two-membered mixed culture of a *Pseudomonas* sp. and *Klebsiella aerogenes* grown with glucose and *p*-hydroxybenzoate as the limiting nutrients. In this case, no interactions other than substrate competition were seen. It was concluded that *K. aerogenes* eliminated the *Pseudomonas* sp. from the culture because the former species could metabolize both substrates simultaneously while the latter only used glucose. The apparent advantage of substrate versatility was studied some years later, within the more general context of versatility vs. specialization.[77,80,128-131] For example, the facultatively autotrophic *Thiobacillus versutus* was chosen as a model since, though very versatile and flexible in its metabolism, it exhibits a relatively low $\mu_{max}$-value under most growth conditions when compared with more specialized species. When studied in mixed chemostat cultures, this *Thiobacillus* sp. was shown to be outcompeted in thiosulfate-limited chemostats by the obligate chemolithoautotroph *Thiobacillus neapolitanus*, and in acetate-limited cultures by a chemoorganotroph.[128] Since *T. versutus* was capable of mixotrophic growth in acetate plus thiosulfate limited chemostats, its competitiveness was also studied under such growth conditions in combination with either of the above mentioned competitors.[128] In mixed cultures with the specialist, coexistence was possible at relatively low acetate reservoir concentrations but at higher acetate concentrations the "specialist" was eliminated from the culture. A comparable result was obtained in mixed cultures with a heterotroph (a *Spirillum* sp.); however, in this case the heterotroph was gradually eliminated with increasing thiosulfate concentrations (see Figure 6). These results are readily explained in terms of mixotrophic potential of *Thiobacillus versutus*. With increasing quantities of the *nonshared*

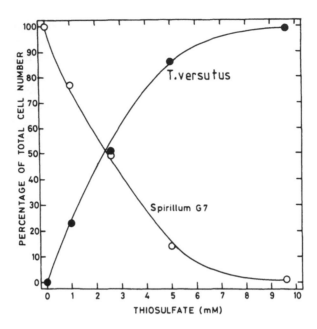

FIGURE 6.    Effect of different concentrations of thiosulfate on the outcome of the competition between *T. versutus* and a heterotrophic *Spirillum* G7, for acetate. the inflowing medium combined acetate (10 m*M* together with increasing concentrations of thiosulfate (0—10 m*M*). The percentage of *T. versutus* (●- - -●) and *Spirillum* G7 (○- - -○) were determined after steady states had been established. (From Gottschal, J. C., De Vries, S., and Kuenen, J. G., *Arch. Microbiol.*, 121, 241, 1979. With permission.)

substrate (acetate in the first case, thiosulfate in the second) the number of "mixotrophs" in the mixed culture will increase and as a result, metabolize an increasing fraction of the *shared* substrate.

Similar observations have been reported for mixed cultures of two clostridial species[129] and for cocultures of two *E. coli* strains.[130] In both cases, one of the competing organisms could metabolize both substrates fed into the culture while the other could only use one of them. The examples illustrate a principle of paramount ecological importance: the presence of one or more additional substrates may cause the "generalist" to outcompete the "specialist" even though the latter clearly has a much higher $\mu_{max}$ on the *shared* substrate.

To further our understanding of the competitive behavior of microbial species under multiple substrate limitation, attempts have been made to describe such cultures mathematically.[132-136] Although this is beyond the scope of the present review, one result, the importance of cell yield, is particularly worth mentioning here. For example, in mixed cultures like those just described above, it can be predicted that the higher the yield of the versatile species on the *nonshared* substrate, the more effectively it competes for the *shared* substrate.[132,136] This is in contrast to a situation when the competition is for a single growth-limiting substrate only where growth yield is unimportant. Another important prediction from such mathematical studies is that in the case of pure substrate competition, a stable mixed culture of n species can only be sustained if the number of growth limiting substrates is ≥ n (see for example Reference 132). However important this principle may be in theory, in practice it will not be easy to find a habitat in which other types of interactions will not occur simultaneously. The chemostat may well be used to study some of these other interactions as discussed below.

The presence of inert particles has been investigated in batch[137-140] and in continuous cultures.[141-143] Maigetter and Pfister[142] found that the addition of kaolinite-clay particles to mixed continuous cultures of *Chromobacterium lividum* and a *Pseudomonas* sp. caused a drop in numbers of the free *C. lividum* cells. While strong aggregation effects were observed with this species, the data do not allow conclusions to be drawn about the possible effect of such particles on substrate competition in this mixed culture. Laanbroek et al.,[143,144] in their study on the competitiveness of different sulfate-reducing bacteria, noted marked stimulation of acetate oxidation by illite clay particles with *Desulfobacter postgatei,* though they saw no effect of illite on competition for limiting sulfate. Clearly more work is needed before any general conclusions can be drawn concerning the nature of the effect of clay particles on microbial competition. However, such an area of investigation is tremendously important considering the ubiquity of natural habitats containing mineral particles.

Another factor which can influence competition in continuous culture is temperature, and this has previously received little attention in mixed culture studies. Perhaps the best studied example concerns the growth of a facultatively psychrophilic *Spirillum* sp. and an obligately psychrophilic *Pseudomonas* sp. at various temperatures when competing for limiting lactate.[145] It was found that at 16°C, the *Spirillum* sp. outcompeted the *Pseudomonas* sp. at all dilution rates tested, while at 2°C the situation was completely reversed. At intermediate temperatures the $\mu$-s relationships were shown to cross. Pirt[16] mentioned another interesting example of the effect of temperature on the outcome of the competition between a strain of *K. aerogenes* and a *Pseudomonas* sp. with glucose and *p*-hydroxybenzoate as growth-limiting substrates, the two species could coexist at 40°C; however, at temperatures below 37°C, *K. aerogenes* outcompeted the *Pseudomonas* sp. completely, because at this temperature it could use glucose and *p*-hydroxybutyrate simultaneously, while at 40°C only the latter substrate was used by *K. aerogenes*. This behavior may well be common but, so far, few examples have been reported.

Numerous workers have reported on the structure of communities in which the members are mutually interdependent. The state of our knowledge in this field has repeatedly been reviewed in recent years.[35,122,146-153] In this chapter, therefore, we will only consider a few examples to illustrate the importance of continuous culture in investigating such communities.

Megee et al.[154] studied mixed cultures of a riboflavin-producing *Saccharomyces cerevisiae* and a riboflavin-requiring *Lactobacillus casei* both in batch and continuous culture. At a relatively low pH value (5.5), stable coexistence of the two species was obtained under glucose-limiting conditions in the absence of exogenous riboflavin at dilution rates below 0.21 per hour. *L. casei* is supplied with riboflavin by the yeast, and both organisms compete for glucose, which results in a pattern of interactions which may be termed commensalism + competition, using a terminology described by Bungay and Bungay.[146] The same experiments were performed at a higher pH (6.5). In the absence of pH regulation, the authors concluded, based on the results of a mathematical description of the culture, that commensalism had now turned into mutualism, since acidification by *L. casei* down to pH 5.5 stimulated growth of the yeast. Furthermore, use of media containing riboflavin once more changed the type of interaction; pure competition for glucose was the result and this caused *L. casei* to outcompete the yeast.

A completely commensal interaction was also reported for mixed continuous cultures of the fresh water alga *Ankistrodesmus braunii* with bacterial species grown under phosphate-limiting conditions.[155] A stable mixed culture was obtained with a *Flavobacterium* sp. growing at the expense of algal exudates. However, cocultivation of the alga with a *Pseudomonas* sp. resulted in washout of the alga which was explained in terms of the pseudomonad outcompeting the alga for limiting phosphate. When the *Pseudomonas* sp. was added to a stable mixed culture of the alga and *Flavobacterium* sp., washout of the alga was observed. Although competition for phosphate may strongly reduce the algal population, it is not clear why complete washout should occur, since growth of the bacteria is dependent on algal excretion products, as illustrated, for example, in the work done on mutualism + competition by Meyer et al.[156] The data of Mayfield and Inniss[155] do not allow clear conclusions on this point since the experiments were curtailed too early. This kind of mixed-culture work with algae and bacteria certainly needs further attention, since it constitutes a major step in nutrient cycling in many aquatic systems.

In a well-studied example of commensalism in mixed-continuous culture, *Propionibacterium shermanii* grew at the expense of lactate produced by *Lactobacillus plantarum* from the primary growth-limiting substrate glucose.[157] At dilution rates below 0.07 per hour stable mixed cultures were thus obtained, and a simple mathematical model accurately described properly some of the growth features observed under steady state and transient conditions. Lactate was also the mediator in a commensal relationship between *Streptococcus mutans* and *Veillonella alcalescens*.[158] Both strains, commonly found in dental plaque, coexisted in mixed continuous cultures supplied with glucose as the limiting nutrient. They were studied in continuous culture since this may simulate some important characteristics of the oral ecosystems.[159-161] In this commensal relationship, lactate was replaced by acetate, propionate, and ethanol, whose lower dissociation constants might reduce demineralization of the enamel by acid.[158] It was shown with *Streptococcus cremoris*, some years later,[162] that removal of lactate by *Pseudomonas stutzeri* under anaerobic conditions in the presence of nitrate established a strong mutualism in lactose-limited mixed chemostat cultures. The rationale here is that by maintaining very low external lactate concentrations, *S. cremoris* is able to gain more energy from the efflux of lactate than at higher external concentrations.[163] Under such growth conditions, *S. cremoris* exhibits a 60 to 70% higher cell yield based on lactose consumed.[162] Such an example of mutualism is sometimes referred to as protocooperation, indicating that this type of interaction is not obligatory for either population.[122] This type of interaction can be established in many different ways, most obviously where one species produces stimulatory, or removes inhibitory, compounds affecting the other. Such interactions might well be widespread in nature; however, few cases have been studied in detail in continuous culture. A good example was reported by Wilkinson et al.[164] who found, using a mixed culture of a methane utilizing pseudomonad and three other species, that a *Hyphomicrobium* sp. grew on and hence removed inhibitory amounts of methanol formed by methane oxidation. Protocooperation between *Streptococcus thermophilus* and *Lactobacillus bulgaricus* was demonstrated during attempts to produce yogurt by continuous cultivation.[165] While both species can grow well on milk, they stimulate each other by producing amino acids (*L. bulgaricus*) and formic acid (*S. thermophilus*). Yet another bacterial interaction, interspecies $H_2$ transfer,[166] is frequently encountered in many anaerobic habitats, and this may in many cases be considered protocooperative. Ianotti et al.[167] were first to investigate this fundamental interaction in continuous culture. *Ruminococcus albus* (a rumencellulolytic species) was grown in glucose-limited chemostats either alone or in coculture with *Vibrio succinogenes*. In monoculture ethanol, acetate, hydrogen, and carbon dioxide were the only end products. In the mixed culture, when fumarate was included in the feed acetate, carbon dioxide and succinate were the only end products. Electrons liberated in the fermentation were combined with protons to form $H_2$, which was subsequently used by *V. succinogenes* to reduce fumarate to succinate. The resultant increase in acetate production allowed *R. albus* to obtain more energy per mole of glucose fermented. In a more recent study, Traore et al.[168] studied $H_2$ transfer between *Desulfovibrio vulgaris* and *Methanosarcina barkeri* in cocultures grown with lactate or pyruvate as the sole substrate in the absence of sulfate. Both in batch and in continuous cultures, hydrogen, produced by *D. vulgaris*, was used by *M. barkeri* to form methane. The methanogen successfully replaced sulfate as the electron acceptor. A major difference between batch and continuous cultures was that free $H_2$ concentration remained extremely low in the latter, while in batch cultures a transient accumulation occurred. Apparently, a tighter coupling between $H_2$ production and consumption existed under substrate limiting conditions. Such $H_2$-transfer reactions, when they occur in nature, will almost certainly be under electron donor limiting conditions as was the case in the continuous culture experiments.

## Predation in Defined Mixed Cultures

Knowledge of the importance of predation on bacterial populations to mineral cycling in natural ecosystems is still far from complete. Many studies in this field have appeared and have been well reviewed in recent years, with particular emphasis on predation by protozoa.[169-172] In this review, we would only like to emphasize the part chemostat studies can play in elucidating the impact of bacterial grazing in nature. Several workers have used continuous culture techniques to study selected predator-prey relationships.[122,173-180] Most of these studies have been discussed extensively in previous reviews[9,122] and therefore we will only mention one example which clearly illustrates the possibilities and limitations of the chemostat approach to the study of this normally complex interaction. Jost et al.[175] performed mixed continuous culture experiments with glucose as carbon and energy source and different combinations of *Azotobacter vinelandii, Escherichia coli,* plus the ciliate *Tetrahymena pyriformis.* With the medium used, the ciliate could only grow with bacteria as a food source. Growth of the ciliate at the expense of *A. vinelandii* resulted in coexistence of both species at three different holding times tested. However, at the two shorter holding times (5.9 and 23 hr), sustained oscillations for both bacterial and ciliate densities and for residual glucose concentrations were observed. At a 40 hr, holding-time-damped oscillations eventually led to stable coexistence. Similar results were found for *T. pyriformis* when grown with *E. coli* as a food source: sustained oscillations were observed at 7.1 and 15.2 hr holding time. These results did not, however, fit a mathematical model based on simple Monod-kinetics, developed for this kind of culture.[174,181] Minimum bacterial densities and residual glucose concentrations were found to be much higher than predicted. This lack of agreement between model and experiments could be largely eliminated by replacing the Monod-type model by a "multi saturation" model.[181] In this it was assumed that the predator became less effective at low prey densities, a conclusion supported by recent evidence.[182,183] Later theoretical work by the same group[184] indicated that incorporation of another factor, wall growth, into the Monod-type model, could equally well describe, in a semiquantitative way, the behavior of the mixed cultures. The constant reinoculation of bacteria from the wall, where the ciliates did not graze, could generate much higher cell densities. In order to help decide between the two alternatives of Jost et al.,[175] the effect of wall growth on the predator-prey system involving *E. coli* was recently tested.[185] By treating the chemostat walls with a bacteriostatic silicone agent, it could be shown that in cases where sustained oscillations of species densities occurred, minimum bacterial densities were several orders of magnitude lower than in chemostats with untreated walls. While these results do not really reveal to what extent wall growth was involved in the work of Jost et al.,[175] it clearly shows the possible impact of even the slightest culture heterogeneity on seemingly simple two-membered microbial systems.

## Complex Communities

It is just this spatial heterogeneity, so clearly present in almost any natural ecosystem, which raises a question mark over the use of simple chemostats for modelling more complex microbial ecosystems. This is probably the reason why so few attempts have been made to use simple, single-stage continuous culture systems for this purpose. Slyter et al.[186,187] maintained a rumen population in continuous culture, and this produced a pattern of fatty acids and bacterial species similar to that observed in the rumen. A major difference, however, was that protozoa initially present in large numbers were lost from the culture. Veilleux and Rowland[188] were only able to maintain in continuous culture representative hindgut microbial populations of two different rats when a second fermenter was connected in series with the first. They concluded that this was due to the complex nature of the medium and the spatial heterogeneity caused by wall growth and cell recycling between the stages. Freter et al.[189] came to similar conclusions while simulating mouse hindgut microbial flora in continuous culture. In an interesting paper by Marsh et al.,[160] chemostat cultivation was described for complex microbial communities established by inoculating glucose- or amino acid-limited continuous cultures with human dental plaque suspensions. At different dilution rates, a range of different communities became established. The authors noted that neither qualitatively nor quantitatively did the community composition ever reproduce the composition of the inoculum, although the overall metabolic activity resembled that thought to occur in vivo. Yet in spite of the differences noted, the chemostat was considered to be an environmentally relevant model for the oral cavity. With due appreciation of the immense difficulties involved in studying the dental plaque community, we believe that this is some-what too optimistic a view. If the community composition in the chemostat does not largely resemble that of the ecosystem under investigation, then we are clearly studying a microbial population which will react differently in response to experimental manipulations. Conclu-sions from such experiments would tell us little or nothing about the natural ecosystem we wish to study.

One is forced to make a choice between experimental conditions which simulate in vivo systems as closely as possible and an approach in which certain fundamental aspects from the ecosystem are singled out and studied in detail. It is of course the combination of these two approaches which is needed to learn to understand the functioning of microbial eco-systems, as this and other chapters in this book clearly stress.

### Controlled Discontinuous Growth

Although continuous culture was designed to create controlled constant growth conditions, it is equally well suited to investigating controlled *dis*continuous growth. In ecological research, this aspect deserves far more attention than it has received so far. With few exceptions (for example, the deep ocean ecosystem), microorganisms in most natural habitats will be subjected to changes in their physico-chemical environment. It seems clear that such changes will have a profound effect on the properties of the microbial communities prevailing in these habitats. There are only a few cases where chemostats have been used to study the effect of short-term environmental changes on simple mixed microbial cultures. One such is from the work of Leegwater[113] where competition between different bacterial species for limiting glucose under conditions of continuous or pulse-wise addition of glucose was described. In a mixed continuous culture of *Klebsiella aerogenes* and a *Bacillus* sp., the latter always outgrew *K. aerogenes* with continuous glucose addition, while with pulsed glucose addition *K. aerogenes* predominated. However, for unexplained reasons, the *Bacillus* sp. was never eliminated completely. Similarly, at dilution rates above 0.2 per hour with a continuous supply of glucose, *K. aerogenes* always outcompeted *Escherichia coli;* however, pulsing the glucose supply reversed this effect. Finally, *Bacillus subtilis* Marburg was shown to outcompete *E. coli* with a continuous supply of glucose (at D = 0.14 per hour) but was itself displaced during pulsed glucose addition. For a full appreciation of these results, experiments on the physiological responses of each individual species should also be con-sidered.[113] It is tempting to conclude from these competition experiments, however, that in general species like *K. aerogenes* and to a lesser extent *E. coli* are better adapted to respond effectively to sudden increases in substrate supply than are the *Bacillus* spp. tested here. However, much more work in the laboratory and in the field is needed before we can relate such properties to the ecological niches of these and comparable species.

A group of microorganisms for which discontinuous substrate supply is particularly ob-vious is the complex community found in dental plaque. Van der Hoeven et al.[190] reported preliminary results for mixed chemostat cultures of different oral *Streptococci* which indicated that pulsed glucose addition, in combination with slightly reduced pH-values, led to the selection of certain strains.

In natural water-bodies, changes in incident light intensities play a key role in the growth of phototrophic microorganisms (see References 87 and 191 to 194 and references therein). Moreover, as primary production is influenced by fluctuations in light-intensity, the entire microbial community in such ecosystems is affected. An understanding of such communities requires the study of phototrophic organisms under various light-regimes, in pure and in mixed cultures. Again, cultivation in continuous culture offers an excellent system for studying this type of discontinuous growth condition. This has indeed been done for many years with pure cultures, as already discussed; on the other hand, only a few studies with mixed cultures in chemostats have been described. A good illustration of the approach was due to Van Gemerden,[191] who showed that with intermittent light-dark illumination, *Chromatium weissii* and *Chromatium vinosum* coexisted in sulfide-limited chemostats, whereas with continuous light *Chr. vinosum* always outgrew the former species. This was explained in terms of the sulfide which accumulated in the dark periods, and was most rapidly oxidized by *Chr. weissii* when the light was turned on, while at the low residual sulfide concentrations found during the remainder of the light period, *Chr. vinosum* was the faster growing species.[191]

The effects of different light-dark regimes on the growth of phototrophic species were also studied in cultures of the cyanobacterium *Oscillatoria agardhii* and the green alga *Scenedesmus protuberans*.[195] It had earlier been shown that *O. agardhii* grew faster than *S. protuberans* under conditions of very low continuous light-irradiance in continuous culture.[87] Loogman showed that *O. aghardii* was also the better competitor when a light-dark regime was applied in which the light periods were short relative to the dark periods (for example 1.5 hr light to 6.5 hr dark). With longer light periods, *S. protuberans* grew faster than the cyanobacterium.

Part of our own work on the ecological niche of "mixotrophic" thiobacilli concerned their metabolic flexibility and reactivity.[131,196,197] It was found that in three-membered mixed cultures of the "mixotrophic" species *T. versutus*, the obligately autotrophic *T. neapolitanus* and a heterotrophic *Spirillum* alternate supply of acetate and thiosulfate resulted in complete washout of *T. versutus*. So, in spite of the fact that this latter species was the only one which could grow on both substrates, its physiologically more specialized competitors were apparently more reactive under such growth conditions. It was concluded[131] that a flexible metabolism as found in *T. versutus* is well adapted to growth with a continuous supply of mixed substrates but that it is incompatible with nutrient conditions in a dynamically changing environment. This conclusion is based on observations that *T. versutus*, given an alternating supply of substrates, rapidly loses the oxidative capacity for a substrate in the period when it is absent. In less flexible species, on the other hand, oxidative capacities remain unaffected for long periods of time even in the absence of one of its substrates.

**Enrichment of Pure and Mixed Cultures**

Laboratory studies in microbial ecology depend strongly on the procedures used to isolate the desired organisms. This is usually done using selective enrichment for the desired physiological type of bacteria or bacterial community. By far the most common technique is batch culture enrichment. In such cultures, while the initial conditions can be well-defined, the microbial activities of the developing population, the environmental conditions in the system, and therefore the type of selection pressure applied are constantly changing. The problem of changing conditions plus the relatively high substrate concentrations required in such cultures can lead to poorly reproducible isolation of organisms adapted to unrealistically high substrate concentrations. Thus, in batch cultures with saturating substrate concentrations, selection pressure works in favor of species exhibiting the highest $\mu_{max}$ values. Enrichment in substrate limited chemostats, on the other hand, features many properties which make it better suited to the selection of microbial species in ecologically oriented research. These and related problems have been discussed extensively by several authors.[10,12,119,121,198-201] We may summarize the major characteristics of chemostat enrichments as follows:

1.   Selection pressure works in favor of those species which grow best at often very low substrate concentrations, at a rate which is below their maximum specific growth rate. Since in natural environments, bacteria generally grow and compete for limiting nutrients at suboptimal growth rates, species isolated in the chemostat may be more representative of environmentally relevant populations.
2.   Enrichment takes place under conditions which can be controlled for long periods of time. Conditions can be constant or exhibit programed changes in, for example, light/dark cycles, pH, aerobic/anaerobic transitions, and variations of the rate, concentration, and nature of substrate supplied. This feature makes the chemostat a very precise selection tool since most growth conditions can be kept constant within narrow limits.
3.   As a consequence of low growth-limiting nutrient concentrations that can be maintained, selection under multiple substrate limitation can be incorporated, a situation likely to prevail in many natural habitats.
4.   Chemostat enrichment is probably the best method for selectively isolating microbial communities. Simple microbial food chains and more complex food webs, which in batch culture would appear largely as successional changes in populations, will in continuous flow systems all be able to grow at the same time.

A few examples which may serve to illustrate the important features of typical chemostat enrichments will now be discussed.

Jannasch[119] was probably the first to use a substrate-limited chemostat to isolate bacteria from natural habitats. Using different reservoir concentrations of the growth-limiting substrate (lactate) and different dilution rates from 0.05 to 0.5 per hour, different species became dominant and were subsequently isolated. A *Spirillum* sp. and a *Pseudomonas* sp., isolated at low and high dilution rate, respectively, were further investigated in mixed and pure cultures. The *Spirillum* sp. was most competitive below dilution rates of 0.55 per hour, while above this value the *Pseudomonas* was the more competitive species. Clearly their $\mu$-s relationships crossed. Similar findings have since been reported for many different species and substrates.[9,35,42,120-125] A detailed description of the isolation of two different species in phosphate-limited continuous cultures when run at relatively high and relatively low dilution rates is a good illustration of this principle[123] (see earlier section on defined mixed cultures). Matin and Veldkamp[42] also isolated a *Spirillum* sp. and a *Pseudomonas* sp. exhibiting crossing $\mu$-s relationships, with the *Spirillum* being the more competitive species at low-dilution rates. *Escherichia coli* was outcompeted when grown in chemostat enrichment cultures run at dilution rates below 0.35 per hour, though it dominated in chemostats run at dilution rates between 0.35 and 0.5 per hour.[202] An important point emerging from this type of enrichment is that apparently many species isolated from the environment are *specialized* in growth at very low nutrient concentrations. These bacterial types may play a crucial role in mineralization in low-nutrient habitats and will be overlooked entirely when only high substrate concentrations are used in isolation procedures.

Yet, another important group of bacteria is easily overlooked. Certain species appear to be specialized for the simultaneous utilization of more than one substrate, present at low concentration. An example of such a "generalist" is the "mixotrophic" *Thiobacillus versutus*. These bacteria never dominate in batch cultures or in single nutrient-limited chemostats, because under such conditions they are always outgrown by heterotrophic or by autotrophic "specialists". Stimulated by our increased understanding of mixed substrate utilization in pure continuous cultures and by the specific behavior of *T. versutus*,[77,128] chemostat enrichments were performed with various mixtures of thiosulfate + acetate as growth limiting substrates. In this way, successful enrichments of versatile thiobacilli were obtained from a number of freshwater samples.[80]

Mixed substrate-limited chemostat enrichment does not always lead to the dominance of a single species. For example, Chian and Mateless[127] repeatedly found stable mixed cultures of two species in enrichments with glucose + butyrate as limiting substrates. Similar results were obtained by Laanbroek[203] with glutamate + aspartate as limiting substrates. Perhaps the fundamental difference here can be attributed to the fact that the *Thiobacillus* enrichments used an "autotrophic" substrate (thiosulfate) and an organic carbon source (acetate). In such a mixture the organic carbon source can have a considerable energy saving effect, since it makes $CO_2$-fixation redundant.[73,75-77] No such effect is expected with two functionally similar substrates unless they are metabolized via pathways with markedly different energetics. In this respect, it would be particularly interesting to set up chemostat enrichments with mixtures of $H_2$ or $NH_4^+$ + organic compound, or for example $H_2S$ + light + organic compound.

Earlier we discussed the use of chemostats in maintaining and studying complex mixed microbial populations as they are found in natural habitats. Chemostat enrichment can also be used successfully when attempting to construct communities which are not necessarily found as such in nature. The microbial degradation of xenobiotics, e.g., certain pesticides and many different toxic wastes, has been investigated in this way. For example, parathion, an organophosphorus insecticide, was shown to be partly degraded by a stable aerobic community of three species which developed over a two year period in a parathion-limited continuous culture.[204] One species, *Pseudomonas stutzeri*, was shown to hydrolyze parathion cometabolically while growing on degradation products of *p*-nitrophenol (*p*-NP), which was itself degraded by *Pseudomonas aeruginosa*. The latter could not hydrolyze parathion but grew well on the hydrolysis product (*p*-NP). The third species, a coryneform, was not further characterized; neither was its role in the degradation process elucidated.

A more complicated, seven-membered, stable community was obtained when enriching in a chemostat for bacteria able to degrade the herbicide dalapon.[205,206] Over the last few years, more examples of community-based breakdown of xenobiotic compounds have been described.[151] Although the type of communities obtained in this way may not always play an important part in natural ecosystems, the way they are organized might nevertheless help us in finding out how other ecologically interesting communities are constructed.

## CONCLUDING REMARKS

The enormous complexity of most natural environments in which bacteria thrive has prompted microbial ecologists to construct experimental model systems for use in studying the behavior of microorganisms under well-controlled conditions. Continuous culture, one example of such a model system, has become extremely popular, among both microbial ecologists and physiologists, in particular, as a means of studying microbial growth under nutrient limiting conditions, a situation likely to prevail in most natural habitats.

The danger with laboratory model systems is undoubtedly that they may be identified with the natural system itself. What we have tried to do in the preceding pages is to present an overview of several different approaches to the use of continuous culture as a tool in studying some of the basic phenomena in microbial ecology. It is to this type of simplification of the "real world" that techniques of continuous culture can be most fruitfully applied. We have also tried to emphasize that the chemostat not only offers a useful means for studying pure cultures but can allow the analysis of interspecies interactions in simple, defined mixed cultures as well. Furthermore, we are convinced that examples in which chemostats have been operated in a controlled *dis*continuous mode illustrate that the chemostat provides a very powerful method for studying the effects of environmental change on the growth of microorganisms.

In recent years several refinements to the basic concept of continuous culture have been developed aiming at a better simulation of inhomogeneities in space and time, effects which so clearly complicate the study of microbes in nature (see elsewhere in this volume). We must never forget, however, that not even the most sophisticated laboratory model system will ever make redundant experimentation *in situ*. It is the continued endeavor to verify whether eco-physiological principles, found in laboratory model systems, really apply to field-situations, which will ultimately let us understand microbial life in nature.

## NOTE ADDED IN PROOF

Literature survey for this manuscript was completed in May 1985.

## ACKNOWLEDGMENTS

We are indebted to M. Broens-Erenstein and M. Pras for invaluable help during the preparation of the manuscript.

## REFERENCES

1. **Wimpenny, J. W. T.,** Spatial order in microbial ecosystems, *Biol. Rev.,* 56, 295, 1981.
2. **Wimpenny, J. W. T., Lovitt, R. W., and Coombs, J. P.,** Laboratory model systems for the investigation of spatially and temporally organized microbial ecosystems, in *Microbes in Their Natural Environments,* 34th Symp. Soc. Gen. Microbiol., Slater, J. H., Whittenbury, R., and Wimpenny, J. W. T., Eds., Cambridge University Press, Cambridge, 1983, 67.
3. **van Es, F. B., Laanbroek, H. J., and Veldkamp, H.,** Microbial ecology: an overview, in *Aspects of Microbial Metabolism and Ecology,* Codd, G. A., Ed., Academic Press, London, 1984, 1.
4. **Monod, J.,** La technique de culture continu; théorie et applications, *Ann. Inst. Pasteur,* Paris, 79, 390, 1950.
5. **Novick, A. and Szilard, L.,** Description of the chemostat, *Science,* 112, 715, 1950.
6. **Novick, A. and Szilard, L.,** Experiments with the chemostat on spontaneous mutation of bacteria, *Proc. Natl. Acad. Sci. U.S.A.,* 36, 708, 1950.
7. **Tempest, D. W.,** The place of continuous culture in microbiological research, *Adv. Microb. Physiol.,* 4, 223, 1970.
8. **Veldkamp, H.,** Continuous culture in microbial physiology and ecology, in *Patterns of Progress,* Cook, J. G., Ed., Meadowfield Press, Shildon, Durham, England, 1976.
9. **Veldkamp, H.,** Ecological studies with the chemostat, *Adv. Microb. Ecol.,* 1, 59, 1977.
10. **Harder, W., Kuenen, J. G., and Matin, A.,** Microbial selection in continuous culture, *J. Appl. Bacteriol.,* 43, 1, 1977.
11. **Bull, A. T. and Brown, C. M.,** Continuous culture applications to microbial biochemistry, in *International Review of Biochemistry, Microbial Biochemistry,* Vol. 21, Quayle, J. R., Ed., University Park Press, Baltimore, 1979, 177.
12. **Kuenen, J. G. and Harder, W.,** Microbial competition in continuous culture, in *Experimental Microbial Ecology,* Burns, R. G. and Slater, J. H., Eds., Blackwell Scientific Publiciations, Oxford, 1982, 342.
13. **Dykhuizen, D. E. and Hartl, D. L.,** Selection in chemostats, *Microbiol. Rev.,* 47, 150, 1983.
14. **Herbert, D., Elsworth, R., and Telling, R. C.,** The continuous culture of bacteria: a theoretical and experimental study, *J. Gen. Microbiol.,* 14, 601, 1956.
15. **Tempest, D. W.,** The continuous cultivation of microorganisms. I. Theory of the chemostat, in *Methods in Microbiology,* Vol. 2, Norris, J. R. and Ribbons, D. W., Eds., Academic Press, New York, 1970, 259.
16. **Pirt, S. J.,** *Principles of Microbe and Cell Cultivation,* Blackwell Scientific Publications, Oxford, 1975.
17. **Bazin, M. J.,** Theory of continuous culture, in *Continuous Cultures of Cells,* Vol. 1, Calcott, P. H., Ed., CRC Press, Boca Raton, Fla., 1981, 27.

18. **Evans, C. G. T., Herbert, D., and Tempest, D. W.,** The continuous cultivation of microorganisms. II. The construction of a chemostat, in *Methods in Microbiology,* Vol. 2, Norris, J. R. and Ribbons, D. W., Eds., Academic Press, New York, 1970, 277.

19. **Calcott, P. H.,** The construction and operation of continuous cultures, in *Continuous Cultures of Cells,* Vol. 1, Calcott, P. H., Ed., CRC Press, Boca Raton, Fla., 1981, 13.

20. **Drew, S. W.,** Liquid culture, in *Manual of Methods for General Bacteriology,* Costilow, R. N., Ed., American Society for Microbiology, Washington, D.C., 1981, 151.

21. **Monod, J.,** *Recherches sur la croissance des cultures bactériennes,* Hermann, Paris, 1942.

22. **Stouthamer, A. H.,** Energetic aspects of the growth of microorganisms, in *Microbial Energetics,* 27th Symp. Soc. Gen. Microbiol., Haddock, R. A. and Hamilton, W. A., Eds., Cambridge University Press, Cambridge, 1977, 285.

23. **Stouthamer, A. H.,** The search for correlation between theoretical and experimental growth yields, in *International Review of Biochemistry, Microbial Biochemistry,* Vol. 21, Quayle, J. R., Ed., University Park Press, Baltimore, 1979, 1.

24. **Hempfling, W. P. and Rice, C. W.,** Microbial bioenergetics during continuous culture, in *Continuous Culture of Cells,* Vol. 2, Calcott, P. H., Ed., CRC Press, Boca Raton, Fla., 1981, 99.

25. **Jannasch, H. W.,** Estimation of bacterial growth rates in natural waters, *J. Bacteriol.,* 99, 156, 1969.

26. **Shilo, M.,** Strategies of adaptation to extreme conditions in aquatic microorganisms, *Naturwissenschaften,* 68, 384, 1980.

27. **Konings, W. N. and Veldkamp, H.,** Energy transduction and solute transport mechanisms in relation to environments occupied by microorganisms, in *Microbes in Their Natural Environments,* 34th Symp. Soc. Gen. Microbiol., Slater, J. H., Whittenbury, R., and Wimpenny, J. W. T., Eds., Cambridge University Press, Cambridge, 1983, 153.

28. **Tempest, D. W. and Neijssel, O. M.,** Eco-physiological aspects of microbial growth in aerobic nutrient-limited environments, in *Adv. Microb. Ecol.,* 2, 105, 1978.

29. **Konings, W. N. and Veldkamp, H.,** Phenotypic responses to environmental change, in *Contemporary Microbial Ecology,* Ellwood, D. C., Hedger, J. N., Latham, M. J., Lynch, J. M., and Slater, J. H., Eds., Academic Press, London, 1980, 161.

30. **Tempest, D. W. and Neijssel, O. M.,** Metabolic compromises involved in the growth of microorganisms in nutrient-limited (chemostat) environments, in *Basic Life Sciences,* Vol. 18, Hollaender, A., Ed., Plenum Press, New York, 1981, 335.

31. **Harder, W. and Dijkhuizen, L.,** Physiological responses to nutrient limitation, *Ann. Rev. Microbiol.,* 37, 1, 1983.

32. **Tempest, D. W., Neijssel, O. M., and Zevenboom, W.,** Properties and performance of microorganisms in laboratory culture: their relevance to growth in natural ecosystems, in *Microbes in Their Natural Environments,* 34th Symp. Soc. Gen. Microbiol., Slater, J. H., Whittenbury, R., and Wimpenny, J. W. T., Eds., Cambridge University Press, Cambridge, 1983, 119.

33. **Harder, W., Dijkhuizen, L., and Veldkamp, H.,** Environmental regulation of microbial metabolism, in *The Microbe 1984: Part II Prokaryotes and Eukaryotes,* 36th Symp. Soc. Gen. Microbiol., Kelly, D. P. and Carr, N. G., Eds., Cambridge University Press, Cambridge, 1984, 51.

34. **Pardee, A. B.,** Response of enzyme synthesis and activity to environment, in *Microbial Reaction to Environment,* 11th Symp. Soc. Gen. Microbiol., Meynell, G. G. and Gooder, H., Eds., Cambridge University Press, Cambridge, 1961, 19.

35. **Veldkamp, H. and Jannasch, H. W.,** Mixed culture studies with the chemostat, *J. Appl. Chem. Biotechnol.,* 22, 105, 1972.

36. **Dean, A. C. R.,** Influence of environment on the control of enzyme synthesis, *J. Appl. Chem. Biotechnol.,* 22, 245, 1972.

37. **Matin, A.,** Microbial regulatory mechanisms at low nutrient concentrations as studied in chemostat, in *Strategies of Microbial Life in Extreme Environments,* Dahlem Conf. Life Sci. Res. Rep. 13, Shilo, M., Ed., Verlag Chemie, New York, 1979, 323.

38. **Matin, A.,** Regulation of enzyme synthesis as studied in continuous culture, in *Continuous Culture of Cells,* Vol. 2, Calcott, P. H., Ed., CRC Press, Boca Raton, Fla., 1981, 69.

39. **Harder, W. and Dijkhuizen, L.,** Strategies of mixed substrate utilization in microorganisms, *Philos. Trans. R. Soc. London Ser. B,* 297, 459, 1982.

40. **Herbert, D.,** The chemical composition of microorganisms as a function of their environment, in *Microbial Reaction to Environment,* 11th Symp. Soc. Gen. Microbiol., Meynell, G. G. and Gooder, H., Eds., Cambridge University Press, Cambridge, 1961, 391.

41. **Tempest, D. W. and Wouters, J. T. M.,** Properties and performance of microorganisms in chemostat culture, *Enzyme Microb. Technol.,* 3, 283, 1981.

42. **Matin, A. and Veldkamp, H.,** Physiological basis of the selective advantage of a *Spirillum* sp. in a carbon-limited environment, *J. Gen. Microbiol.,* 105, 187, 1978.

43. **Poindexter, J. S.**, Oligotrophy, *Adv. Microb. Ecol.*, 5, 63, 1981.

44. **Luscombe, B. M. and Gray, T. R. G.**, Effect of varying growth rate on the morphology of *Arthrobacter*, *J. Gen. Microbiol.*, 69, 433, 1971.

45. **Luscombe, B. M. and Gray, T. R. G.**, Characteristics of *Arthrobacter* grown in continuous culture, *J. Gen. Microbiol.*, 82, 213, 1974.

46. **Clark, J. B.**, Sphere-rod transitions in *Arthrobacter*, in *Developmental Biology of Prokaryotes*, Parish, J. H., Ed., Blackwell Scientific Publications, Oxford, 1979, 73.

47. **Dawes, E. A., Midgley, M., and Whiting, P. H.**, Control of transport systems for glucose, gluconate and 2-oxo-gluconate and of glucose metabolism in *Pseudomonas aeruginosa*, in *Continuous Culture*, Vol. 6, *Applications and New Fields*, Dean, A. C. R., Ellwood, D. C., Evans, C. G. T., and Melling, J., Eds., Ellis Horwood, Chichester, U.K., 1976, 195.

48. **Dawes, E. A.**, Carbon metabolism, in *Continuous Culture of Cells*, Vol. 2, Calcott, P. H., Ed., CRC Press, Boca Raton, Fla., 1981, 1.

49. **Neijssel, O. M., Hueting, S., Crabbendam, K. J., and Tempest, D. W.**, Dual pathways of glycerol assimilation in *Klebsiella aerogenes* NCIB 418. Their role and possible functional significance, *Arch. Microbiol.*, 104, 83, 1975.

50. **Brown, C. M., MacDonald-Brown, D. S., and Meers, J. L.**, Physiological aspects of microbial inorganic nitrogen metabolism, *Adv. Microb. Physiol.*, 11, 1, 1974.

51. **Brown, C. M.**, Nitrogen metabolism in bacteria and fungi, in *Continuous Culture*, Vol. 6, *Applications and New Fields*, Dean, A. C. R., Ellwood, D. C., Evans, C. G. T., and Melling, J., Eds., Ellis Horwood, Chichester, U.K., 1976, 170.

52. **Dawes, E. A. and Senior, P. J.**, The role and regulation of energy reserve polymers in micro-organisms, *Adv. Microb. Physiol.*, 10, 135, 1973.

53. **Sutherland, I. W.**, Biosynthesis of microbial exopolysaccharides, *Adv. Microb. Physiol.*, 23, 79, 1982.

54. **Neijssel, O. M. and Tempest, D. W.**, The physiology of metabolite overproduction, in *Microbiol Technology*, 29th Symp. Soc. Gen. Microb., Bull, A. T., Ellwood, D. C., and Ratledge, C., Eds., Cambridge University Press, Cambridge, 1979, 53.

55. **Ellwood, D. C. and Tempest, D. W.**, Effects of environment on bacterial wall content and composition, *Adv. Microb. Physiol.*, 7, 83, 1972.

56. **Ellwood, D. C. and Robinson, A.**, Bacterial envelope structure and macromolecular composition, in *Continuous Culture of Cells*, Vol. 2, Calcott, P. H., Ed., CRC Press, Boca Raton, Fla, 1981, 39.

57. **Minnikin, D. E. and Abdolrahimzadeh, H.**, The replacement of phosphatidylethanolamine and acidic phospholipids by an ornithine-amide lipid and a minor phosphorus-free lipid in *Pseudomonas fluorescens* NCMB 129, *FEBS Lett.* 43, 257, 1974.

58. **Harder, W. and Dijkhuizen, L.**, Mixed substrate utilization in microorganisms, in *Continuous Culture*, Vol. 6, *Application and New Fields*, Dean, A. C. R., Ellwood, D. C., Evans, C. T. G., and Melling, J., Eds., Ellis Harwood, Chichester, U.K., 1976, 297.

59. **Magasanik, B.**, Catabolite repression, *Cold Spring Harb. Symp. Quant. Biol.*, 26, 249, 1961.

60. **Magasanik, B.**, Genetic control of nitrogen assimilation in bacteria, *Ann. Rev. Genet.*, 16, 135, 1982.

61. **Bader, F. G., Meyer, J. S., Frederickson, A. G., and Tsuchiya, H. M.**, Comments on microbial growth rate, *Biotechnol. Bioeng.*, 17, 279, 1975.

62. **Bader, F. B.**, Kinetics of double-substrate limited growth, in *Microbial Population Dynamics*, Bazin, M. J., Ed., CRC Press, Boca Raton, Fla., 1982, 1.

63. **Egli, T., Lindley, N. D., and Quayle, J. R.**, Regulation of enzyme synthesis and variations of residual methanol concentration during carbon-limited growth of *Kloeckera* sp. 2201 on mixtures of methanol and glucose, *J. Gen. Microbiol.*, 129, 1269, 1983.

64. **Gottschal, J. C.**, Mixed substrate utilization by mixed cultures, in *Bacteria in Nature*, Vol. 2, Leadbetter, E. R. and Poindexter, J. S., Eds., Plenum Press, New York, 1986, 261.

65. **Clark, B. and Holms, W. H.**, Control of sequential utilization of glucose and fructose by *Escherichia coli*, *J. Gen. Microbiol.*, 95, 191, 1976.

66. **Mateles, R. I., Chian, S. K., and Silver, R.**, Continuous culture on mixed substrates, in *Microbial Physiology and Continuous Culture*, Powell, E. O., Evans, C. G. T., Strange, R. E., and Tempest, D. W., Eds., Her Majesty's Stationery Office, London, 1967, 232.

67. **Silver, R. S. and Mateles, R. I.**, Control of mixed substrate utilization in continuous cultures of *Escherichia coli*, *J. Bacteriol.*, 97, 535, 1969.

68. **Smith, M. E. and Bull, A. T.**, Studies of the utilization of coconut water waste for the production of the food yeast *Saccharomyces fragilis*, *J. Appl. Bacteriol.*, 41, 81, 1976.

69. **Dijkhuizen, L., van der Werf, B., and Harder, W.**, Metabolic regulation in *Pseudomonas oxalaticus* OX1. Diauxic growth on mixtures of oxalate and formate or acetate, *Arch. Microbiol.*, 124, 261, 1980.

70. **Gottschal, J. C. and Kuenen, J. G.**, Physiological and ecological significance of facultative chemolithotrophy and mixotrophy in chemolithotrophic bacteria, in *Microbial Growth on $C_1$ Compounds*, Dalton, H., Ed., Heyden, London, 1981, 92.

71. **Egli T. and Harder, W.**, Growth of methylotrophs on mixed substrates, in *Microbial Growth on C₁ Compounds*, Crawford, R. L. and Hanson, R. S., Eds., American Society for Microbiology, Washington, D.C., 1984, 330.

72. **Dijkhuizen, L. and Harder, W.**, Current views on the regulation of autotrophic carbon dioxide fixation via the Calvin cycle in bacteria, *Antonie van Leeuwenhoek*, 50, 473, 1984.

73. **Dijkhuizen, L. and Harder, W.**, Regulation of autotrophic and heterotrophic metabolism in *Pseudomonas oxalaticus* OX1. Growth on mixtures of oxalate and formate in continuous culture, *Arch. Microbiol.*, 123, 55, 1979.

74. **Dijkhuizen, L., Knight, M., and Harder, W.**, Metabolic regulation in *Pseudomonas oxalaticus* OX1. Autotrophic and heterotrophic growth on mixed substrates, *Arch. Microbiol.*, 116, 77, 1978.

75. **Dijkhuizen, L. and Harder, W.**, Regulation of autotrophic and heterotrophic metabolism in *Pseudomonas oxalaticus*. OX1. Growth on mixtures of acetate and formate in continuous culture, *Arch. Microbiol.*, 123, 47, 1979.

76. **Van Verseveld, H. W., Boon, J. P., and Stouthamer, A. H.**, Growth yields and the efficiency of oxidative phosphorylation of *Paracoccus denitrificans* during two-(carbon) substrate-limited growth, *Arch. Microbiol.*, 121, 213, 1979.

77. **Gottschal, J. C. and Kuenen, J. G.**, Mixotrophic growth of *Thiobacillus* A2 on acetate and thiosulphate as growth limiting substrates in the chemostat, *Arch. Microbiol.*, 126, 33, 1980.

78. **Harrison, A. P.**, Genomic and physiological comparisons between hetrerotrophic thiobacilli and *Acidiphilium cryptum*, *Thiobacillus versutus* sp.nov., and *Thiobacillus acidophilus* nom.rev., *Int. J. Syst. Bacteriol.*, 33, 21, 1983.

79. **Smith, A. L. and Kelly, D. P.**, Competition in the chemostat between an obligately and a facultatively chemostat between an obligately and a facultatively chemolithotrophic *Thiobacillus*, *J. Gen. Microbiol.*, 115, 377, 1979.

80. **Gottschal, J. D. and Kuenen, J. G.**, Selective enrichment of facultatively chemolithotrophic thiobacilli and related organisms in continuous culture, *FEMS Microbiol. Lett.*, 7, 241, 1980.

81. **Sinclair, C. G. and Ryder, D. N.**, Models for the continuous culture of microorganisms under both oxygen and carbon limiting conditions, *Biotechnol. Bioeng.*, 17, 375, 1975.

82. **Lee, A. L., Ataai, M. M., and Shuler, M. L.**, Double-substrate-limited growth of *Escherichia coli*, *Biotechnol. Bioeng.*, 26, 1398, 1984.

83. **Cooney, C. L., Koplov, H. M., and Häggström, M.**, Transient phenomena in continuous culture, in *Continuous Cultures of Cells*, Vol. 1, Calcott, P. H., Ed., CRC Press, Boca Raton, Fla., 1981, 143.

84. **Harrison, D. E. F. and Topiwala, H. H.**, Transient and oscillatory states of continuous culture, *Adv. Biochem. Eng.*, 3, 167, 1974.

85. **Harrison, D. E. F.**, The regulation of respiration rate in growing bacteria, *Adv. Microb. Physiol.*, 14, 243, 1976.

86. **Morris, J. G.**, Changes in oxygen tension and the microbial metabolism of organic carbon, in *Aspects of Microbial Metabolism and Ecology*, Codd, G. A., Ed., Academic Press, London, 1984, 59.

87. **Liere, L. van**, On *Oscillatoria agardhii gomont*, experimental ecology and physiology of a nuisance bloom-forming cyanobacterium, Ph.D. thesis, University of Amsterdam, The Netherlands, 1979.

88. **Gibson, C. E. and Jewson, D. H.**, The utilization of light by microorganisms, in *Aspects of Microbial Metabolism and Ecology*, Codd, G. A., Ed., Academic Press, London, 1984, 97.

89. **Topiwala, H. H. and Sinclair, C. G.**, Temperature relationship in continuous culture, *Biotechnol. Bioeng.*, 13, 795, 1971.

90. **Gaudy, A. F.**, The transient response to pH and temperature shock loading of fermentation systems, *Biotechnol. Bioeng.*, 17, 1051, 1975.

91. **Hanson, R. S.**, The physiology and diversity of bacterial endospores, in *Developmental Biology of Prokaryotes*, Parish, J. H., Ed., Blackwell Scientific, Oxford, 1979, 37.

92. **Dow, C. S., Whittenbury, R., and Carr, N. G.**, The 'shut down' or 'growth precursor' cell. An adaptation for survival in a potentially hostile environment, in *Microbes in Their Natural Environments*, 34th Symp. Soc. Gen. Microbiol., Slater, J. H., Whittenbury, R., and Wimpenny, J. W. T., Eds., Cambridge University Press, Cambridge, 1983, 187.

93. **Dawes, E. A. and Mandelstam, J.**, Sporulation of *Bacillus subtilis* in continuous culture, *J. Bacteriol.*, 103, 529, 1970.

94. **Strange, R. E.**, Microbial response to mild stress, in *Patterns of Progress*, Cook, J. G., Ed., Meadowfield Press, Shildon Co., Durham, 1976.

95. **Meiberg, J. B. M.**, Metabolism of methylated amines in *Hyphomicrobium* spp., Ph.D. thesis, University of Groningen, The Netherlands, 1979.

96. **Matin, A., Veldhuis, C., Stegeman, V., and Veenhuis, M.**, Selective advantage of a *Spirillum* sp. in a carbon-limited environment. Accumulation of poly-β-hydroxybutyric acid and its role in starvation, *J. Gen. Microbiol.*, 112, 349, 1979.

97. **Lawrence, A.**, Microbial diversity, a consequence of the aquatic environment, Ph.D. thesis, University of Warwick, U.K., 1978.

98. **Konings, W. N. and Michels, P. A. M.**, Electron-transfer-driven solute translocation across bacterial membranes, in *Diversity of Bacterial Respiratory Systems*, Vol. 1, Knowles, C. J., Ed., CRC Press, Boca Raton, Fla., 1980, 33.

99. **Otto, R., ten Brink, B., Veldkamp, H., and Konings, W. N.**, The relation between growth rate and electrochemical proton gradient of *Streptococcus cremoris, FEMS Microbiol. Lett.*, 16, 69, 1983.

100. **Otto, R., Klont, B., ten Brink, B., and Konings, W. N.**, The phosphate potential, adenylate energy charge and proton motive force in growing cells of *Streptococcus cremoris, Arch. Microbiol.*, 139, 338, 1984.

100a. **Otto, R.**, Unpublished; as cited in **Konings, W. N. and Veld Kamp, H.**, *Microbes in Their Natural Environments*, 34th Symp. Soc. Gen. Microbiol., Slater, J. H., Whittenbury, R. and Wimpenny, J. W. T., Eds., Cambridge University Press, Cambridge, 1983, 153.

101. **Koch, A. L.**, Microbial growth in low concentrations of nutrients, in *Strategies of Microbial Life in Extreme Environments*, Dahlem Conf. Life Sci. Res. Rep. 13, Shilo, M., Ed., Verlag Chemie, New York, 1979, 261.

102. **Singh, U. N.**, Adaptation in micro-organisms: variation in macromolecular composition with growth rate, *J. Theor. Biol.*, 59, 107, 1976.

103. **Singh, U. N.**, Adaptation in micro-organisms II: Transient behavior following shift-up and shift-down, *J. Theor. Biol.*, 72, 459, 1978.

104. **Barford, J. P., Panment, N. B., and Hall, R. J.**, Lag phases and transients, in *Microbial Population Dynamics*, Bazin, M. J., Ed., CRC Press, Boca Raton, Fla., 1982, 55.

105. **Pickett, A. M.**, Growth in a changing environment, in *Microbial Population Dynamics*, Bazin, M. J., Ed., CRC Press, Boca Raton, Fla., 1982, 91.

106. **Daigger, G. T. and Grady, C. P. L., Jr.**, An assessment of the role of physiological adaptation in the transient response of bacterial cultures, *Biotechnol. Bioeng.*, 24, 1427, 1982.

107. **Maaløe, O. and Kjeldgaard, N. O.**, *Control of Macromolecular Synthesis*, W. A. Benjamin, New York, 1966.

108. **Koch, A. L.**, How bacteria face depression, recession and derepression, *Perspect. Biol. Med.*, 20, 44, 1976.

109. **Koch, A. L.**, The inefficiency of ribosomes functioning in *Escherichia coli* growing at moderate rates, *J. Gen. Microbiol.*, 116, 165, 1980.

110. **Maaløe, O.**, Regulation of protein synthesizing machinery, ribosomes, IRNA, factors, etc., in *Biology Control and Development*, Goldberger R., Ed., Plenum Press, New York, 1979, 520.

111. **Nierlich, D. P.**, Regulation of bacterial growth, RNA and protein synthesis, *Ann. Rev. Microbiol.*, 32, 393, 1978.

112. **Harvey, R. J.**, Metabolic regulation in glucose-limited chemostat cultures of *Escherichia coli, J. Bacteriol.*, 104, 698, 1970.

113. **Leegwater, M. P. M.**, Microbial reactivity: its relevance to growth in natural and artificial environments, Ph.D. thesis, University of Amsterdam, The Netherlands, 1983.

114. **Höfle, M. G.**, Transient response of glucose-limited cultures of *Cytophaga johnsonae* to nutrient excess and starvation, *Appl. Environ. Microbiol.*, 47, 356, 1984.

115. **Höfle, M. G.**, Glucose uptake of *Cytophaga johnsonae* studied in batch and continuous culture, *Arch. Microbiol.*, 133, 289, 1982.

116. **O'Brien, R. W., Neijssel, O. M., and Tempest, D. W.**, Glucose phosphoenolpyruvate phosphotransferase activity and glucose uptake rate of *Klebsiella aerogenes* growing in chemostat culture, *J. Gen. Microbiol.*, 116, 305, 1980.

117. **Gause, G. F.**, *The Struggle for Existence*, Williams and Wilkins, Baltimore, 1934.

118. **Powell, E. O.**, Criteria for the growth of contaminants and mutants in continuous culture, *J. Gen. Microbiol.* 18, 259, 1958.

119. **Jannasch, H. W.**, Enrichment of aquatic bacteria in continuous culture, *Arch. Mikrobiol.*, 59, 165, 1967.

120. **Jannasch, H. W. and Mateles, R. I.**, Experimental bacterial ecology studied in continuous culture, *Adv. Microb. Physiol.* 11, 165, 1974.

121. **Veldkamp, H.**, Mixed culture studies with the chemostat, in *Continuous Culture*, Vol. 6, *Applications and New Fields*, Dean, A. C. R., Ellwood, D. C., Evans, C. G. T., and Melling, J., Eds., Ellis Horwood, Chichester, U.K., 1976, 315.

122. **Frederickson, A. G.**, Behaviour of mixed cultures of microorganisms, *Ann. Rev. Microbiol.*, 31, 63, 1977.

123. **Kuenen, J. G., Boonstra, J., Schröder, H. G. J., and Veldkamp, H.**, Competition for inorganic substrates among chemoorganotrophic and chemolithotrophic bacteria, *Microb. Ecol.*, 3, 119, 1977.

124. **Laanbroek, H. J., Geerligs, H. J., Peynenburg, A. A. C. M., and Siesling, J.,** Competition for L-lactate between *Desulfovibrio, Veillonella,* and *Acetobacterium* species isolated from anaerobic intertidal sediments, *Microb. Ecol.,* 9, 341, 1983.

125. **Veldkamp, H., Van Gemerden, H., Harder, W., and Laanbroek, H. J.,** Microbial competition, in *Current Perspectives in Microbial Ecology,* Klug, M. J. and Reddy, C. A., Eds., American Society for Microbiology, Washington, D.C., 1984, 279.

126. **Winogradsky, S.,** *Microbiologie du sol,* Oeuvres complètes, Masson, Paris, 1949.

127. **Chian, S. K. and Mateles, R. I.,** Growth of mixed cultures on mixed substrates. I. Continuous Culture, *Appl. Microbiol.,* 16, 1337, 1968.

128. **Gottschal, J. C., De Vries, S., and Kuenen, J. G.,** Competition between the facultatively chemolithotrophic *Thiobacillus* A2, an obligately chemolithotrophic *Thiobacillus* and a heterotrophic *Spirillum* for inorganic and organic substrates, *Arch. Microbiol.,* 121, 241, 1979.

129. **Laanbroek, H. J., Smit, A. J., Klein Nulend, G., and Veldkamp, H.,** Competition for L-glutamate between specialized and versatile Clostridium species, *Arch. Microbiol.,* 120, 61, 1979.

130. **Dykhuizen, D. and Davies, M.,** An experimental model: bacterial specialists and generalists competing in chemostats, *Ecology,* 61, 1213, 1980.

131. **Beudeker, R. F., Gottschal, J. C., and Kuenen, J. G.,** Reactivity versus flexibility in thiobacilli, *Antonie van Leeuwenhoek,* 48, 39, 1982.

132. **Taylor, P. A. and Williams, P. J. LeB.,** Theoretical studies on the co-existence of competing species under continuous flow conditions, *Can. J. Microbiol.,* 21, 90, 1975.

133. **Bazin, M. J., Saunders, P. T., and Prosser, J. I.,** Models of microbial interactions in the soil, *Crit. Rev. Microbiol.,* 4, 463, 1976.

134. **Tilman, D.,** Resource competition between planktonic algae: an experimental and theoretical approach, *Ecology,* 58, 338, 1977.

135. **Yoon, H., Klinzing, G., and Blanch, H. W.,** Competition for mixed substrates by microbial populations, *Biotechnol. Bioeng.,* 19, 1193, 1977.

136. **Gottschal, J. C. and Thingstad, T. F.,** Mathematical description of competition between two and three bacterial species under dual substrate limitation in the chemostat: a comparison with experimental data, *Biotechnol. Bioeng.,* 24, 1403, 1982.

137. **Zobell, C. E.,** The effect of solid surfaces upon bacterial activity, *J. Bacteriol.,* 46, 37, 1943.

138. **Jannasch, H.,** Studies on planktonic bacteria by means of a direct membrane filter method, *J. Gen. Microbiol.,* 18, 609, 1958.

139. **Stotzky, G.,** Influence of clay minerals on microorganisms. II. Effect of various clay species, homoionic clays, and other particles on bacteria, *Can. J. Microbiol.,* 12, 831, 1966.

140. **Kjelleberg, S.,** Effects of interfaces on survival mechanisms of copiotrophic bacteria in low-nutrient habitats, in *Current Perspectives in Microbial Ecology,* Klug, M. J. and Reddy, C. A., Eds., American Society for Microbiology, Washington, D.C., 1984, 151.

141. **Button, D. K.,** Effect of clay on the availability of dilute organic nutrients to steady-state heterotrophic populations, *Limnol. Oceanogr.,* 14, 95, 1969.

142. **Maigetter, R. Z. and Pfister, R. M.,** A mixed bacterial population in a continuous culture with and without koalinite, *Can. J. Microbiol.,* 21, 173, 1975.

143. **Laanbroek, H. J., Geerligs, H. J., Sijtsma, L., and Veldkamp, H.,** Competition for sulfate and ethanol among *Desulfobacter, Desulfobulbus* and *Desulfovibrio* species isolated from intertidal sediments, *Appl. Environ. Microbiol.,* 47, 329, 1984.

144. **Laanbroek, H. J. and Pfennig, N.,** Influence of clay (illite) on substrate utilization by sulfate-reducing bacteria, *Arch. Microbiol.,* 134, 161, 1983.

145. **Harder, W. and Veldkamp, H.,** Competition of marine psychrophilic bacteria at low temperatures, *Antonie van Leeuwenhoek,* 37, 51, 1971.

146. **Bungay, H. R., III and Bungay, M. L.,** Microbial interactions in continuous culture, *Adv. Appl. Microbiol.,* 10, 269, 1968.

147. **Meers, J. L.,** Growth of bacteria in mixed cultures, *Crit. Rev. Microbiol.,* 2, 139, 1973.

148. **Harrison, D. E. F.,** Mixed cultures in industrial fermentation processes, *Adv. Appl. Microbiol.,* 24, 129, 1978.

149. **Slater, J. H. and Bull, A. T.,** Interactions between microbial populations, in *Companion to Microbiology,* Bull, A. T. and Meadow, P. M., Eds., Longman, London, 1978, 181.

150. **Miura, Y., Tanaka, H., and Okazaki, M.,** Stability analysis of commensal and mutual relations with competitive assimilation in continuous mixed culture, *Biotechnol. Bioeng.,* 22, 929, 1980.

151. **Slater, J. H.,** Mixed cultures and microbial communities, in *Mixed Culture Fermentations,* Bushell, M. E. and Slater, J. H., Eds., Academic Press, London, 1981, 1.

152. **Bull, A. T., and Slater, J. H.,** Microbial interactions and community structure, in *Microbial Interactions and Communities,* Vol. 1, Bull, A. T. and Slater, J. H., Eds., Academic Press, London, 1982, 13.

153. **Linton, J. D. and Drozd, J. W.**, Microbial interactions and communities in biotechnology, in *Microbial Interactions and Communities*, Vol. 1, Bull, A. T. and Slater, J. H., Eds., Academic Press, London, 1982, 357.

154. **Megee, R. D. III, Drake, J. F., Fredrickson, A. G., and Tsuchiya, H. M.**, Studies in intermicrobial symbiosis. *Saccharomyces cerevisiae* and *Lactobacillus casei, Can. J. Microbiol.*, 18, 1733, 1972.

155. **Mayfield, C. I. and Inniss, W. E.**, Interactions between freshwater bacteria and *Ankistrodesmus braunii* in batch and continuous culture, *Microb. Ecol.*, 4, 331, 1978.

156. **Meyer, J. S., Tsuchiya, H. M., and Fredrickson, A. G.**, Dynamics of mixed populations having complementary metabolism, *Biotechnol. Bioeng.*, 17, 1065, 1975.

157. **Lee, I. H., Fredrickson, A. G., and Tsuchiya, H. M.**, Dynamics of mixed cultures of *Lactobacillus plantarum* and *Propionibacterium shermanii, Biotechnol. Bioeng.*, 18, 513, 1976.

158. **Mikx, F. H. M. and Van der Hoeven, J. S.**, Symbiosis of *Streptococcus mutans* and *Veillonella alcalescens* in mixed continuous cultures, *Arch. Oral Biol.*, 20, 407, 1975.

159. **Ellwood, D. C. and Hunter, J. R.**, The mouth as a chemostat, in *Continuous Culture*, Vol. 6, *Applications and New Fields*, Dean, A. C. R., Ellwood, D. C., Evans, C. G. T., and Melling, J., Eds., Ellis Horwood, Chichester, U.K., 1976, 270.

160. **Marsh, P. D., Hunter, J. R., Bowden, G. H., Hamilton, J. R., McKee, A. S., Hardie, J. M., and Ellwood, D. C.**, The influence of growth rate and nutrient limitation on the microbial composition and biochemical properties of a mixed culture of oral bacteria grown in a chemostat, *J. Gen. Microbiol.* 129, 755, 1983.

161. **Van der Hoeven, J. C. and de Jong, M. H.**, Continuous culture studies and their relation to the in vivo behaviour of oral bacteria, in *Continuous Culture*, Vol. 8, *Biotechnology, Medicine and Environment*, Dean, A. C. R., Ellwood, D. C., and Evans, C. G. T., Eds., Ellis Horwood, Chichester, U.K., 1984, 89.

162. **Otto, R., Hugenholtz, J., Konings, W. N., and Veldkamp, H.**, Increase of molar growth yield of *Streptococcus cremoris* for lactose as a consequence of lactate consumption by *Pseudomonas stutzeri* in mixed culture, *FEMS Microbiol. Lett.*, 9, 85, 1980.

163. **ten Brink, B.**, The generation of metabolic energy in bacteria: the energy recycling model, Ph.D. thesis, University of Groningen, The Netherlands, 1984.

164. **Wilkinson, T. G., Topiwala, H. H., and Hamer, G.**, Interactions in a mixed bacterial population growing on methane in continuous culture, *Biotechnol. Bioeng.*, 16, 41, 1974.

165. **Driessen, F. M.**, Protocooperation of yogurt bacteria in continuous cultures, in *Mixed Culture Fermentations*, Bushell, M. E. and Slater, J. H., Eds., Academic Press, London, 1981, 99.

166. **Bryant, M. P., Wolin, E. A., Wolin, M. J., and Wolfe, R. S.**, *M. omelianski*, a symbiotic association of two species of bacteria, *Arch. Microbiol.*, 59, 20, 1967.

167. **Ianotti, E. L., Kafkewitz, D., Wolin, M. J., and Bryant, M. P.**, Glucose fermentation products of *Ruminococcus albus* grown in continuous culture with *Vibrio succinogenes, J. Bacteriol.*, 114, 1231, 1973.

168. **Traore, A. S., Fardeau, M. L., Hatchikian, C. E., LeGall, J., and Belaich, J. P.**, Energetics of growth of a defined mixed culture of *Desulfovibrio vulgaris* and *Methanosarcina barkeri:* interspecies hydrogen transfer in batch and continuous cultures, *Appl. Environ. Microbiol.*, 46, 1152, 1983.

169. **Fenchel, T.**, The ecology of marine microbenthos, *Ophelia*, 6, 1, 1969.

170. **Fenchel, T., and Harrison, P.**, The significance of bacterial grazing and mineral cycling for the decomposition of particulate detritus, in *The Role of Terrestrial and Aquatic Organisms in the Composition Processes*, Anderson, J. M., Ed., Blackwell Scientific Publications, Oxford, 1976, 285.

171. **Finlay, B.**, Procedures for the isolation, cultivation and identification of protozoa, in *Experimental Microbial Ecology*, Burns, R. G. and Slater, J. H., Eds., Blackwell Scientific Publications, Oxford, 1982, 44.

172. **Bark, A. W. and Watts, J. M.**, A comparison of the growth characteristics and spatial distribution of hypolimnetic ciliates in a small lake and an artificial lake ecosystem, *J. Gen. Microbiol.*, 130, 3113, 1984.

173. **Sterkin, V. E. and Samoilenko, V. A.**, Analysis of the kinetics of interaction between parasite and host in continuous culture, *Microbiologiya*, 41, 408, 1972.

174. **Tsuchiya, H. M., Drake, J. F., Jost, J. L., and Fredrickson, A. G.**, Predator-prey interactions of *Dictyostelium discoideum* and *Escherichia coli* in continuous culture, *J. Bacteriol.*, 110, 1147, 1972.

175. **Jost, J. L., Drake, J. F., Fredrickson, A. G., and Tsuchiya, H. M.**, Interactions of *Tetrahymena pyriformis, Escherichia coli, Azotobacter vinelandii* and glucose in a minimal medium, *J. Bacteriol.*, 113, 834, 1973.

176. **Bader, F. G., Tsuchiya, H. M., and Fredrickson, A. G.**, Grazing of ciliates on blue-green algae: effects of light shock on the grazing relation and on the algal population, *Biotechnol. Bioeng.*, 18, 333, 1976.

177. **Bader, F. G., Tsuchiya, H. M., and Fredrickson, A. G.**, Grazing of ciliates on blue-green algae: effects of ciliate encystment and related phenomena, *Biotechnol. Bioeng.*, 18, 311, 1976.

178. **Dent, V. E., Bazin, M. J., and Saunders, P. T.**, Behaviour of *Dictyostelium discoideum amoebae* and *Escherichia coli* grown together in chemostat culture, *Arch. Microbiol.*, 109, 187, 1976.

179. **Varon, M.,** Selection of predation-resistant bacteria in continuous culture, *Nature,* 277, 386, 1979.

180. **Crowley, P. H., Straley, S. C., Craig, R. J., Culin, J. D., Fu, Y. T., Hayden, T. L., Robinson, T. A., and Straley, J. P.,** A model of prey bacteria, predator bacteria, and bacteriophage in continuous culture, *J. Theor. Biol.,* 86, 377, 1980.

181. **Jost, J. L., Drake, J. F., Tsuchiya, H. M., and Fredrickson, A. G.,** Microbial food chains and food webs, *J. Theor. Biol.,* 41, 461, 1973.

182. **Habte, M. and Alexander, A.,** Protozoan density and the coexistence of protozoan predators and bacterial prey, *Ecology,* 59, 140, 1978.

183. **Watson, P. J., Ohtaguchi, K., and Frederickson, A. G.,** Kinetics of growth of the ciliate *Tetrahymena pyriformis* on *Escherichia coli, J. Gen. Microbiol.,* 122, 323, 1981.

184. **Bonomi, A. and Fredrickson, A. G.,** Protozoan feeding and bacterial wall growth, *Biotechnol. Bioeng.,* 18, 239, 1976.

185. **Ratman, D. A., Pavlon, S., and Fredrickson, A. G.,** Effects of attachment of bacteria to chemostat walls in a microbial predator-prey relationship, *Biotechnol. Bioeng.,* 24, 2675, 1982.

186. **Slyter, L. L., Nelson, W. O., and Wolin, M. J.,** Modifications of a device for maintenance or the rumen microbial population in continuous culture, *Appl. Microbiol.,* 12, 373, 1964.

187. **Slyter, L. L., Bryant, M. P., and Wolin, M. J.,** Effect of pH on population and fermentation in a continuously cultured rumen ecosystem, *Appl. Microbiol.,* 14, 573, 1966.

188. **Veilleux, B. G. and Rowland, I.,** Simulation of the rat intestinal ecosystem using a two-stage continuous culture system, *J. Gen. Microbiol.,* 123, 103, 1983.

189. **Freter, R., Stauffer, E., Cleven, D., Holdeman, L. V., and Moore, W. E. C.,** Continuous flow cultures as in vitro models of the ecology of large intestinal flora, *Infect. Immun.,* 39, 666, 1983.

190. **van der Hoeven, J. S., de Jong, M. H., Camp, P. J. M., and van den Kieboom, C. W. A.,** Competition between oral *Streptococcus* species in the chemostat under alternating conditions of glucose-limitation and excess, *F.E.M.S. Microbiol. Ecol.,* 31, 373, 1985.

191. **van Gemerden, H.,** Coexistence of organisms competing for the same substrate: an example among the purple sulfur bacteria, *Microb. Ecol.,* 1, 104, 1974.

192. **Rhee, G. Y.,** Continuous culture in phytoplankton ecology, in *Advances in Aquatic Microbiology,* Vol. 2, Droop, M. R. and Jannasch, H. W., Eds., Academic Press, London, 1980, 151.

193. **Rhee, G. Y., Gotham, I. G., and Chisholm, S. W.,** Use of cyclostat cultures to study phytoplankton ecology, in *Continuous Culture of Cells,* Calcott, P. H., Ed., CRC Press, Fla., 1981, 159.

194. **van Gemerden, H. and Beeftink, H. H.,** Ecology of phototrophic bacteria, in *The Phototrophic Bacteria: Anaerobic Life in the Light. Studies in Microbiology,* Vol. 4, Ormerod, J. G., Ed., Blackwell Scientific, Oxford, England, 1983, 146.

195. **Loogman, J. G.,** Influence of photoperiodicity on algal growth kinetics, Ph.D. thesis, University of Amsterdam, The Netherlands, 1982.

196. **Gottschal, J. C., Pol, A., and Kuenen, J. G.,** Metabolic flexibility of *Thiobacillus* A2 during substrate transitions in the chemostat, *Arch. Microbiol.,* 129, 23, 1981.

197. **Gottschal, J. C., Nanninga, H. J., and Kuenen, J. G.,** Growth of *Thiobacillus* A2 under alternating growth conditions in the chemostat, *J. Gen. Microbiol.,* 126, 85, 1981.

198. **Schlegel, H. G. and Jannasch, H. W.,** Enrichment cultures, *Ann. Rev. Microbiol.,* 21, 49, 1967.

199. **Brown, C. M., Ellwood, D. C., and Hunter, J. R.,** Enrichments in a chemostat, in *Techniques for the Study of Mixed Populations,* Lovelock, D. W. and Davies, R., Eds., Academic Press, London, 1978, 213.

200. **Slater, J. H. and Hartman, D. J.,** Microbial ecology in the laboratory: experimental systems, in *Experimental Microbial Ecology,* Burns, R. G. and Slater, J. H., Eds., Blackwell Scientific Publications, London, 1982, 255.

201. **Parkes, R. J.,** Methods for enriching, isolating, and analysing microbial communities in laboratory systems, in *Microbial Interactions and Communities,* Vol. 1, Bull, A. T. and Slater, J. H., Eds., Academic Press, London, 1982, 45.

202. **Jannasch, H. W.,** Competitive elimination of *Enterobacteriaceae* from sea water, *Appl. Microbiol.,* 16, 1616, 1968.

203. **Laanbroek, H. J.,** Ecology and physiology of L-aspartate- and L-glutamate fermenting bacteria, Ph.D. thesis, University of Groningen, The Netherlands, 1978.

204. **Daughton, C. G. and Hsieh, D. P. H.,** Parathion utilization by bacterial symbionts in a chemostat, *Appl. Environ. Microbiol.,* 34, 175, 1977.

205. **Senior, E., Bull, A. T., and Slater, J. H.,** Enzyme evolution in a microbial community growing on the herbicide Dalapon, *Nature,* London, 263, 476, 1976.

206. **Senior, E.,** Characterization of a microbial association growing on the herbicide Dalapon, Ph.D. thesis, University of Kent, U.K., 1977.

Chapter 3

# MULTISTAGE CHEMOSTATS AND OTHER MODELS FOR STUDYING ANOXIC ECOSYSTEMS

**R. John Parkes and Eric Senior**

## CHARACTERISTICS OF ANOXIC ECOSYSTEMS

Anaerobic ecosystems are of widespread environmental importance. Such habitats are not restricted to well-studied examples like the gastrointestinal tracts of animals, sewage digesters, eutrophic lakes, waterlogged soils, and lower layers of aquatic sediments. Jørgensen[1] has calculated that in the sediments of coastal areas, anaerobic sulfate-reduction accounted for approximately 50% of the total degradation of organic matter, and in a freshwater lake sediment, anaerobic nitrate reduction and methanogenesis accounted for between 30 and 50% of the organic carbon mineralized.[2] In addition, the presence of anaerobic microniches within generally aerobic environments considerably extends the environmental importance of anaerobic metabolism.[3]

The characteristics of anaerobic metabolism are quite different from those of aerobic metabolism. Complete mineralization of organic compounds to $CO_2$, often by the activity of a single bacterial species, is common in the presence of oxygen but complete degradation of organic matter under anaerobic conditions characteristically requires the cooperative metabolism of a mixed population of bacteria, each of which contributes a partial oxidation of the organic compound, until complete mineralization to $CO_2$ or $CH_4$ is achieved. These anaerobic populations can use a variety of inorganic electron acceptors, often in a sequence (Figure 1), which reflects the amount of energy that can be liberated from a common electron donor.[4] Most of these bacteria specialize in reducing a specific electron acceptor. For example, sulfate-reducing bacteria use sulfate, while methanogenic bacteria use carbon dioxide (Figure 2). Therefore, in structured ecosystems such as sediments and landfill, the sequence of electron acceptor utilization is usually mirrored by discrete changes in bacterial populations (Figure 3). The thermodynamic advantage due to the use of a more oxidized electron acceptor is often reflected in a kinetic advantage to that group of bacteria. Thus, though sulfate-reducing and methanogenic bacteria can grow under the same conditions,[5] if they are using the same limiting substrate ($H_2$ or acetate), the sulfate-reducing bacteria will outcompete the methanogen since it has a higher affinity for these substrates.[6,7]

Fermentative bacteria (Figure 2) require no external electron acceptor and therefore are not dependent on gradients of electron acceptors. However, the interrelationship between the fermentative bacteria and those that catabolize fermentation products such as $H_2$ and acetate are very close (Figure 2). Often the fermentative reactions are only thermodynamically possible at very low concentrations of $H_2$, and therefore fermenters and $H_2$ utilizers like the *Syntrophomonas wolfei* system[8] are very closely coupled. Even if the reaction is not inhibited by high concentrations of $H_2$, $H_2$ concentration often influences the products of the reaction. At low $H_2$ concentrations, more oxidized substrates (particularly acetate) are produced and more of the energy of the reaction is conserved in the form of ATP.[9] A similar increase in energy conservation is produced if other fermentation products, such as lactate are maintained at low levels.[10] Sulfate-reducing and methanogenic bacteria therefore play a central controlling role in anaerobic decomposition of organic matter by other anaerobes, by removing toxic metabolites, directing electron flow to limited reduced end products, and enhancing growth rates and yields, or by supplying essential growth factors.[11] Anaerobic bacteria often derive only a little energy from their reactions, and a large amount of the potential energy

RESPIRATION TYPE

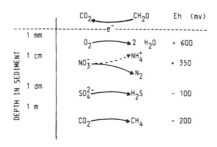

FIGURE 1.    Theoretical distribution of the main types of respiration in a marine sediment, together with approximate Eh values required for the initiation of the reaction.

is retained in reduced end products, such as $H_2S$ and $CH_4$. As a result of this, carbon and energy flow in anaerobic systems is not so tightly coupled as in aerobic environments and the system is often energy limited. Reduced end products can diffuse from the anoxic environment into the aerobic environment where they can be efficiently exploited by chemoautotrophs and photoautotrophs and therefore represent an important energy link between aerobic and anaerobic systems.[12]

Anaerobic systems can generally be considered as obligately heterogeneous communities characterized by mixed microbial populations (Figure 2), with very strong interdependence, acting on mixed substrates, strongly influenced, both positively and negatively, by chemical gradients, in particular of electron acceptors and donors and $H_2$ concentrations. What is more, such systems are often energy limited. All these considerations should be taken into account in the design of laboratory models to study anaerobic systems.

This heterogeneity in anaerobic communities is often enhanced by the presence of surfaces such as sediment and soil particles and fecal pellets. Surfaces tend to concentrate both inorganic and organic compounds and thus stimulate bacterial attachment and growth. Surface growth is often characterized by the production of a mucilagenous exudate which may serve to accumulate nutrients and prevent diffusion of exoenzymes and the products of their action away from the cell surface. Although this microbial film community is initially very active, as community biomass increases, activity per unit biomass decreases, presumably due to the bacteria lower in the film becoming diffusion limited for some essential substrate.[13] Therefore, the development of microbial films on surfaces leads in the long term to regulation and stabilization of both community biomass and metabolism, which may be crucial for the development of the close associations which are essential for efficient metabolism under anaerobic conditions (see Figure 2 and previous discussion). The metabolism of the microbial film can also provide anaerobic microenvironments within an aerobic system.[3] If the particles themselves become consolidated, as often occurs in sediments and soils, they provide positional stability for the microbial biomass so that a structured community develops which can display the theoretical hierarchy of bacterial activities (Figure 1) both in space (within microbial films on particles) and time (i.e., sediment/soil depth, Figure 3).

The presence of consolidated particles also tends to select for microbial communities that contain filamentous organisms, which exhibit chemotaxis, and are capable of gliding. Such communities are very different from open water communities which tend to be dominated by unicelluar coccoid, vibrioid, spirilloid, and more or less rod-shaped microbes.[14] These filamentous organisms, together with other motile bacteria, exploit the steep chemical gradients that often exist within anaerobic sediments and will migrate to optimum chemical gradients over several millimeters.[12] In marine sediments, the filamentous gliding habit is

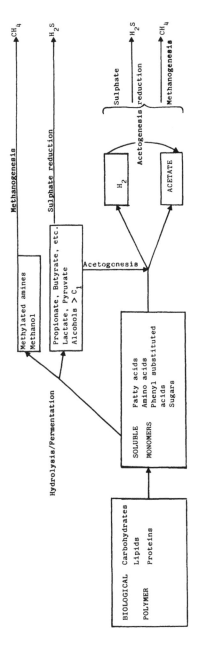

FIGURE 2. A scheme of carbon flow through an anaerobic ecosystem showing some of the functional metabolic groups (bold type) and their interrelationships.

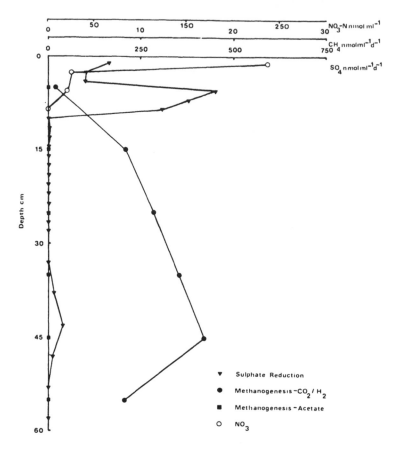

FIGURE 3.    An example of bacterial community structure in an organically polluted marine sediment.

also exhibited by some heterotrophic sulfate-reducing bacteria.[15] Chemoautotrophic bacteria are also important in anoxic water columns, but there the chemocline extends over several meters.[12] In both environments, the chemoautotrophs play an important role in both energy and sulfur cycling. However, organisms which exhibit chemotaxis may not be successfully isolated in normal homogeneous enrichment conditions and require heterogeneous enrichment, including gradients of growth conditions and often a stagnant aqueous solution that allows free movement, for effective isolation.[14]

The presence of clay particles can also have a direct positive effect on the metabolism of some anaerobic bacteria. This was demonstrated for certain sulfate-reducing bacteria and was thought to be of ecological importance in energy-limited sediments.[16]

## SOME EXAMPLES OF COMMUNITY STRUCTURE IN ANOXIC ECOSYSTEMS

### Marine Sediments

In marine sediments, oxygen is normally limited to the top 3 to 5 mm depth, with occasional penetration to deeper layers as a result of bioturbation by benthic macrofauna.[17] It is therefore not surprising that anaerobic metabolism plays a dominant role in the degradation of organic matter within these sediments. In coastal sediments, anaerobic sulfate reduction alone accounts for 50% of the degradation of organic matter,[1] reflecting the ready supply of sulfate from seawater. In certain sediments, the theoretical sequence of electron acceptor utilization (Figure 1) is found (Figure 3), whereby changes in bacterial activity reflect limitation due

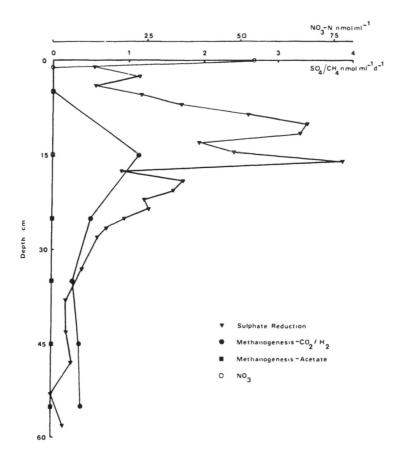

FIGURE 4. An example of bacterial community structure in a nonpolluted marine sediment.

to removal of the required electron acceptors. In these circumstances, sulfate-reduction is still quantitatively the most important anaerobic process[18-22] but it is limited to the top layer of the sediment due to the rapid removal of sulfate, thus enabling methanogenesis to develop at lower depths in the sediment (Figure 3). Although the maximum activities of sulphate-reduction and methanogenesis are spatially separated, the two processes are not mutually exclusive, despite the theoretical ability of the sulfate-reducing bacteria to outcompete the methanogens for similar substrates.[6,7] In other sediments, the maximum activities of sulfate reduction and methanogenesis occur together (Figure 4, see also the Creek site in Senior et al.[20]), even though sulfate-reduction is still quantitatively the more important process. The reasons for these different patterns in community structure are unclear, but may be related to the availability of organic carbon in the different sites. The sediment exhibiting spatial separation of methanogenesis and sulfate-reduction (Figure 3) was polluted with waste from an alginate factory, but the sediment showing no such separation was from an unpolluted site in the same area (Figure 4, see Parkes and Taylor[22] for further details). The situation may also indicate the presence of substrates for methanogenesis which are not competed for by sulfate-reducing bacteria (Figure 2, also References 21 and 23 to 25). If intertidal sediments are subjected to very high loadings of both nutrients and readily degradable organic carbon, methanogenesis can become the dominant process.[26] These and other interactions are shown schematically in Figure 2.

Acetate has been shown to be the most important substrate for sulfate-reducing bacteria within marine sediments, accounting for approximately 50 to 60% of all sulfate-reduction,

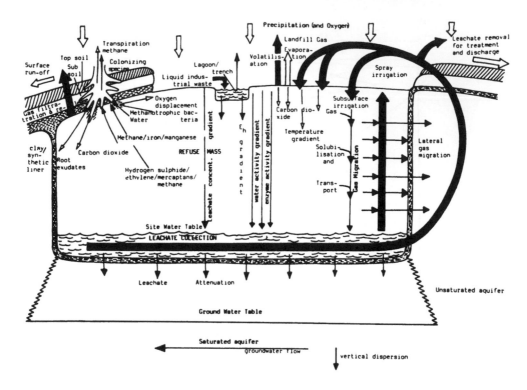

FIGURE 5.    Diagrammatic representation of landfill ecosystems; see Senior and Balba[36] for further details.

followed by propionate and butyrate (10 to 20%) and $H_2$ (5 to 10%).[21,27,28] This acetate seems to be metabolized by a different group of sulfate-reducing bacteria than those metabolizing $H_2$.[29] *Desulfobacter* sp. appears to be responsible for acetate utilization, while $H_2$ is probably utilized by *Desulfobulbus* sp.[30] More detailed account of microbial interactions within marine sediments is given in the recent review by Nedwell.[31]

## Landfill

In the landfill ecosystem, bacterial species are often confronted by a complex mixture of organic substrates (often in nonlimiting concentrations) and electron acceptors, and thus species diversity is high, especially in recently emplaced refuse. In most natural anoxic habitats, the aqueous phase is usually oligotrophic,[32] but this is unlikely to be the case in landfill, particularly in the early stages of refuse decomposition. Initially, the biodegradation of refuse components is dominated by periods of intense aerobic metabolism which result in the depletion of molecular oxygen and ultimately in the dominance of anoxic processes which are functionally similar to other anoxic systems (Figure 2). During this period, the microorganisms, through their own metabolism, provoke extensive shifts in the chemical environment which in turn mediates rapid phenotypic changes among species.[33]

As a consequence of the methods of refuse emplacement, stratification in landfill is initially more likely to be temporal than vertical. However, vertical stratification does develop with time as the concentrations of exogenous electron acceptors become limited, although the picture is complicated by the presence of horizontal and temporal gradients (Figure 5). The system is characterized by gradients of multidirectional temperature, gas, liquid ($a_w$), Eh, pH, enzyme activity,[34] and vectoral solute flow, and thus this spatial heterogeneity should be incorporated into any laboratory model used to study landfill. Site conditions, refuse emplacement strategies, and reclamation practices are *major factors in controlling* conditions prevailing within the landfill system, particularly, pH, Eh, water content, and gas partial pressures. Other important control factors are molecular properties such as water solubility, lipid/water partition coefficient, volatility, molecular size, charge, shape, and functional group presence,[35] as well as microbial attachment to organic and inorganic surfaces, microbe-microbe interactions, bidirectional diffusions across oxic-anoxic interfaces, overlapping niches and interfaces such as solid-liquid, solid-gas, and liquid-gas.[32] Further details of the complex interactions within landfill ecosystems can be found in the recent review by Senior and Balba.[36]

It is obvious from the descriptions of community interactions in both marine sediments and landfill, that there is great similarity between the two environments and that they share certain attributes common to all anoxic systems (see earlier section on characteristics of anoxic ecosystems and Figure 2). There are, however, differences in the terminal steps of organic carbon oxidation, since marine sediments are dominated by sulfate reduction, whereas landfills tend to be dominated by methanogenesis. This means that hydrogen will be a more important intermediate and that acetogenic bacteria (Figure 2) will play a much more important role in carbon mineralization in landfill than in marine sediments. In this respect, the landfill system will tend to be more similar to freshwater than to marine sediments (see Banat and Nedwell).[37] There are also differences in the influence of feedback controls and toxic effects in sediments and landfill systems, with landfill systems being much more affected by such factors than sediment systems.[36]

## THE REQUIREMENT FOR MODELS OF ANOXIC ECOSYSTEMS

It is clear from what has been discussed so far that anaerobic systems are obligatory heterogeneous ecosystems with mixed closely interacting bacterial communities, acting on mixed substrates and strongly influenced by chemical gradients of both inorganic and organic compounds. Often such systems are further complicated by the presence of inorganic particles, which not only have a strong direct effect on bacterial metabolism, but also allow the development of structured communities, both in space and time. Investigations of such complex communities *in situ* are made more difficult by all the problems normally associated with field work. These include disturbance of the system by the sampling procedure, changes in samples due to transportation, analytical problems in a heterogeneous sample where compounds of interest are often present in trace amounts, natural variation from sample to sample, and constantly changing environmental conditions. In addition, there is the added problem of sampling and handling anoxic systems in an oxic environment. The subtle interplay between abiotic and biotic factors in anaerobic ecosystems makes it difficult to be certain that the processes being studied are exclusively biological. Another problem common to all field work is that often due to ignorance or to logistical limitations not all the controlling factors in the environment can be studied simultaneously so that although correlations may be obtained at the end of the study, direct cause and effect and the role of unmeasured parameters are not always clear.

In addition to the above there are some specific problems with *in situ* investigations of anaerobic systems which makes the use of laboratory model systems particularly attractive and these will be discussed briefly. Obtaining undisturbed cores of coastal marine sediments can present difficulties, and a soft-landing, hydraulically damped coring device[38] is much more suitable than the gravity cores which are often used.[39] After sampling, subsequent handling must exclude oxygen, for example by using Hungate's gassing technique,[40] and care has to be taken to minimize sediment disturbance, since even mixing the sediment or diluting it to produce a slurry can cause changes in sediment activity.[41] It is difficult to obtain accurate estimates of the contribution to the total biomass of individual bacterial populations: for example, viable counts of sulfate-reducing bacteria tend to give population estimates that are approximately 1000-fold too low.[42] Partly, for this reason, research into community interactions has concentrated on the biochemical activities of individual populations and, although interesting information has been obtained (see earlier section on examples of community structure in anoxic ecosystems and Figures 3 and 4 for example), these techniques do not provide information about the biomass,[43] growth yields, or growth rates of individual populations. This is an area where model systems should prove invaluable.

Recent data indicate that there are some basic problems with using activity measurements based on $^{14}C$-radiotracer techniques, which may lead to considerable overestimate of *in situ* activities.[44-49] This discrepancy is probably due to the presence of multiple substrate pools within pore water, which have different biological availabilities, coupled with the equilibration of the $^{14}C$-label across the pools,[50,51] Again, this problem ought to be amenable to a model system approach. The use of a continuous culture laboratory model would also allow dynamic interactions, including carbon flow, between different bacterial populations to be studied without the presence of the vast majority of organic matter, which remains in the sediment but is largely unavailable to microorganisms.[42]

The chemical determination of microbial biomass and community structure, especially the analysis of cell wall constituents,[22,52] seems a very promising technique, but more information needs to be obtained about the degradation rates of the specific biomarkers under anaerobic conditions, so that biomarkers can be more realistically interpreted as "living" or "dead" biomass. A multiple chemostat seems an ideal tool for such an investigation.

There is thus an obvious need to complement basic field investigation with experimental work in the laboratory, which can be conducted under defined and controlled conditions. It must be stressed, however, that these laboratory systems must be ecologically relevant if realistic information is to be obtained. The multiple chemostat system has certain attributes which make it an attractive tool for studying anaerobic systems and these will now be discussed.

## CHARACTERISTICS OF MULTISTAGE CHEMOSTATS AND THEIR APPLICABILITY TO ANOXIC SYSTEMS

A multistage chemostat is a continuous flow culture system incorporating a number of linked vessels in series (Figure 6A). Such a system seems an ideal model for studying anoxic systems, since anaerobic decomposition proceeds via the sequential action of different bacterial types (Figure 2) and there is thus the potential for isolating each sequential event within individual vessels of the system. In some respects, the multistage system resembles batch enrichment cultures, which produce a spectrum of physiological states arranged in temporal order since growth of the organisms is influenced by the past history of the culture. In a multiple chemostat system, however, the spectrum of physiological types is spatially arranged through each of the different vessels, and since it is an open system, distribution of the various physiological types within it will be time independent and therefore easier to study than an equivalent batch system. In a multistage chemostat, stepped changes in solute concentrations are produced rather than true gradients;[53] therefore spatial resolution tends to be poor, but this may not be a disadvantage when it is used as a model for anaerobic systems where steep chemical gradients are the norm, for example, in sediments, landfill, chemoclines of stratified water bodies,[12] microbial films on surfaces and microenvironments, etc. Another advantage of the system is that it tends to lead to the enrichment of more ecologically relevant bacterial communities, since except for the first vessel, enrichment conditions within each vessel will be controlled by the metabolism of the bacterial communities in the previous vessel, in a similar manner to changes in conditions in *in situ* structured ecosystems.[54] The unidirectional flow of organic substrates and electron acceptors in a multistage chemostat is similar to the situation in structured ecosystems such as sediments

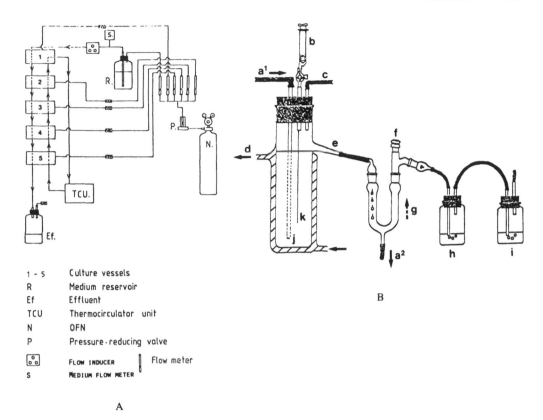

FIGURE 6. (A) Five vessel anoxic landfill model. (B) Single vessel of the five-vessel anoxic landfill model. (From Coutts, D. A. P., Senior, E., and Balba, M. T. M., *J. Appl. Bacteriol.*, 62, 253, 1987. With permission.)

and microbial films on inert particles, although unfortunately unidirectional flow of reduced metabolic products in the model is not a faithful reflection of natural environments where these compounds can diffuse upwards to be reoxidized.[28,42] The multistage system, just like normal chemostats, can be operated under defined conditions in each vessel, but in addition, separate inputs into the various vessels can also be made, for example to simulate ground water containing sulfate entering a sulfate-limited sediment.

A feature of structured microbial communities is that bacterial biomass often maintains a relatively constant position while solutes flow past. Theoretically, both the biomass and solutes will have the same residence time in a multiple-stage chemostat. In practice, however, when such systems are used to model anaerobic processes, low dilution rates (for example, D = 0.006 per hour, Parkes and Taylor[54]) and very low or zero stirring speeds quickly allow wall growth to develop. There is therefore often a significant portion of the biomass which is retained within individual vessels of the system while the solute flows past. Such heterogeneity can deliberately be increased by including more sites for surface growth such as glass beads, soil, or sediment particles, but unfortunately, as the heterogeneity increases, the system becomes less amenable to analysis by simple steady-state kinetics.[55]

If all the vessels of the system are the same volume, the dilution rate of individual vessels is the same, but wash out, if it occurs, will proceed sequentially down the vessel array. The system as a whole will also have approximately the same residence time as a single vessel of the same total volume, but the residence time for the individual vessels will be equal to the residence time for the total system divided by the number of vessels in the system. The flow rate of the system therefore has to be carefully selected to prevent the wash out of individual vessels. The working volume of each vessel can be varied, and thus the dilution rate and therefore the residence time of individual vessels may be selected. This is a most useful attribute, since the dilution rate at points within the system may be decreased to accommodate slow-growing organisms, for example, methanogens or bacteria catalyzing a reaction which provides little energy for growth.

In common with chemostats in general, a multiple chemostat system can be run under a choice of limiting conditions, such as energy, carbon, nitrogen, or sulfate, thus making it a very useful tool for studying factors controlling community interactions and structure within anaerobic systems, as well as the flow of carbon in such systems. Consideration, however, must be given in such a heterogeneous system to the possibility that although the whole system may be characterized by one type of limitation, an individual component or components of the system may be controlled by different limitations (see, for example, Wilkinson et al.[56]).

A multistage chemostat system should also provide information about the yields of anaerobes under various limiting conditions or when competitive or synergistic interactions are taking place. However, it can not be too strongly emphasized that media composition and conditions within each vessel must be carefully selected if environmentally relevant data are to be obtained.

## EXAMPLES OF THE USE OF MULTISTAGE CHEMOSTATS TO STUDY ANOXIC ECOSYSTEMS

### Landfill

In one of our laboratories (Senior) a multistage continuous culture system has been developed to study both the processes and the diversity of the decomposer species responsible for the methanogenic fermentation of hexanoic acid, which has been shown to be present in high concentrations (5770 mg/$\ell$) in leachate of young landfills[36] (Figure 6A and B and Figure 7).

The five individual, unstirred, water-jacketed (d) vessels were each of 700 m$\ell$ working volume and were maintained at a constant temperature of 25°C by means of a Churchill 05-CTCV thermocirculator (Churchill Instrument Co., Uxbridge, U.K.). The influent medium (a1) was introduced into the first chemostat by a Watson Marlow MHRE/2 flow inducer (Watson Marlow Ltd., Falmouth, U.K.) to give a constant dilution rate of 0.03 per hour. The resulting effluent (e) was discharged via the angled overflow tube and was introduced into the base of the second chemostat by means of a centrally-located tube (j). The vessel linkage was repeated throughout the array, to ensure that the bulk flow characteristics approximated a continuous but segmented plug flow with an overall dilution rate of 0.006 per hour and effluent from the fifth chemostat was collected in a 20-$\ell$ reservoir.

Anaerobiosis was maintained throughout by individual chemostat overpressures (0.1 atm) of oxygen-free nitrogen (c) and was facilitated by the use of oxygen-impermeable butyl rubber tubing. This arrangement, however, prevented both the vertical migration of species and upwards diffusion of gases. Separation of the gas and liquid effluent streams were effected by means of the U-tube configurations which were linked in series to pressure heads of zinc acetate (1% w/v) (h) and barium hydroxide (18 g/$\ell$) (i) to trap hydrogen sulfide and carbon dioxide, respectively, thus enabling individual mass balances to be constructed for each chemostat. Samples for culture analyses were withdrawn into sterile syringes (b), via the three-way valves attached to the fixed needles (k), whereas samples for gas analysis were aseptically removed into gas syringes through the U-tube suba seals (f).

Figure 8 (A, B and C) shows the "steady-state" concentrations of methane and the intermediate fatty acids in the individual chemostats in the presence of influent concentrations of 10 (8A), 5 (8B), and 1.4 m$M$ sulfate (8C).[102] As maximum concentrations of both methane and acetate were observed in the first vessel of the system, spatial separation of the component

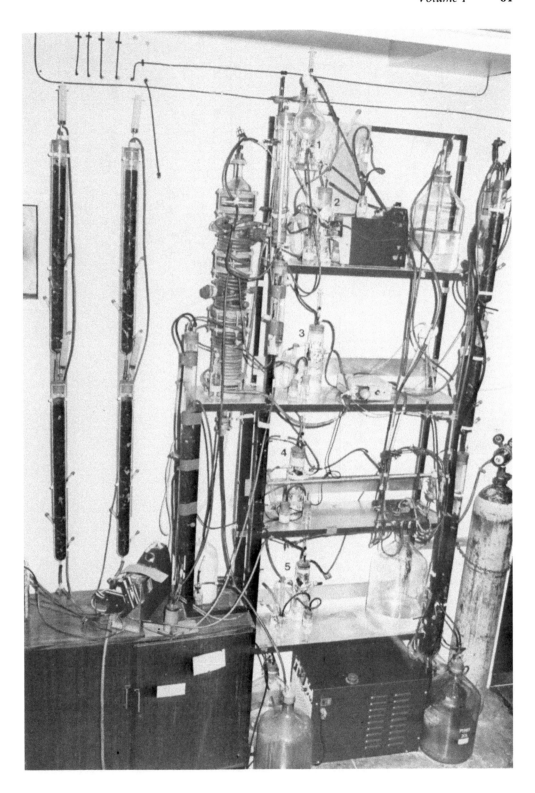

FIGURE 7. Five-vessel anoxic landfill model, 1 to 5 are the growth vessels.

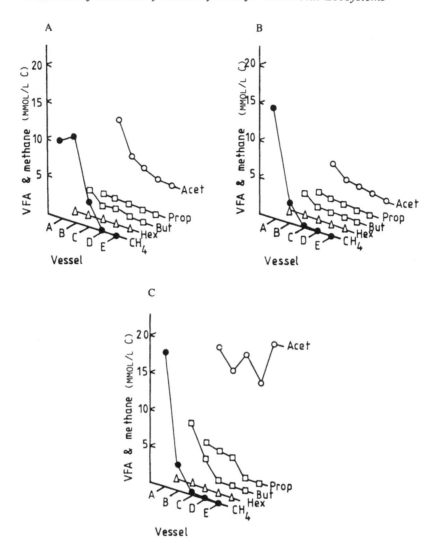

FIGURE 8.   Changes in concentrations of metabolic intermediates during the anoxic catabolism of hexanoic acid (5 m$M$) in a five-vessel landfill model in the presence of (A) 10, (B) 5, and (C) 1.4 m$M$ sulfate.

species of the microbial association was not accomplished and, thus, identification of the exact interfaces could not be resolved. As a consequence, an extra dimension, nonconstant dilution rate was added to a second multistage model (Figure 9). In this array, the working volumes of three individual chemostats were increased from 310 m$\ell$ in Vessel 1, to 700 m$\ell$ in Vessel 2, and finally to 1600 m$\ell$ in Vessel 3, with the result that the slower-growing methanogens were spatially displaced from the first stage due to the higher dilution rates in this vessel (Figure 10 and Reference 102).

**Bacterial Enrichment**

The majority of enrichment and isolation methods developed to date have concentrated on aerobic species and many of these methods, which have been reviewed by Parkes,[55] often select against some of the more abundant types of microorganisms present in natural ecosystems.[57] This is particularly true when closed culture *enrichments* are used which result in the isolation of much less representative populations than do chemostat methods.[54,58]

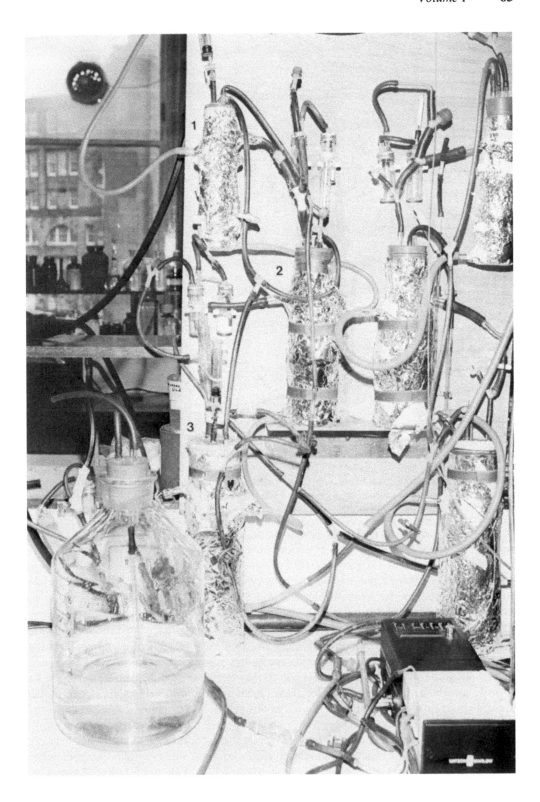

FIGURE 9. Three-vessel anoxic landfill model with decreasing dilution rate, 1 to 3 are the growth vessels.

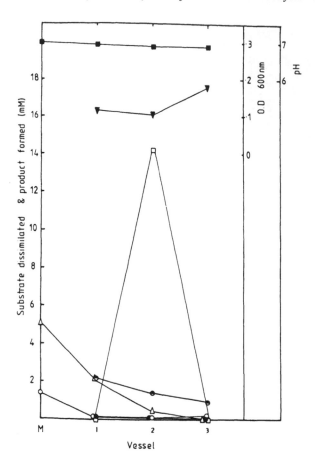

FIGURE 10.  Changes in concentrations of residual sulfate (○ — ○) hydrogen sulfide (● — ●), total HS⁻ (◓ — ◓), Methane (□ — □), pH (■ — ■), residual hexanoic acid (△ — △), and optical density (▼ — ▼) during the anoxic catabolism of hexanoic acid (5 m*M*) in the presence of 1.4 m*M* sulfate in a three-vessel landfill model subject to decreasing dilution rate. M = concentrations in input medium.

However, there are some environments where batch culture enrichment may be an appropriate enrichment procedure, for example, landfill and discontinuous input of organic matter to carbon-limited soils or sediments.[55]

The gradients that develop within a multivessel chemostat system lead to different selection pressures in each of the component vessels and thus may result in the enrichment and isolation of cultures which are representative of the ecosystem under study. Parkes and Taylor[54] adopted this approach to isolate bacterial associations of different respiratory types (aerobic, anaerobic, facultatively aerobic, and facultatively anaerobic bacteria) from marine sediment. Four linked chemostats were used in this study (Figure 11). Each vessel consisted of a Quickfit® vessel (1 ℓ) and top, with stainless steel tubing for temperature control, gassing, sampling, medium outflow, and a magnetic stirrer for mixing. Connections between vessels were made with flexible neoprene tubing. Gas flow was controlled by rotameters and medium flow by peristaltic pumps (Watson Marlow Ltd., Falmouth, U.K.). Gases were sterilized by filtration (Microflow Ltd., Hants, U.K.) and medium by autoclaving (120°C).

Medium composition (in g/ℓ) was glucose, 1; cellulose, 0.1; yeast extract, 0.5; $NH_4NO_3$, 0.001; $NH_4Cl$, 0.5; $K_2HPO_4$, 1.4; $KH_2PO_4$, 0.35; EDTA disodium salt, 0.067; $MgCl_2 \cdot 6H_2O$, 7; $FeSO_4 \cdot 7H_2O$, 0.05; NaCl, 20.1; resazurin, 1 mℓ of a 0.1% w/v solution, and 1

FIGURE 11.   Four-vessel sediment model. A, medium pump. B medium in, C medium out; 1 to 4 are the growth vessels.

m$\ell$ of trace elements solution. Trace elements solution contained (in g/$\ell$): NaHCO$_3$, 2; MnSO$_4$ · 4H$_2$O, 12; ZnSO$_4$ · 7H$_2$O, 7.2; (NH$_4$)$_6$Mo$_7$O$_{24}$ · 4H$_2$O, 0.04; CuSO$_4$ · 5H$_2$O, 1.23; Al$_2$(SO$_4$)$_3$ · 16H$_2$O, 0.2; CoCl$_2$ · 6H$_2$O, 2.4; FeSO$_4$ · 7H$_2$O, 7; H$_3$BO$_4$, 0.31; CaCl$_2$, 2.2, and EDTA disodium salt, 7.5. The pH of the medium was adjusted to 7.2 with 6-$N$ NaOH prior to autoclaving.

Aerobic medium was pumped into Vessel 1 at a rate of 3 m$\ell$/hr. This vessel also received, from a second reservoir, Na$_2$SO$_4$ (2.15 g/$\ell$), at a rate of 2 m$\ell$/hr in order to provide a sulfate concentration of approximately 6 m$M$. The combined flow rate into the multiple-vessel system was 5 m$\ell$/hr giving a dilution rate of 0.008 per hour in Vessel 1 (650 m$\ell$) and 0.006 per hr in the remaining three vessels (900 m$\ell$ working volume). Nitrogen (oxygen-free) was passed over the headspace of all the vessels at a rate of 60 m$\ell$/min and air was pumped into the culture of Vessel 1 at a rate of 1 m$\ell$/min. Vessel 1 was autoclaved with 650 m$\ell$ of complete medium (including 6 m$M$ SO$_4$) and allowed to cool in air. Vessel 2 was autoclaved with 900 m$\ell$ of complete medium but cooled under nitrogen. The other vessels were autoclaved without medium. The vessels were all connected and flow from both reservoirs started, then Vessels 1 and 2 were inoculated each with 2 m$\ell$ of sediment. Vessels 3 and 4 were allowed to fill up slowly from the previous vessels and, when at working volumes, were also inoculated with sediment. This was in order to try and overcome possible substrate-accelerated death of bacterial isolates.

Confirmation that realistic populations had been enriched was obtained by the close correlation between the lipid fatty acid distribution of a mixed culture of sulfate-reducing bacteria isolated from the multistage system and the lipid distribution of the zone of maximum sulfate-reduction activity within the sediment. Both distributions had significant concentrations of C14:0, iso and anteiso C15:0, C16:0, C16:1$\omega$7, and C18:1$\omega$7. This was in contrast to the lipid fatty acid profile of *Desulfovibrio desulfuricans*, a sulfate-reducing bacterium obtained by batch enrichment, which was dominated by iso C17:1$\omega$ 7 fatty acids and correlated poorly with the sediment.[54]

### Bacterial Interactions within Marine Sediments

Senior et al.[59] and Thompson et al.[60] reported using a five-vessel multistage chemostat system to study the anaerobic decomposition of glucose and benzoate. A carbon-sulfate ratio was selected so that there was insufficient sulfate present to allow complete oxidation of the organic substrate through sulfate-reduction. Dilution rates were slow (D = 0.0018 per hour for the whole system) and each vessel had the same culture volume and hence the same dilution rate (0.009 per hour). Under these conditions spatial separation of sulfate-reduction, methanogenesis, and acetogenesis occurred along the array of vessels in a manner similar to the vertical spatial separation found in natural sediments, where organic carbon is also in excess (e.g., Figure 3). After a long period of stabilization (110 days), the systems attained "steady states", which allowed carbon balances to be measured and thus both rates of carbon degradation and bacterial biomass production to be quantified.

### Cyclic Oxidation and Reduction

Wimpenny et al. (this volume) used a multistage model to examine cyclic oxidation and reduction of both nitrogen and sulfur compounds. However, to facilitate such studies provision must be made for bidirectional flows. Similar models were also used by Herbert (Chapter 5, this volume) to examine the interactions between *Clostridium butyricum*, *Desulfovibrio* sp., and *Chromatium* sp., which resulted in a functional sulfur cycle.

### Phase Separations

Separation of functional groups of anaerobic bacteria by the use of multivessel models, some of which incorporated tubular reactors, have been achieved by a number of workers[61,62] as a prerequisite for process optimization. Ghose and Bhadra[61] used a five-stage open culture system, which consisted of four linked digesters, each of 2ℓ capacity, and a fifth of 10 ℓ volume, to examine energy recovery in the biomethanation of cow dung. By separating the acidogenic phases from biomethanation in the 10 ℓ vessel, volatile solids reduction, methane yield and residual solids NPK value all increased by approximately 50% compared with a single-stage batch culture control and the energy recovery doubled.

### Digestive Systems

The rumen, with its heterogeneous content involving as many as 20 species of protozoa and 100 species of bacteria,[63] poses a considerable challenge for ecosystem modeling, particularly when food constituents are degraded at different rates and recalcitrant molecules have different residence times.

Czerkawski and Breckenridge[65] developed a relatively simple but extremely effective model rumen (Rusitec) to examine the catabolisms of various dietary components. In this model, a perspex vessel of 1 ℓ working volume is used, within which two or four[64] nylon-gauze bags, containing various components of solid digesta, are placed. The bags are moved up and down 50 to 60 mm at a rate of 8 c/min in the presence of artificial saliva. Although this model is not a true continuous culture system, it has some aspects of an open system and allowed Czerkawski[64] to develop a conceptual compartmentation model of three, four, and five units.

## OTHER LABORATORY MODELS FOR STUDYING ANOXIC SYSTEMS

Although both the isolation and study of anaerobic bacteria and anaerobic communities are still dominated by the use of batch growth systems, different model systems, especially homogeneous single-stage chemostats, are increasingly being used. Some of these will be briefly reviewed as systems for studying anaerobic communities that contrast with the multistage chemostat.

Homogeneous single-stage chemostats have been used for bacterial enrichment,[66,67] for studying the growth of monocultures of bacteria,[68-70] especially rumen bacteria,[71-75] to provide competition between monocultures of bacteria,[58,76-78] as well as coculture studies.[79,80] Two-stage systems have been used to study bacterial competition,[5,81] anaerobic digestion,[62] and as a simple gut ecosystem simulation model.[82] In addition, semicontinuous systems[83] have also been developed.

Some heterogeneous models have also been used but much less extensively than the homogeneous variety. These include gel-stabilized systems (Wimpenny et al.,[53] Macfarlane et al.,[84] see also Chapter 1, Volume II) and batch and continuous culture in the presence of particles.[16,78]

In aquatic sediments, sediment slurry systems have been extensively used,[21,23,24,25,27,30,37,85] and although slurrying sediments can affect the rates of processes,[41] the pathway of carbon flow seems unaffected.[28]

Models based essentially on lysimeters[86] have been used to study landfill systems and can include provision for the control of refuse ingress, composition, particle size, density, degree of decomposition, temperature, degree of refuse saturation, moisture movement, and leachate recirculation. They may also be equipped with gas probes, thermocouples, tensiometers, neutron probe access tubes, and sampling ports. Such systems have been used to examine gas production,[87-92] leachate generation,[86,93-95] attenuation,[96] and codisposal[97] in the laboratory, in semiscale and *in situ*. Most of these studies, however, have focused on terminal products and not on fundamental microbiology. Similarly, studies with laboratory percolation columns,[98-100] which result in vertical stratification of microbes and their metabolites, of codisposal and leachate attenuation, have been somewhat lacking in definition with respect to fundamental microbiology and biochemistry.

Finally, mathematical models have been developed to simulate contaminant transport from a landfill.[101]

# REFERENCES

1. **Jørgensen, B. B.**, Mineralization of organic matter in the sea bed — the role of sulphate-reduction, *Nature*, 296, 643, 1982.
2. **Jones, J. G. and Simon, B. M.**, Differences in microbial decomposition processes in profundal and littoral lake sediments, *J. Gen. Microbiol.*, 123, 297, 1981.
3. **Jørgensen, B. B.**, Bacterial sulfate reduction within reduced micro niches of oxidised marine sediments, *Mar. Biol.*, 41, 7, 1977.
4. **Thauer, R. K., Jungermann, K., and Decker, K.**, Energy conservation in chemotrophic anaerobic bacteria, *Bacteriol. Rev.*, 41, 100, 1977.
5. **Cappenberg, T. E.**, A study of mixed continuous cultures of sulfate-reducing and methane producing bacteria, *Microb. Ecol.*, 2, 60, 1975.
6. **Kristjansson, J. K., Schönheit, P., and Thauer, R. K.**, Different $K_s$ values for hydrogen of methanogenic bacteria and sulfate reducing bacteria: an explanation for the apparent inhibition of methanogenesis by sulfate, *Arch. Microbiol.*, 131, 278, 1982.
7. **Schönheit, P., Kristjansson, J. K., and Thauer, R. K.**, Kinetic mechanism for the ability of sulfate reducers to out-compete methanogens for acetate, *Arch. Microbiol*, 132, 285, 1982.
8. **McInerney, M. J., Bryant, M. P., Hespell, R. B., and Costerton, J. W.**, *Syntrophomonas wolfei* gen. nov. sp. nov., an anaerobic syntrophic fatty acid-oxidizing bacterium, *Appl. Environ. Microbiol.*, 41, 1029, 1981.
9. **Tewes, F. J. and Thauer, R. K.**, Regulation of ATP-synthesis in glucose fermenting bacteria involved in interspecies hydrogen transfer, in *Anaerobes and Anaerobic Infections*, Gottschalk, G., Pfennig, N., and Werner, H., Eds., Fischer Verlag, Stuttgart, 1980, 97.
10. **Thauer, R. K. and Morris, J. G.**, Metabolism of chemotrophic anaerobes: old views and new aspects, in *The Microbe 1984 II: Prokaryotes and Eukaryotes*, Kelly, D. P. and Carr, N. G., Eds., Cambridge University Press, Cambridge, 1984, 123.

11. **Zeikus, J. G.,** Metabolic communication between biodegradative populations in nature, in *Microbes in their Natural Environments,* Slater, J. H., Whittenbury, R., and Wimpenny, J. W. T., Eds., Cambridge University Press, Cambridge, 1983, 423.

12. **Jørgensen, B. B.,** Ecology of the bacteria of the sulphur cycle with special reference to anoxic-oxic interface environments, *Philos. Trans. R. Soc. Lond., Ser. B,* 298, 543, 1982.

13. **Hargrave, B. T. and Phillips, G. A.,** Oxygen uptake of microbial communities on solid surfaces, in *Aquatic Microbial Communities,* Cairns, J., Jr., Ed., Garland Publishing Inc., New York, 1977, chap. 17.

14. **Pfennig, N.,** Microbial behaviour in natural environments, in *The Microbe 1984 II: Prokaryotes and Eukaryotes,* Kelly, D. P. and Carr, N. G., Eds., Cambridge University Press, Cambridge, 1984, 23.

15. **Widdel, F., Kohring, G. W., and Mayer, F.,** Studies on dissimilatory sulfate-reducing bacteria that decompose fatty acids. III. Characterization of the filamentous gliding *Desulfonema limicola* gen. nov. sp. nov., and *Desulfonema magnum* sp. nov., *Arch. Microbiol.,* 134, 286, 1983.

16. **Laanbroek, H. J. and Geerligs, H. J.,** Influence of clay particles (illite) on substrate utilization by sulfate-reducing bacteria, *Arch. Microbiol.,* 134, 161, 1983.

17. **Revsbech, N. P., Sørensen, J., and Blackburn, T. H.,** Distribution of oxygen in marine sediments measured with microelectrodes, *Limnol. Oceanogr.,* 25, 403, 1980.

18. **Sørensen, J., Jørgensen, B. B., and Revsbech, N. P.,** A comparison of oxygen, nitrate and sulfate respiration in coastal marine sediments, *Microb. Ecol.,* 5, 105, 1979.

19. **Winfrey, M. R., Marty, D. G., Bianchi, J. M., and Ward, D. M.,** Vertical distribution of sulfate reduction, methane production, and bacteria in marine sediments, *Geomicrobiol. J.,* 2, 341, 1981.

20. **Senior, E., Lindström, E. B., Banat, I. M., and Nedwell, D. B.,** Sulfate reduction and methanogenesis in the sediment of a saltmarsh on the east coast of the United Kingdom, *Appl. Environ. Microbiol.,* 43, 987, 1982.

21. **Winfrey, M. R. and Ward, D. M.,** Substrates for sulfate reduction and methane production in intertidal sediments, *Appl. Environ. Microbiol.,* 45, 193, 1983.

22. **Parkes, R. J. and Taylor, J.,** Characterization of microbial populations in polluted marine sediments, *J. Appl Bacteriol.,* Suppl. 155S, 1985.

23. **King, G. M., Klug, M. J., and Lovley, D. R.,** Metabolism of acetate, methanol, and methylated amines in intertidal sediments of Lowes Cove, Maine, *Appl. Environ. Microbiol.,* 45, 1848, 1983.

24. **Banat, I. M., Nedwell, D. B., and Balba, M. T.,** Stimulation of methanogenesis by slurries of saltmarsh sediment after the addition of molybdate to inhibit sulphate-reducing bacteria, *J. Gen. Microbiol.,* 129, 123, 1983.

25. **King, G. M.,** Metabolism of trimethylamine, choline, and glycine betaine by sulfate-reducing and methanogenic bacteria in marine sediments, *Appl. Environ. Microbiol.,* 48, 719, 1984.

26. **Mountfort, D. O. and Asher, R.,** Role of sulfate reduction versus methanogenesis in terminal carbon flow in polluted intertidal sediment of Waimea Inlet, Nelson, New Zealand, *Appl. Environ. Microbiol.,* 42, 252, 1981.

27. **Sørensen, J., Christensen, D., and Jørgensen, B. B.,** Volatile fatty acids and hydrogen as substrates for sulfate-reducing bacteria in anaerobic marine sediment, *Appl. Environ. Microbiol.,* 42, 5, 1981.

28. **Christensen, D.,** Determination of substrates oxidized by sulfate reduction in intact cores of marine sediments, *Limnol. Oceanogr.,* 29, 189, 1984.

29. **Banat, I. M., Lindström, E. B., Nedwell, D. B. and Balba, M. T.,** Evidence for coexistence of two distinct functional groups of sulfate-reducing bacteria in salt marsh sediment, *Appl. Environ. Microbiol.,* 42, 985, 1981.

30. **Taylor J. and Parkes, R. J.,** Identifying different populations of sulphate-reducing bacteria within marine sediment systems, using fatty acid biomarkers, *J. Gen. Microbiol.,* 131, 631, 1985.

31. **Nedwell, D. B.,** The input and mineralization of organic carbon in anaerobic aquatic sediments, in *Advances in Microbial Ecology,* Vol. 7, Marshall, K. C., Ed., Plenum Press, New York, 1984, 93.

32. **Marshall, K. C.,** Reactions of organisms, ions and macromolecules at interfaces, in *Contemporary Microbial Ecology,* Ellwood, D. C., Hedger, J. N., Latham, M. J., Lynch, J. M., and Slater, J. H., Eds. Academic Press, 1980, 93.

33. **Tempest, D. W., Neijssel, O. M., and Zevenboom, W.,** Properties and performance of microorganisms in laboratory culture: their relevance to growth in natural ecosystems, in *Microbes in their Natural Environments,* Society for General Microbiology Symp. 34, Slater, J. H., Whittenbury, R., and Wimpenny, J. W. T., Eds., Cambridge University Press, Cambridge, 1983, 119.

34. **Rees, J. F. and Grainger, J. M.,** Rubbish dump or fermenter? Prospects for the control of refuse fermentation to methane in landfills, *Process Biochem.,* Nov./Dec., 41, 1982.

35. **Fewson, C. A.,** Biodegradation of aromatics with industrial relevance, in *Microbial Degradation of Xenobiotics and Recalcitrant Compounds,* Federation of European Microbiological Societies Symp. 12, Leisinger, T., Hütter, R., Cook, A. M., and Nüesh, J., Eds., Academic Press, Orlando, Fla., 1981, 141.

36. **Senior, E. and Balba, M. T. M.,** Landfill Biotechnology, in *Bioenvironmental Systems,* Vol. II., Wise, D. L., Ed., CRC Press, Boca Raton, Fla., 1987, 17.

37. **Banat, I. M. and Nedwell, D. B.**, Mechanisms of turnover of $C_2$-$C_4$ fatty acids in high-sulphate and low-sulphate anaerobic sediments, *FEMS. Microbiol. Lett.*, 17, 107, 1983.

38. **Craib, J. S.**, A sampler for taking short undisturbed marine cores, *J. Cons. Int. Explor. Mer.*, 30, 34, 1965.

39. **Baxter, M. S., Farmer, J. G., McKinley, I. G., Swan, D. S., and Jack, W.**, Evidence of the unsuitability of gravity coring for collecting sediment in pollution and sedimentation rate studies, *Environ. Sci. Technol.*, 15, 843, 1981.

40. **Hungate, R. E.**, A roll tube method for cultivation of strict anaerobes, in *Methods in Microbiology*, Vol. 3B, Norris, J. R. and Ribbons, D. W., Eds., Academic Press, London, 1969, 117.

41. **Jørgensen, B. B.**, A comparison of methods for quantification of bacterial sulfate reduction in coastal marine sediments. I. Measurement with radiotracer techniques, *Geomicrobiol. J.*, 1, 11, 1978.

42. **Jørgensen, B. B.**, A comparison of methods for the quantification of bacterial sulfate reduction in coastal marine sediments. III. Estimation from chemical and bacteriological field data, *Geomicrobiol. J.*, 1, 49, 1978.

43. **Meyer-Reil, L. A., Dawson, R., Liebezeit, G., and Tiedge, H.**, Fluctuations and interactions of bacterial activity in sandy beach sediments and overlying waters, *Mar. Biol.*, 48, 161, 1978.

44. **Burnison, B. K. and Morita, R. Y.**, Heterotrophic potential for amino acid uptake in a naturally eutrophic lake, *Appl. Microbiol.*, 27, 488, 1974.

45. **Dawson, R. and Gocke, K.**, Heterotrophic activity in comparison to the free amino acid concentrations in Baltic seawater samples, *Oceanol. Acta.*, 1, 45, 1978.

46. **Christensen, D. and Blackburn, T. H.**, Turnover of tracer ($^{14}C$, $^3H$ labelled) analine in inshore marine sediments, *Mar. Biol.*, 58, 97, 1980.

47. **Gocke, K., Dawson, R., and Liebezeit, G.**, Availability of dissolved free glucose to heterotrophic microorganisms, *Mar. Biol.*, 62, 209, 1981.

48. **Christensen, D. and Blackburn, T. H.**, Turnover of $^{14}C$-labelled acetate in marine sediments, *Mar. Biol.*, 71, 113, 1982.

49. **Shaw, D. G., Alperin, M. J., Reeburgh, W. S., and McIntosh, D. J.**, Biogeochemistry of acetate in anoxic sediments of Skan Bay, Alaska, *Geochim. Cosmochim. Acta.*, 48, 1819, 1984.

50. **Parkes, R. J. and Taylor, J.**, Demonstration, using *Desulfobacter* sp., of two pools of acetate with different biological availabilities in marine pore water, *Mar. Biol.*, 83, 271, 1984.

51. **Jones, J. G. and Simon, B. M.**, Measurement of microbial turnover of carbon in anoxic freshwater sediments: cautionary comments, *J. Microbiol. Methods*, 3, 47, 1984.

52. **White, D. C.**, Analysis of microorganisms in terms of quantity and activity in natural environments, in *Microbes in their Natural Environments*, Slater, J. H., Whittenbury, R. and Wimpenny, J. W. T., Eds., Cambridge University Press, Cambridge, 1983, 37.

53. **Wimpenny, J. W. T., Lovitt, R. W., and Coombs, J. P.**, Laboratory model systems for the investigation of spatially and temporally organised microbial ecosystems, in *Microbes in their Natural Environments*, Slater, J. H., Whittenbury, R., and Wimpenny, J. W. T., Eds., Cambridge University Press, Cambridge, 1983, 67.

54. **Parkes, R. J. and Taylor, J.**, The relationship between fatty acid distributions and bacterial respiratory types in contemporary marine sediments, *Estuarine Coastal Mar. Sci.*, 16, 173, 1983.

55. **Parkes, R. J.**, Methods for enriching, isolating, and analysing microbial communities in laboratory systems, in *Microbial Interactions and Communities*, Vol. 1, Bull, A. T. and Slater, J. H., Eds., Academic Press, London, 1982, chap. 3.

56. **Wilkinson, T. G., Topiwala, H. H., and Hamer, G.**, Interactions in a mixed bacterial population growing on methane in continuous culture, *Biotech. Bioeng.*, 16, 41, 1974.

57. **Harder, W. and Dijkhuizen, L.**, Mixed substrate utilization, in *Continuous Culture: Applications and New Fields*, Dean, A. C. R., Ellwood, D. C., Evans, C. G. T., and Melling, J., Eds., Ellis Horwood, Chichester, U.K., 1976, 297.

58. **Kuenen, J. G. and Robertson, L. A.**, Competition among chemolithotrophic bacteria under aerobic and anaerobic conditions, in *Current Perspectives in Microbial Ecology*, Klug, M. J. and Reddy, C. A., Eds., American Society for Microbiology, Washington, D.C., 1984, 306.

59. **Senior, E., Balba, M. T. M., Lindström, E. B., and Nedwell, D. B.**, Analysis of interacting anaerobic microbial associations by use of multi-stage open culture systems, *Soc. Gen. Microbiol. Quarterly*, 8, 40, 1980.

60. **Thompson, L. A., Nedwell, D. B., Balba, M. T. M., Banat, I. M., and Senior, E.**, The use of multiple-vessel, open-flow systems to investigate carbon flow in anaerobic micobial communities, *Microb. Ecol.*, 9, 189, 1983.

61. **Ghose, T. K. and Bhadra, A.**, Maximisation of energy recovery in biomethanation process, *Process Biochem.*, Oct./Nov., 23, 1981.

62. **Hobson, P. N., Summers, R., and Harries, C.,** Single- and multi-stage fermenters for treatment of agricultural wastes, in *Microbiological Methods for Environmental Biotechnology,* Grainger, J. M. and Lynch, J. M., Eds., Academic Press, Orlando, Fla., 1984, 119.

63. **Hungate, R. E.,** *The Rumen and Its Microbes,* Academic Press, Orlando, Fla., 1966.

64. **Czerkawski, J. W.,** Microbial fermentation in the rumen, *Proc. Nutr. Soc.,* 43, 101, 1984.

65. **Czerkawski, J. W. and Breckenbridge, G.,** Design and development of a long-term rumen simulation technique (Rusitec), *B. J. Nutr.,* 38, 371, 1977.

66. **Keith, S. M., Herbert, R. A., and Harfoot, C. G.,** Isolation of new types of sulphate-reducing bacteria from estuarine and marine sediments using chemostat enrichments, *J. Appl. Bacteriol.,* 53, 29, 1982.

67. **Laanbroek, H. J., Abee, T., and Voogd, I. L.,** Alcohol conversions by *Desulfobulbus propionicus* Lindhorst in the presence and absence of sulfate and hydrogen, *Arch. Microbiol.,* 133, 178, 1982.

68. **Hallberg, R. O.,** An apparatus for the continuous cultivation of sulfate-reducing bacteria and its application to geomicrobiological purposes, *Antonie van Leeuwenhoek,* 36, 241, 1970.

69. **Vosjan, J. H.,** Respiration and fermentation of the sulphate-reducing bacterium *Desulfovibrio desulfuricans* in a continuous culture, *Pl. Soil.,* 43, 141, 1975.

70. **Häggström, M. H. and Anderson, E.,** Production of L and D-lactate by *Pediococcus pentosaceus* in batch and steady-state culture, *J. Appl. Bacteriol.,* 52, 119, 1982.

71. **Hobson, P. N.,** Continuous culture of some anaerobic and facultatively anaerobic rumen bacteria, *J. Gen. Microbiol.,* 38, 167, 1965.

72. **Wallace, R. J.,** Control of lactate production by *Selenomonas ruminantium:* homotropic activation of lactate dehydrogenase by pyruvate, *J. Gen. Microbiol.,* 107, 45, 1978.

73. **Russell, J. B. and Baldwin, R. L.,** Comparison of maintenance energy expenditures and growth yields among several rumen bacteria grown on continuous culture, *Appl. Environ. Microbiol.,* 37, 531, 1979.

74. **Mink, R. W., Patterson, J. A., and Hespell, R. B.,** Changes in viability, cell composition, and enzyme levels during starvation of continuously cultures (ammonia-limited) *Selenomonas ruminantium, Appl. Environ. Microbiol.,* 44, 913, 1982.

75. **Silley, P. and Armstrong, D. G.,** Changes in metabolism of the rumen bacterium *Streptococcus bovis* H13/1 resulting from alteration in dilution rate and glucose supply per unit time, *J. Appl. Bacteriol.,* 57, 345, 1984.

76. **Laanbroek, H. J., Geerligs, H. J., Peijnenburg, A. A. C. M., and Siesling, J.,** Competition for L-lactate between *Desulfovibrio, Veillonella* and *Acetobacterium* species isolated from anaerobic intertidal sediments, *Microb. Ecol.,* 9, 341, 1983.

77. **Veldkamp, H., Van Gemerden, H., Harder, W., and Laanbroek, H. J.,** Competition among bacteria: an overview, in *Current Perspective in Microbial Ecology,* Klug, M. J. and Reddy, C. A., Eds., American Society for Microbiology, Washington, D.C., 1984, 279.

78. **Laanbroek, H. J., Geerligs, H. J., Sijtsma, L., and Veldkamp, H.,** Competition for sulfate and ethanol among *Desulfobacter, Desulfobulbus,* and *Desulfovibrio* species isolated from intertidal sediments, *Appl. Environ. Microbiol.,* 47, 329, 1984.

79. **Allison, M. J. and Reddy, C. A.,** Adaptations of gastrointestinal bacteria in response to changes in dietary oxalate and nitrate, in *Current Perspectives in Microbiol Ecology,* Klug, M. J. and Reddy, C. A., Eds., American Society for Microbiology, Washington, D.C., 1984, 248.

80. **Russel, J. B. and Allen, M. S.,** Physiological basis for interactions among rumen bacteria: *Streptococcus bovis* and *Megasphaera elsdenii* as a model, in *Current Perspectives in Microbiol Ecology,* Klug, M. J. and Reddy, C. A., Eds., American Society for Microbiology, Washington, D.C., 1984, 239.

81. **Parker, R. B.,** Continuous-culture system for ecological studies of microorganisms, *Biotechnol. Bioeng.,* 8, 473, 1966.

82. **Veilleux, B. G. and Rowland, I.,** Simulation of the rat intestinal ecosystem using a two stage continuous culture system, *J. Gen. Microbiol.,* 123, 103, 1981.

83. **Miller, T. L. and Wolin, M. J.,** Fermentation by the human large intestine microbial community in a *in vitro* semicontinuous culture system, *Appl. Environ. Microbiol.,* 42, 400, 1981.

84. **Macfarlane, G. T., Russ, M. A., Keith, S. M., and Herbert, R. A.,** Simulation of microbial processes in estaurine sediments using gel-stabilized systems, *J. Gen. Microbiol.,* 130, 2927, 1984.

85. **Laanbroek, H. J. and Pfennig, N.,** Oxidation of short-chain fatty acids by sulfate-reducing bacteria in freshwater and in marine sediments, *Arch. Microbiol.,* 128, 330, 1981.

86. **Raveh, A. and Avnimelch, Y.,** Leaching of pollutants from sanitary landfill models, *J. Water Pollut. Control Fed.,* 51, 2705, 1979.

87. **Tanaka, M., Ikeguchi, T., Hanashima, M., Yamasaki, K., and Matsufuji, Y.,** Experimental studies of leachate and gas generation resulting from the use of wastes in land reclamation, in *Reclamation 83 Papers,* Industrial Seminars Ltd., Tunbridge Wells, U.K., 1983, 353.

88. **Games, L. M. and Haynes, J. M.,** On the mechanisms of carbon dioxide and methane production in natural anaerobic environments, *Proc. 2nd Int. Symp. Environ. Biogeochem.,*

89. **Baker, J. M., Peters, C. J., Perry, R., and Knight, C. P. V.,** Odour problems associated with solid waste disposal, *Public Health Eng.,* 12, 115, 1984.
90. **Klink, R. E. and Ham, R. K.,** Effects of moisture movement on methane production in solid waste landfill samples, *Resour. Recov. Conserv.,* 8, 29, 1982.
91. **DeWalle, F. B., Chian, E. S. K., and Hammerberg, E.,** Gas production from solid waste in landfills, *J. Environ. Eng. Div.,* EE3, 415, 1978.
92. **Ham, R. K.,** Predicting gas generation from landfills, *Waste Age,* Nov., 50, 1979.
93. **Qasim, S. R. and Burchinal, J. C.,** Leaching from simulated landfills, *J. Water Pollut. Control Fed.,* 42, 371, 1970.
94. **Tittlebaum, M. E.,** Organic carbon content stabilisation through landfill leachate recirculation, *J. Water Pollut. Control Fed.,* 54, 428, 1982.
95. **Pohland, F. G.,** Landfill stabilisation with leachate recycle, Annual progress Report, EPA, No. EP-00658, Environmental Protection Agency, Washington, D.C., 1972.
96. **Campbell, D. J. V., Parker, A., Rees, J. F., and Ross, C. A. M.,** Attenuation of potential pollutants in landfill leachate by Lower Greensand, *Waste Manage. Res.,* 1, 31, 1983.
97. **Schneider, von W., Gorbauch, H., and Rump, H. H.,** Studies on the behaviour of pollutants in refuse-filled lysimeters, *Forum Stadte-Hygiene,* 32, 200, 1981.
98. **Josephson, J.,** Immobilisation and leachability of hazardous wastes, *Environ. Sci. Technol.,* 16, 219A, 1982.
99. **Loch, J. P. G., Lagas, P., and Haring, B. J. A. M.,** Behaviour of heavy metals in soil beneath a landfill, *Sci. Total Environ.,* 21, 203, 1981.
100. **Senior, E. and Balba, M. T. M.,** The use of single-stage and multi-stage fermenters to study the metabolism of xenobiotic and naturally-occurring molecules by interacting microbial associations, in *Microbiological Methods for Environmental Biotechnology,* Grainger, J. M. and Lynch, J. M., Eds., Academic Press, Orlando, Fla., 1984, 275.
101. **Dasgupta, D., Sengupta, S., Wong, K. V., and Nemerow, N.,** Two-dimensional time dependent simulation of contaminant transport from a landfill, *Appl. Math. Modelling,* 8, 203, 1984.
102. **Coutts, D. A. P., Senior, E., and Balba, M. T. M.,** Multi-stage chemostat investigations of interspecies interactions in a hexanoate-catabolising microbial association isolated from anoxic landfill, *J. Appl. Bacteriol.,* 62, 251, 1987.

Chapter 4

BIDIRECTIONALLY LINKED CONTINUOUS CULTURE: THE GRADOSTAT

**Julian W. T. Wimpenny**

## INTRODUCTION

One great advantage of homogeneous liquid culture devices like chemostats is that they are open systems capable of reaching steady states. The latter are invaluable since they can reveal unequivocal responses to environmental manipulations. Temporal but not spatial heterogeneity can be examined in the single-stage chemostat. Good examples of time-dependent changes using the chemostat are the responses of photosynthetic bacteria to cycles of light intensity[1] or to cycles of nutrient concentration.[2] Spatial heterogeneity needs either a tube reactor or some interlinked multivessel array. These will be discussed.

### Multistage Continuous-Culture Systems

Multistage continuous-culture systems can incorporate spatial and temporal heterogeneity though flow is in one direction only, and therefore reciprocal interactions between neighboring vessels is impossible. The kinetics of multistage continuous-culture systems has been discussed by Herbert.[3] Such systems have been used to observe the sequential degradation of complex wastes such as coke-oven effluents[4] and diesel oil,[5] and in addition to investigate nitrification and denitrification in effluent streams.[6]

Nedwell and his colleagues[7,8] have established a sediment model system consisting of five vessels linked together in series and is aimed at reproducing the sequential degradation of substrates such as benzoate and glucose using sulfate and carbon dioxide as electron acceptors. Similar systems have been constructed to observe the degradation of paper and paper-mill effluents[9] (and see Parkes and Senior, this volume).

### The Introduction of Bidirectionality

Margalef[10] suggested the need for bidirectionally linked multistage systems in an interesting and thought-provoking article entitled "Laboratory Analogues of Estuarine Plankton Systems". Some of the comments and predictions made by the author are worth reading. Talking about multistage systems he says: "Any system of chemostats affords an ideal way for mapping time series into space; going down the row of flasks is equivalent to progressing along an ecological succession". An interesting example of just such an experiment was published earlier by the Czech authors Slezak and Sikyta[11] who set up a multistage continuous-flow system to model a batch culture in which each of the growth stages was represented by a single vessel. The nonsteady state time series of a microbial batch culture was thus mapped onto a steady state spatial array.

Margalef devised systems in which "a certain amount of diffusion and contamination 'upstream' has been permitted as in natural systems". These were the seeds that lead in the end to the development of bidirectionally linked multistage systems. Margalef went much further conceptually than simple bidirectionality. He speculated on three dimensional arrays, though he felt that they were too complex to establish practically. He imagined a ten by ten by ten array of vessels, for example, all linked together with rubber tubing and pumps.

"The adequately programed pumps could forward the liquid in either direction and by this mechanism, the effects of turbulent mixing could be simulated by pumping fluid from one vessel to another, stirring, pumping the same volume back again to the first and so on."

Such techniques, he considered, could give "a dynamic insight on how spatial patterns develop".

It is clear from these short excerpts that Margalef had made the critical jump from simple single-stage chemostats via multidimensional arrays, straight through to a sense that spatial heterogeneity was a prerequisite for pattern formation. Sadly, this work was largely ignored for some time after this.

Multistage systems have been examined from another point of view, this time in assessing survival of species. Stephanopoulos and Fredrickson[12] have discussed the principles of competitive exclusion which indicate that in a uniform environment, if two species are competing for a single nutrient, one must be eliminated. Coexistence is possible if each is using a different nutrient and, in addition, under certain other circumstances. There are still problems with the competitive exclusion principle, since examples exist where numerous species appear to coexist apparently competing for a very small range of nutrients. Temporal and spatial heterogeneity may explain such coexistence. Stephanopoulos and his colleagues have analyzed both situations. Spatial heterogeneity was invoked by considering a two-vessel system where the output from one vessel fed a second whose output was partially recycled to the first. Under these conditions, it was possible to find two species which could coexist when grown on the same substrate, each growing faster in a different vessel. Jager[21] investigating the mathematics of the gradostat came to similar conclusions.

A bidirectionally linked model system was developed by Cooper and Copeland[13] based on some of the ideas of Margalef discussed above and used also to investigate estuarine ecosystems. It consisted of a series of five 9-gal plastic containers. Each container was connected to its neighbor by two glass tubing couplers. Reservoirs were located at each end of the system and medium could overflow from each end. The containers were gently mixed and transfer of solutes was by diffusion through the glass coupling tubes. It is clear that actual transfer rates were dependent not only on the flow rates of medium through the system, but on the dimensions of the coupling tubes and on the stirring rate.

Cooper and Copeland used this system to investigate the microbial flora of an estuarine ecosystem. In this case, the gradient established was one of salt concentration. Freshwater flowed down through the system while saltwater was fed in from the second reservoir up the array of vessels. Salinity gradients are conservative in that the organisms present in the model do not metabolize NaCl, so that the gradient is unaffected by cell concentration. In operation, the system became colonized by the same microflora as was present at different positions in the estuary itself. While the results were qualitatively the same, they were not quantitatively similar to patterns in the natural habitat.

## THE GRADOSTAT

The Cooper-Copeland model originated as an ecologist's answer to the problem of modeling salinity gradients in a natural ecosystem; however, we arrived at a similar solution from the point of view of microbial physiologists versed in the art of continuous-culture techniques. Robert Lovitt and I have developed over the last few years a bidirectional compound chemostat which we named the "gradostat".[14-16] The perhaps infelicitous title arose after considering names of some of the other continuous-culture equipment capable of operating under steady-state conditions. Consider, for example, the chemostat itself which operates at a constant chemical composition, the turbidostat whose absorbance is held steady, and perhaps more esoteric instruments like the cyclostat which imposes a constant cycle of chemical composition on an otherwise steady state. A system that generates a constant gradient of one solute or other can only in the face of this competition be called a gradostat! Two versions of the gradostat have finally been developed: the first, in which separate fermenter vessels are connected together by pumped tubes or by overflows; the second, reverting to diffusion coupling as originally described by Cooper and Copeland.

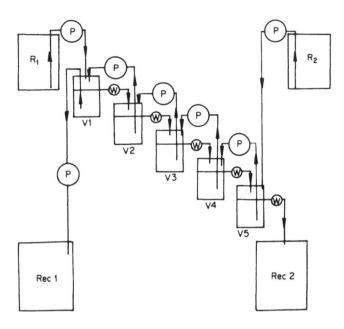

FIGURE 1. A diagramatic representation of the TC-gradostat: V1 to V5 are fermenter vessels; R1 and R2 are the two medium reservoirs feeding each end of the array; Rec1 and Rec2 are receivers; P indicates tubing pumped lines; W are overflow weirs.

## The Tubing-Coupled (TC) Gradostat

The TC gradostat (Figures 1 and 2) consists of a series of bidirectionally linked fermentation vessels. We have found in practice that five vessels is a suitable number to use. An odd number is sensible since it provides a "middle". More than five vessels is unwieldy while fewer than five reduces the resolution of the system. The vessels are linked together so that culture is transferred between neighboring vessels in two directions at the same time. Solutes are fed into the system from reservoirs located at each end of the array, and there are outlets for culture from each of the end vessels. A moment of reflection indicates that two different solutes (dyes for example) which are present in each reservoir will be distributed as opposing stepped gradients across the whole array. As will be shown later, under steady-state conditions these solute gradients will be linear from each source to each sink so long as the volumes of each vessel and the flow rates in each direction are the same. The presence of cells acting as sinks for substrates or as sources for products alter the distribution patterns.

### Construction of the Gradostat

Each vessel consists of a single fermentation unit. We have used simple 1 $\ell$ all-glass fermentation jars for four out of five vessels and employed a small commercial bench top chemostat as the fifth vessel to obtain reasonable oxygen transfer rates where aerobiosis was required at one of the ends of the array. Each of the glass vessels is equipped with magnetic stirring bars and with a glass weir to allow liquid transfer from vessel to vessel downwards in one direction. All fermenters have sampling ports and temperature sensors and are connected together with butyl rubber or with tygon tubing.

Temperature and agitation rate are controlled in each of the glass vessels by simple fermenter base units constructed in our electronics workshops. The vessels are arranged on a dexion slotted angle framework which provides a series of platforms each higher than the next. Medium flows by gravity from the highest to the lowest vessel over the weirs. The outlets of the latter enter the next lowest vessel below its liquid surface to ensure that

FIGURE 2.    The TC-gradostat in operation.

gas is not transferred up the series from the lowest vessel which is normally aerated. Culture is pumped in the upwards direction, and fresh medium and effluent are all transferred by a multichannel tubing pump. Fresh medium is held in sterile 20 ℓ glass aspirator bottles while the outlets are connected to unsterile glass bottles containing disinfectant. The gradostat is sterilized dry, by autoclaving.

The gradostat is not difficult to operate, although the numerous flow lines needed can lead to something that resembles a heap of spaghetti in the region of the multichannel tubing pump! After sterilization, the equipment is connected upon its stand under a flow of sterile gas, and it is then filled with either inoculated culture or with sterile medium by syphoning it in from the highest vessel and allowing it to flow through the system over the weirs. Where sterile medium is used, the system is inoculated by injecting 5 m$\ell$ of inoculum into each vessel. Temperature control, stirring, and agitation are started and the system allowed to grow and in the end to reach a steady state. Samples are removed through each sampling line.

*Transfer of Materials in the Gradostat*

We are grateful to Mr. D. J. Harries (Department of Pure Mathematics, University College, Cardiff) for his help in solving the transfer equations set out below. These calculations refer solely to the transfer of material which could be any solute or a particle such as a bacterium suspended in an aqueous medium. They only assume that the material is evenly distributed in the fluid and that it is not changing in concentration at all.

To determine the concentration $x_n$, of a solute in the nth vessel, the following differential equation applies:

$$V \frac{dx}{dt} n = ux_{n-1} + vx_{n+1} - (u + v)_{xn} \quad (n = 1, 2 \dots N) \tag{1}$$

where u and v are media flow rates through the vessels and V is the volume of the vessel.

If $x_0 = a$ and $x_{n+1} = b$ which are the concentrations of a different solute in each reservoir, and the initial conditions are such that $x_n = 0$ at time $t = 0 (n = 1, 2 \dots N)$, then solving the differential

$$x_n = \left(\frac{u}{v}\right)^{n/2} \frac{1}{N+1} \sum_{k=1}^{N} \frac{C_k}{\alpha_k} (e^{\alpha k t} - 1)\sin\left(\frac{nk\pi}{N+1}\right) \tag{2}$$

where:

$$\alpha_k = 2\lambda\left(\cos\frac{k\pi}{N+1}\right) + \mu$$

$$C_k = 2\lambda\left[a + b\left(\frac{v}{u}\right)^{N+1/2}(-1)^{k+1}\right]\sin\left(\frac{k\pi}{N+1}\right)$$

$$\lambda = -\frac{\sqrt{u+v}}{V} \quad \text{and} \quad \mu = \frac{-u+v}{V}$$

Equation 2 defines the concentrations of a solute in a particular vessel at any time after pumping commences and is clearly complex, though useful in that it handles the nonsteady state solution. As t tends to infinity, the system tends to a steady state given by

$$x_n = E(u/v)^n + F \tag{3}$$

where $E = (a - b)/\{1 - (u/v)^{N+1}\}$ and $F = \{b - a(u/v)^{N+1}\}$. Where the flow rates u and v are equal, this simplifies to

$$X_n = A + (n + 1)B \tag{4}$$

where $A = \{(N + 2)a - b\}/(N + 1)$ and $B = (b - a)/(N + 1)$.

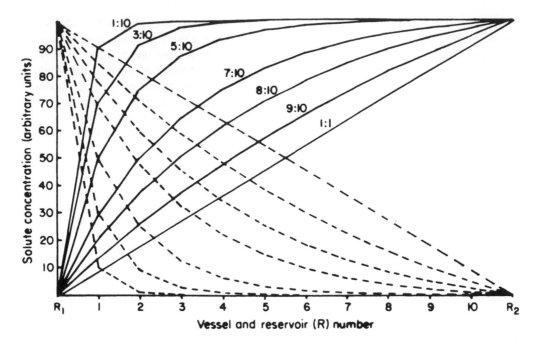

FIGURE 3.  Computer simulation of the steady state distribution of two solutes under conditions of different opposing solute flow rates. Flow rates range from 1:10 to 1:1. It should be noted that such flow regimes are only possible practically in a system which is pumped in each direction (From Wimpenny, J. W. T. and Lovitt, R. W., *Microbiological Methods for Environmental Biotechnology*, Grainger, J. M. and Lynch, J. M., Eds., Academic Press, London, 1984, 293. With permission.)

The equations defining solute transfer are complex, and numerical solutions using computer simulations were developed in my laboratory by S. Jaffe. One program called "gradsim" allows considerable flexibility in investigating the theoretical properties of the gradostat in the presence or absence of microorganisms. A series of simulations were run to determine the effects on solute concentration when the proportions of flow in each direction were varied from one (equal flow rates in both directions) to ten. The results (Figure 3) show that under steady-state conditions, only where the ratio is unity are solutes distributed linearly from sources at each end of the array to sinks at the opposite end. The steady-state solute concentrations at other values show an interesting range. Although unequal flow has not been tested experimentally, it is clear that useful experiments, perhaps in challenging an ecosystem with a xenobiotic, could result from the use of a gradostat in this way. Alterations in proportional flow require slight changes in the construction of the gradostat. Thus the weir system would need to be replaced with a series of pumped lines. One possible consequence is that slight variations in pumping rates due to variations in tube geometry may lead to changes in culture volume in each vessel if care is not taken.

The distribution equations were verified experimentally using methylene blue as a dye marker. Each vessel and one of the reservoirs contained distilled water. The second reservoir contained the dye and the concentration of the latter was determined in each vessel as a function of time after pumping started. Pumping rates and the volumes of each vessel were carefully measured and the slight variations noted almost certainly explain the very slight differences between the theoretical and observed results (Figure 4). It was concluded from these experiments that the system operates as theory dictates in both steady and nonsteady state cases.

*Gradsim* was used to simulate the residence time of particles in the gradostat. Residence times are the mean time that particles can be expected to remain in a flow system such as

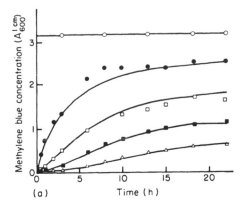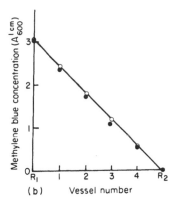

FIGURE 4. Dye distribution in a four-vessel gradostat. (a) As a function of time. Simulated values (continuous lines) and observed results (symbols). Vessels: 1, ●; 2, □; 3, ■; 4, △. (b) Steady state conditions: theoretical (○) and observed (●) values. (From Wimpenny, J. W. T. and Lovitt, R. W., *Microbiological Methods for Environmental Biotechnology*, Grainger, J. M. and Lynch, J. M., Eds., Academic Press, London, 1984, 293. With permission.)

the chemostat before they are washed out of the vessel and are expressed as the reciprocal of the dilution rate that is the volume of the vessel divided by the flow rate through it. The simulation program was flexible enough to allow predictions of residence times not only in the gradostat but also in a chemostat and in a multistage continuous-culture system. The results (Figure 5) show considerable differences. In each case, the dilution rate for the system as a whole is the same. Each vessel in each system contains the same initial concentration of particles which are assumed not to be growing. Individual vessels in a multistage array wash out sequentially as particle-free medium is pumped into the system. The residence time for the system as a whole is approximately the same as for a single vessel having the same total volume, as predicted by Herbert.[3] The situation is quite different in the gradostat where the residence time of any one vessel depends on its position in the array. Residence times of vessels near the periphery of the gradostat are shorter than residence times of the more central vessels. Once more, the average residence time for the gradostat as a whole is approximately the same as that for a single vessel of equivalent volume. This pattern of behavior has important consequences for biological experiments where organisms may be growing near the periphery of the system. This will be discussed in more detail later.

*Growth Experiments with the Gradostat*

A simple experiment is to grow a bacterium in opposing gradients of two essential nutrients fed from opposite ends of the array. *Paracoccus denitrificans* was chosen in view of its ability to grow on a variety of substrates aerobically or anaerobically. Counter gradients of succinate and nitrate were selected for investigation. Preliminary experiments showed that if the concentrations were balanced adequately, the organism grew in the center vessel in a five-vessel array (Figure 6). Cells were distributed away from this position approximately linearly towards each outlet, while growth in vessel three provided a sink for substrates from each reservoir. Results of these experiments suggested that growth was not exactly balanced in the center vessel, but that some growth also took place in the second vessel.

Since the majority of microbial habitats are spatially organized and cells are growing, often at very low concentrations of solute molecules, it is important to determine the kinetic parameters responsible for the precise spatial location of particular groups of organisms. The gradostat is not capable of the same spatial resolution as seen in a natural diffusion coupled system, for example in gel-stabilized models. This is because the gradients are

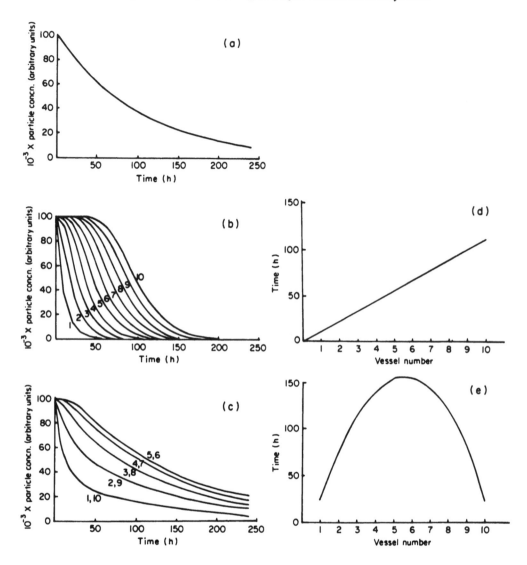

FIGURE 5. Washout curves from computer simulations of (a) a single stage chemostat, (b) a ten-vessel multistage chemostat, (c) a ten-vessel multistage gradostat. Each system contained initially uniformly distributed "particles" which were then washed out by a flow of particle free liquid. Dilution rate was 0.01 per hour for each complete system. Time is for a 40% reduction in particle concentration in (d) the multistage chemostat and (e) the gradostat. (From Wimpenny, J. W. T. and Lovitt, R. W., *Microbiological Methods for Environmental Biotechnology*, Grainger, J. M. and Lynch, J. M., Eds., Academic Press, London, 1984, 293. With permission.)

"discretized", so that resolution is limited to the number of vessels in the array. In the experimental gradostat, five vessels are used, while up to ten can be incorporated into the *gradsim* program. Within the limits imposed by this lack of spatial resolution, certain information on key spatial determinants can be obtained using computer simulations.

The basic strategy we decided to use for these simulations is that the organism is growing in ten vessels in opposing gradients of two essential nutrients. Various different kinetic constants are included in the simulation while all other parameters are held constant. The effects of growth yield were investigated first by holding yield on one substrate constant and varying the other. When the growth yield values are equal, growth occurs in the center of the array. Since the simulations are based on the use of ten vessels, this means that most growth takes place in Vessels 5 and 6. As Figure 7 illustrates, alterations in yield coefficient

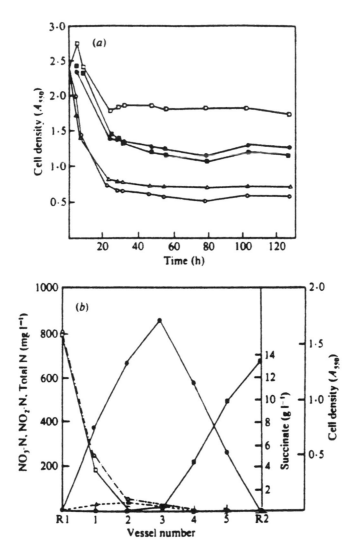

FIGURE 6. Growth of *Paracoccus denitrificans* in a five-vessel gradostat in opposing gradients of nitrate and succinate. (From Lovitt, R. W. and Wimpenny, J. W. T., *J. Gen Microbiol.*, 127, 261, 1981. With permission.)

change the position of maximal growth: the higher the yield on a given substrate the further from the source of that substrate can growth take place.

Similar simulations were carried out where the affinity of the organism for each substrate was varied over quite wide limits. The affinity coefficient, Ks, is the substrate concentration at which growth rates are half of their maximum value[17,18] and tend to be very low (of the order of micromolar). Unexpectedly, wide differences in Ks had only a minor effect on the position of growth. This was probably because the cells were highly substrate limited in all but the center vessels, so that substrate availability rather than affinity was the limiting factor. Once more, this is an important observation, since opposing solute gradients at highly limiting concentrations must be very common in nature.

Growth rate can be altered in the gradostat just as it is in a chemostat, that is by altering the dilution rate. The consequences of changes in dilution rate in the gradostat are again surprising, however, and perhaps relevant to growth in certain natural ecosystems. As the

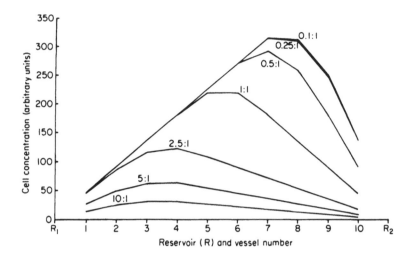

FIGURE 7.   The effect of growth-yield coefficients on predicted growth behavior in a ten-vessel gradostat. Growth was assumed to be limited by two essential nutrients, one from each of the two reservoirs. Ratios of the amount of substrate needed to synthesize one unit of cells varies from 10:1 to 0.1:1. (From Wimpenny, J. W. T. and Lovitt, R. W., *Microbiological Methods for Environmental Biotechnology*, Grainger, J. M. and Lynch, J. M., Eds., Academic Press, London, 1984, 293. With permission.)

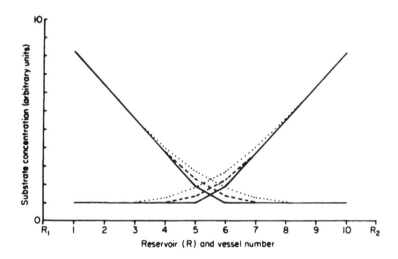

FIGURE 8.   A computer simulation of the effects of various dilution (growth) rates on the steady state concentrations of two essential nutrients flowing in opposite directions from the two reservoirs. Dilution rates in mℓ/hr: ———, 200; - - -, 800; . . . , 1000. (From Wimpenny, J. W. T. and Lovitt, R. W., *Microbiological Methods for Environmental Biotechnology*, Grainger, J. M. and Lynch, J. M., Eds., Academic Press, London, 1984, 293. With permission.)

dilution rate rises, so growth in the central region in the grandostat becomes unrestricted by substrate concentration. Thus, in the simulations, the concentrations of both substrates rise to significant levels near the center of the system (Figure 8). A further rise in dilution rate causes this region of unrestricted growth to spread towards the ends of the system. Only if dilution rate is significantly increased does the cell population at last wash out of the gradostat. This dilution rate expressed on a *per vessel* basis is much higher than the maximum growth

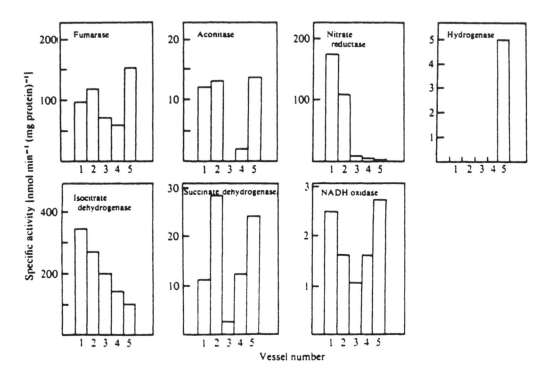

FIGURE 9. Some enzyme levels in cells of *Escherichia coli* grown in a five-vessel gradostat in counter gradients of oxygen + nitrate and glucose. The basal salts medium also contained casamino acids (Oxoid®). (From Lovitt, R. W. and Wimpenny, J. W. T., *J. Gen. Microbiol.*, 127, 61, 1981. With permission.)

rate of the organism. The effect described is related to the spreads in residence times in the gradostat discussed below.

Betty Tang[19] has analyzed the effect of dilution rate on growth position in the gradostat mathematically. She has shown that the position of maximum growth on a single nutrient entering from one reservoir need not be in the vessel adjacent to that reservoir. Increasing flow moves the maximum nearer to the middle of the system. Once more, this must be attributed to the higher washout rates in the vessels at the two ends of the array.

It has been suggested by Kuenen[20] that gradients of essential nutrients are not possible in the gradostat, since if the nutrient is limiting, it cannot spread beyond the region of growth. It is worth pointing out that this is actually true for any spatially organized ecosystem, since if nutrients are diffusing vectorially from a source to a sink they are not likely to penetrate beyond the sink if sink demand exceeds supply. In addition, the simulations just described suggest that gradients are possible over the whole gradostat if dilution rates are fast enough and growth is no longer nutrient limited.

So far, most of the discussion has centered on the theoretical characteristics of the gradostat. A number of laboratory experiments are described below to try and illustrate the value of this model to investigations of the microbiology of gradient systems.

The widespread existence of structured ecosystems where organisms are growing in solute gradients suggests that metabolic versatility has great survival value. Many bacteria are facultative anaerobes which can grow aerobically or anaerobically on a variety of substrates. Many are capable of using inorganic electron acceptors such as nitrate in the absence of oxygen. One such organism is *Escherichia coli*, and this was chosen for growth in the gradostat in opposing gradients of oxygen plus nitrate and glucose. The basal medium was a complex one containing casamino acids and a mineral salts mixture. Although cells, together with culture medium, are transferred at equal rates across the gradostat array, many adaptive changes are seen in cells removed from each of the five vessels (Figure 9). Only cells from the most anaerobic vessel possessed the enzyme hydrogenase which is a good marker for strict anaerobiosis. Cells from the aerobic end of the array possessed nitrate reductase. This enzyme is normally repressed and inhibited by oxygen, suggesting that conditions were oxygen limited at the oxidizing end of the system. Energy charge measurements indicated that cells near the center of the array were either starving or inhibited by the high levels of nitrite which had accumulated. Incorporation of a "transfer" zone, in which cells cannot grow or are actually subject to toxic agents, is itself an interesting property of the gradostat and again distinguishes it from its better known relative the single stage chemostat. It is not possible to model starvation under steady state conditions in the chemostat, since by definition the system must be growing to operate. Once more, there must be numerous natural habitats containing groups of organisms which cannot grow because they lack one essential nutrient, but through which other nutrients diffuse to regions where growth is possible. Such transfer zones are easily incorporated into gradostat experiments.

In a number of experiments which include competitve interactions and enrichments from natural ecosystems, it was noted that the cell population fell *exponentially* from its growth point to other regions in the gradostat. The distribution predicted from Equation 4 given earlier demonstrate that cell (or solute) concentrations should fall in a *linear* fashion from sources to sinks. The experimental results therefore indicate that antagonism and growth inhibition are possible in the gradostat.

The gradostat provides an important tool for investigating competition between organisms. Since there are usually five vessels in the system, there exist five potential niches which could theoretically be occupied by at least five separate species if the nutrient conditions were right. It might be expected that if one organism was occupying some of the space in a gradostat, the addition of a second to the system could either displace the first to another vessel or lead to its complete disappearance from the system. We have seen that *E. coli* can adapt very adequately to a range of environmental factors in the gradostat. What happens if it meets other species? The facultative anaerobe was grown on a casamino acids yeast-extract salts medium in counter gradients of oxygen and glucose in the gradostat. Growth (Figure 10) took place throughout the system. Once a steady state had been reached, a culture of *Pseudomonas aeruginosa* was added. This grew in the aerobic vessels and displaced *E. coli* from this end. The latter continued to happily proliferate at the anaerobic end until another culture, this time of *Clostridium acetobutylicum*, was injected into the system. This now outgrew *E. coli* in the anaerobic region and the latter was reduced to almost insignificant numbers by the time that the experiment was stopped.

The gradostat is well suited to the establishment of enrichment cultures to do particular jobs, since it provides a family of habitats each having a separate physico-chemical identity. One such enrichment culture was established to find a group of organisms capable of oxidizing and reducing inorganic sulfur compounds. The basal medium was a salts solution, and lactate was fed in from the reservoir at the reducing end, while sulfate plus nitrate were added to the system from the oxidizing end. An inoculum derived from water at the base of an oil storage tank was used to seed the gradostat. After a period, the system developed with extensive wall growth. The culture was terminated and the contents of each vessel used to inoculate the gradostat. This time, wall growth was at a minimum and evidence was collected that a functional sulfur cycle was operating. Thus, (Figure 11a) there were obligately an-aerobic sulfate reducing bacteria isolated from the lactate end and sulfide oxidizing nitrate reducers from the oxidizing end. The distribution of sulfur compounds across the array gave evidence that sulfur cycling was indeed taking place.

As with the chemostat, measurements of events taking place between steady states can be informative. We have already described the distribution of dye markers before steady state values were reached: experiments with *Paracoccous denitrificans,* both in the laboratory and through computer simulations, illustrated transients between steady states and showed, in general, good agreement.

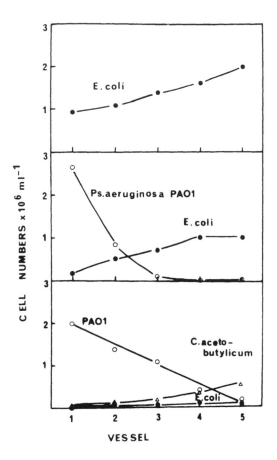

FIGURE 10.    Competition in the gradostat. (a) Growth of *E. coli* alone
in a casamino acids-containing medium in countergradients of oxygen and
glucose. (b) The addition of *Pseudomonas aeruginosa* to the system. (c)
The addition of *Clostridium acetobutylicum* to the same system.

The lactate/sulfate enrichment culture just described showed another example of transient behavior when concentrations of lactate were increased while those of sulfate plus nitrate were reduced. The results (Figure 11b) show that concentrations of lactate in Vessel 5 rise and of sulfate in Vessel 1 fall as might be expected. There is a large transient rise in sulfide concentration in Vessel 4 while a smaller fall in sulfide is recorded in Vessel 5 before the new steady state is reached.

As part of an investigation into the growth of mixed cultures of strict aerobes with anaerobes,[22] *P. dentrificans* was grown in a single-stage chemostat or in a gradostat with *Desulfovibrio desulfuricans*. When grown together in the chemostat, the fate of each species depended on the oxygen level in the gas mixture. At a critical oxygen tension, it was possible to establish an oscillating state where first one species then the other predominated. A stable steady-state population was never observed. When the same species were grown in sulfate plus oxygen vs. lactate counter tradients in the gradostat, the species segregated spatially with the aerobe growing nearest the oxidizing end of the array. Position, especially of the sulfate reducer, was dependent on the relative concentrations of lactate and sulfate (Figure 12).

## The Diffusion-Coupled (DC) Gradostat

Connecting a group of vessels together with tubing and pumps, though effective, is sometimes untidy and bulky. We decided to examine the feasibitility of a gradostat in which

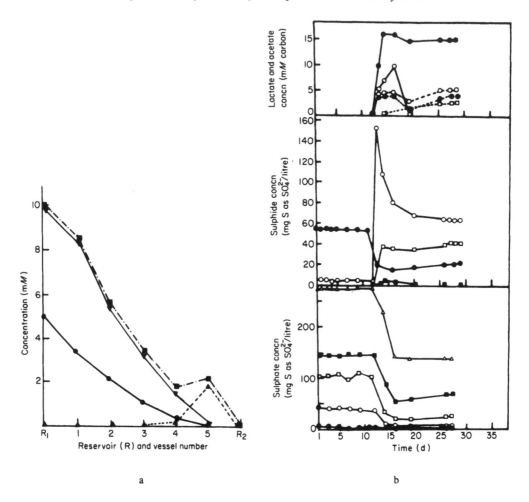

FIGURE 11.    Growth of an enrichment culture in a five-vessel gradostat in opposing gradients of lactate and sulfate plus nitrate. R1 contained 10 m$M$ sodium sulfate and 5 m$M$ sodium nitrate; R2 contained 27 m$M$ sodium lactate. (a) Steady state values: total sulfur, ■; sulfide, ▲; sulfate, ▼; nitrate, ●. (b) Transient values when the reservoir concentration was changed as follows: sodium lactate 33 m$M$; sodium sulfate 5 m$M$; sodium nitrate 1 m$M$. Vessels 1, △; 2, ■; 3, □; 4, ○; 5, ●. Top: ———, lactate; - - -, acetate. (From Wimpenny, J. W. T. and Lovitt, R. W., *Microbiological Methods for Environmental Biotechnology,* Grainger, J. M. and Lynch, J. M., Eds., Academic Press, London, 1984, 293. With permission.)

vessels were connected by diffusion or mechanically assisted diffusion, as originally described by Cooper and Copeland.[13] A number of versions of such a gradostat have been constructed. These range from a simple system of linked bottles, to a sophisticated machine constructed from industrial glassware and stainless steel.

*A Simple Glass Bottle Gradostat*

A simple gradostat was constructed from borosilicate milk dilution bottles. These have a square profile which allows them to be packed together closely. An array of six bottles was built, each connected by a flow-coupling device which enters the vessel through holes drilled in opposite sides of the bottle. After the holes were drilled, the rough edges were fire polished and the whole bottle annealed to relieve any stresses. The array was agitated on a New Brunswick orbital shaker, and exchange between vessels was based on variable-rate rotary mixing. Considerable time was spent in developing the exchange device. Initially, each vessel was connected by a short length of stainless steel hypodermic tubing. Exchange was a linear function of tubing length until the tube was so long that virtually no exchange took place.

The system finally selected consisted of a glass tube containing a glass or stainless steel ball free to move backwards and forwards. Two short lengths of stainless steel enter this chamber through butyl rubber plugs. Gyrotatory motion causes the ball to oscillate backwards and forwards, acting like a loosely coupled pump. The importance of loose coupling must be stressed. If the ball were a very good fit it would simply pump liquid from each vessel alternately into and out of the glass tube with little or no mixing. The degree of mixing produced will depend on the gap between the ball and the glass tube wall. Such a simple gradostat seems suitable for long-term enrichment techniques, but in its present form, does not incorporate good mixing. Exchange rates have been monitored using methylene blue dye transfer and the machine has been used to cultivate anaerobic sulfate reducers together with sulfur oxidizing photosynthetic species.

The square bottles employed make the possibility of a two-dimensional system easy to envisage. All that would be necessary would be to connect bottles together across all four of their sides.

## The Stirred DC-Gradostat

This is a system based on 100-mm glass tubing. A gradostat (Figure 13 a,b,c) was built in collaboration with H. Gest and J. Hayes of the University of Indiana, Bloomington. Industrial borosilicate glass tubing 100 mm in diameter, having 7 mm thick walls, was cut into 75 mm sections. The ends of each section were ground smooth with 150 and 400 mesh carborundum powder to remove irregularities left after sawing the sections. Each section was drilled on opposite sides with 40 and 25 mm diameter holes. End plates and spacer plates were constructed as indicated from 316 grade stainless steel. Shaft bushes and exchange ports were constructed from Delrin plastic. Butyl rubber (3 mm) was used to fabricate seals between all plates and the glass sections. Turbines were centered in each vessel and the shaft turned through a single bearing (LH Fermentation Ltd., Slough, U.K.) by a variable-speed electric motor.

Transfer dynamics were impossible to predict and were determined empirically using dye transfer rates. Once more, methylene blue was used in these experiments, and Figure 14 shows that transfer between vessels in the latest DC-gradostat is a function of stirring rate up to about 1000 rpm. Above this value, exchange rate was constant or fell slightly. This fall may be due to decoupling of the turbine from liquid flow at higher speeds.

The mass transfer rate, $k_La$, was determined as a function of stirring rate by the sulfite method. Although only a few points were determined, $k_La$ values were satisfactory at around 600 per hour at the higher agitation rates (Figure 14). Although not a central aspect of the development of this equipment, it is clear just by inspection of the DC-gradostat in operation that horizontal rotation in the unbaffled vessels leads to excellent mixing and $k_La$ values which approach those seen in the conventional laboratory stirred fermenter.

*Methylophilus methylotrophus* was established in the DC-gradostat in a methanol gradient entering the system from the left-hand reservoir. Each of the five vessels was aerated individually. The experiment was continued for 120 hr by which time a steady state had been reached. Total and viable counts were determined for cells from each vessel (Figure 15). Highest cell numbers and highest overall viability were seen in Vessel 1 nearest the methanol source. Methanol is detectable only in Vessels 1 and 2, and then only in very small amounts, suggesting that the organisms were growing in excess methanol in the first, possibly also in the second vessels. It is clear that the viable fraction falls towards the opposite end of the system where no methanol is available, suggesting that starvation is leading to death of a fraction of the cells. The gradostat provides an opportunity to investigate steady-state starvation in cell populations where growth, as in this case, occurs in one or few vessels only.

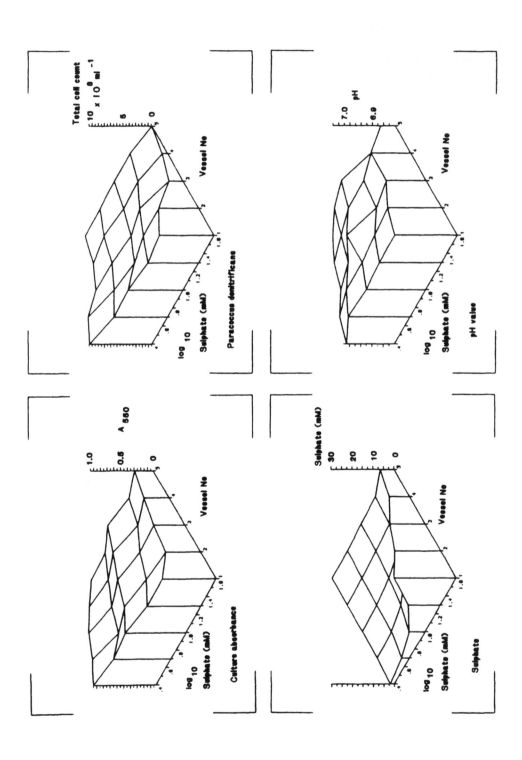

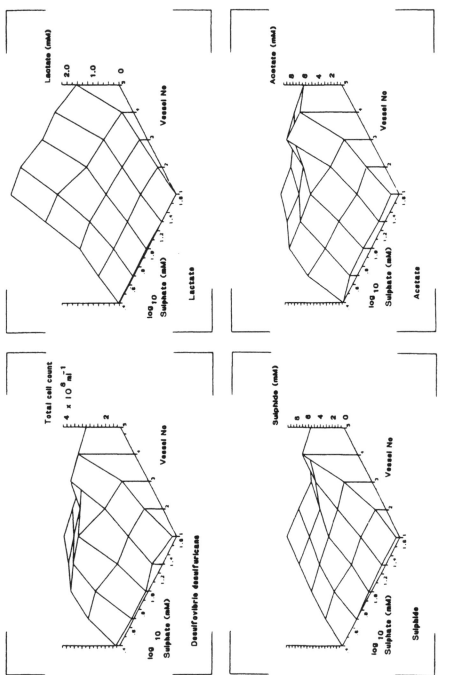

FIGURE 12. Growth of *Paracoccus denitrificans* and *Desulfovibrio desulfuricans* in countergradients of lactate and sulfate plus oxygen. The experiment was run at five different sulfate concentrations from 2.5 to 40 m*M*. Results for each of the variables indicated in the figure are plotted as a three-dimensional surface of activity *vs.* gradostat vessel number and log$_{10}$ sulfate concentration. (Courtesy of H. Abdollahi.)

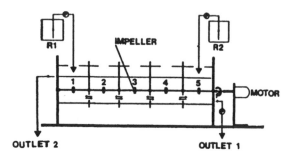

a

FIGURE 13.    The DC-gradostat. (a) Diagram of the whole system. (b) Photograph of the assembled gradostat. (c) Photograph of component parts of the gradostat.

The same system has been used to separate two species of photosynthetic bacteria in a gradient of NaCl concentration. The organisms *Rhodopseudomonas marina/agilis* and *Rhodobacter capsulata* were selected for their difference in salt sensitivity and because they contain different photosynthetic pigments which can be easily detected spectrophotometrically. NaCl (3%) was included in the medium in the left-hand reservoir, while the medium in the right-hand reservoir was left untreated. As the results (Figure 16) show, both organisms grew well in the gradostat, the *Rhodopseudomonas marina/agilis* at high salt concentrations, the *Rhodobacter capsulata* at low concentrations. These results could easily be confirmed by inspection alone since *Rhodopseudomonas marina/agilis* is a bright purple-pink color, while *Rhodobacter capsulata* has a dark greeny-brown appearance and colors in the DC-gradostat reflected the apparent distribution. A more quantitative assessment of distribution was obtained by measuring the ratios of the pigment absorption peaks at 882 and at 858 nm, which are characteristic for *Rhodopseudomonas marina/agilis* and *Rhodobacter capsulata*, respectively. This ratio changed from 1.48 to 0.67 over the array of vessels from the high salt to the low salt regions in the gradostat. Most growth, and the high alkalinity that results from photooxidation of the malate salts, appeared near the center of the array. This is a reflection of the position-dependent retention time in the gradostat alluded to earlier.

There are a number of points that arise out of the DC-gradostat configuration. In contrast to the original tube-coupled device, each chamber is in direct contact with its neighbors. This poses two related problems. Any changes in hydrostatic pressure lead to a pumping action from one chamber to the next. This can be disastrous when it occurs. Fortunately, it is easy to resolve. The exit gas manifolds must be connected together with wide bore tubing. If the latter are too narrow, then during sparging, condensate can bridge the tube, which can itself cause a small rise in pressure in the relevant chamber. The second problem concerns the impeller geometry. The impeller should be located centrally and every effort made to ensure that the flow it induces is symmetrical in the vessel. If it is not, this can also lead to pumping between chambers.

A third problem with direct coupling is caused by the size of the transfer port. This is small enough that it can occasionally become blocked, depending on the type and density of organisms in the system. All these problems are eliminated if the system is modified slightly to ensure that there is at no time a free connection between neighboring vessels. Though this modification has not yet been tested, it is clear that it ought to work and will therefore be described (Figure 17)! Transfer ports are enlarged. This will reduce the possibility of blockages forming. Mounted on the axle is a spring-loaded *disc-shaped plastic shutter.* As it rotates, a window cut in the disc exposes the transfer port for a moment. Medium in the relevant vessel fills the depth of the port. On the other side of the separator plate is

FIGURE 13b

FIGURE 13c

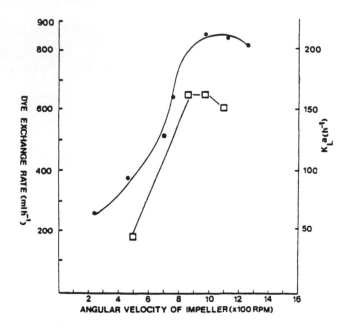

FIGURE 14.    Dye transfer (●) and oxygen transfer (□) rates as functions of rotation speed in the DC-gradostat.

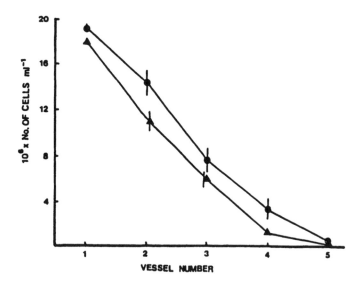

FIGURE 15.    Growth of *Methylophilus methylotrophus* in a methanol gradient in the DC-gradostat. ●, total cell counts; ▲, viable counts.

another spring-loaded disc, also rotating, also fitted with a window which is out of phase with the window on the first disc. As this second window passes the transport port, medium from the opposite side is exchanged for fresh medium from its own side. Each chamber is therefore linked by a smaller chamber whose volume depends on the diameter of the transfer port and its total depth. Transfer rates ought to be rotation-rate independent, apart, that is, from the speedier equilibration due to higher turbulence at faster angular velocities.

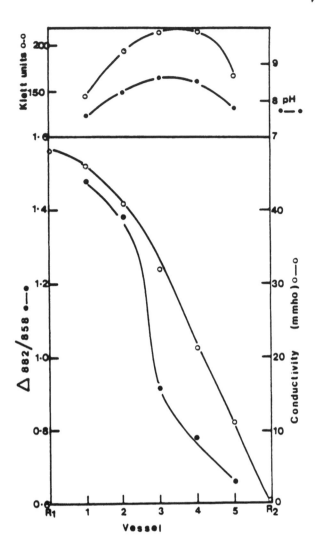

FIGURE 16.  Growth of *Rhodopseudomonas marina/agilis* and *Rhodobacter capsulata* in a sodium chloride gradient in the gradostat. The growth medium contained salt, vitamins, casamino acids, and sodium malate as the major carbon source.

## A Shear-Coupled (SC) Gradostat

A third design for the gradostat has been recently developed based loosely on a concept already briefly described.[23] This system was designed to attempt to increase the spatial resolution of the gradostat while retaining its bidirectionality and its open conformation. The original idea was to use a precision glass tube equipped with a rotating shaft on which was located a number of accurately machined discs. The space between these discs formed each vessel, while the gap between the wall and the disc made an exchange "port". This design was never build since it was felt that exchange would be too quick in such a system. It was decided that the principle was still sound if the discs were made "thick" enough to form a longer path between vessels. A simple model was constructed consisting of a perspex tube and a close-fitting perspex rod. The rod was machined in ten places to cut out chambers which would make up the vessels. Simple bearings were fitted at each end and sampling ports located at the two ends of the system. The shaft was rotated with a variable-speed motor. Initial experiments with dye markers proved highly satisfactory. Examination of the

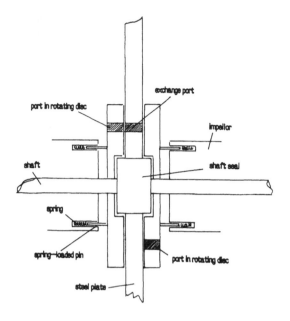

FIGURE 17.    Design for shutter system isolating neighboring vessels, yet still permitting exchange between them.

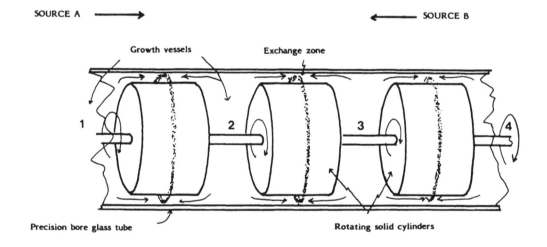

**SHEAR-COUPLED GRADOSTAT**

FIGURE 18.    Design for a shear-coupled gradostat.

thick disc during operation indicates a sharp boundary near the center of the disc where the high dye concentration from one vessel meets the low concentration in its neighbor. The two concentrations mix at this point at a rate which is some function of the rotational shear field. A gradient developed across the array at a rate which was suitable for many biological experiments. At the time of this writing, a full-scale system is still under construction (Figure 18).

# FUTURE DEVELOPMENTS OF THE GRADOSTAT

## Applications in Microbiology

It seems likely that the gradostat will have a number of useful applications in pure and applied microbiological research, A few suggestions are included below:

### Genetic Applications

Some configurations of the gradostat can generate gradients of any desired shape, suggesting that it has valuable properties in genetic experiments designed to investigate the selection of mutant strains resistant to particular agents. For instance, if the ratio of flow rates in either direction is greater than unity, high concentrations of the agent will be present only in the first vessel, providing a strong selective challenge, while the remaining vessels act as "sanctuaries" for sensitive cells.

It ought to be possible to devise experiments using the gradostat, where plasmids specifying particular characteristics can be transferred from one organism to another under specific environmental conditions. Experiments underway at the moment will hopefully throw light on survival of strains possessing antibiotic resistance plasmids and cured strains lacking resistance when the two cell types are grown together in an antibiotic gradient.

### Enrichment

As described earlier, the gradostat is an invaluable tool for the isolation of enrichment cultures where microbial consortia can interact even though some of the constituent species require mutually exclusive habitats. The sulfur cycle is just one example of a system where obligate anaerobes interact with obligate aerobes via inorganic sulfur compounds. Such an interaction would not be possible in single-stage homogeneous systems. Another good example is the nitrogen cycle, where nitrification and nitrate reduction require the presence and absence of oxygen, respectively.

Enrichments for such natural cyclic systems could then provide a useful experimental system for investigating the effects of xenobiotics on microbial ecosystems, a subject that is of great importance environmentally and one which is badly in need of more reliable experimental results.

Such investigation are only one example of the more general possibilities inherent in establishing long-term stable microbial communities and investigating conditions that perturb them.

### Physiological Adaptations

Our experiments with *E. coli* indicate the value of the gradostat in investigating the physiological responses of a single species to changes in environmental makeup. Numerous other examples could be treated in exactly the same way.

An important area here is the survival of organisms in habitats where essential nutrients are either absent or present in vanishingly small concentrations. Starvation is probably commoner in Nature than a condition of well-fed abundance and high growth rates. As has been stressed earlier, the gradostat incorporates "transfer" zones where cells are essentially nongrowing due to the absence of at least one essential nutrient. On the other hand, certain nutrients pass through transfer zones to other areas where growth may be possible. The survival of organisms in transfer zones is an important topic which really demands further research, and one in which the gradostat seems to have potential.

### Competition Studies

The competition experiment described in this chapter is one of a family of interesting experiments that are possible in the gradostat. The experiment showed that a "generalist"

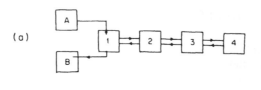

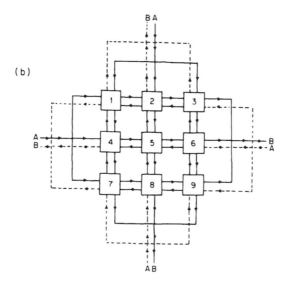

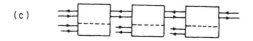

FIGURE 19.    Possible alternative gradostat configurations; (a) the single ended system; (b) two-dimensional systems; (c) separating cells from solutes within each vessel by using membrane filters (cf Herbert, this volume).

organism like *E. coli* can occupy a wide range of habitats; so long as it is in sole possession of the system, the addition of "specialists", in this case obligate aerobes and anaerobes, lead to its exclusion. This experiment adds weight to the prevailing view that specialists will outgrow generalists in their own specific habitat while the generalist can win where other options are open to it. Numerous experiments of this type can be devised to test the hypothesis further.

Similarly, other forms of competition can be investigated, for example amensalism and even predator-prey relationships.

**Other Possible Gradostat Conformations**

It is possible to connect the gradostat in a wide range of different configurations (Figure 19). The only rules are that volumes should remain constant in each vessel and that net input should equal net output. There would be no problem in having a net flow from left to right say, as in a multistage chemostat while exchange took place between each of the vessels. It would also be possible to have material entering both end vessels, or even into all the vessels and leaving the system from just a single vessel. Such possibilities will almost certainly have some parallel in natural ecosystems.

## The Single-Ended System

The principle of bidirectional exchange, since it represents the fundamental property of diffusion, can be applied to double-ended systems where nutrient is entering the array from both ends, but it can equally apply to a system where nutrient is entering from one end only (Figure 19a). A couple of natural examples will make this clearer. The bacterial colony is an interesting collection of organisms growing in a rather flat pattern on the surface of an agar plate. Incubated aerobically, oxygen enters the system from above while nutrients enter from below. Similarly, a narrow stratum in a marine sediment containing sulfur oxidizing bacteria are being fed with sulfide from below and oxygen from above. Both systems can be represented by double-ended gradostat experiments. Bacterial film, however, forms on the surface of inanimate or at any rate impermeable objects in most cases. These can include ship hulls, rocks, steel structures in the sea, and also the surfaces of human teeth. Here solute exchange is through one surface only. Elimination of the reservoir and receiver from one end of the gradostat establishes it as a laboratory analogue of this situation.

## Two-Dimensional Gradostats

It is theoretically possible to construct a two-dimensional gradostat by connecting stirred vessels in two directions at the same time. The simplest two-dimensional array consists of four vessels (Figure 19b), and the system would get unmanageable if much larger arrays were needed. It ought to be possible to construct a version of the gradostat using interconnected square bottles as suggested earlier, though this has not yet been attempted. At any rate, another dimension in space increases the degree of freedom of any one experiment, though it also increases its complexity!

A possible disadvantage of the gradostat is that solutes and organisms travel together throughout the system as a uniform suspension obeying the distribution equations described earlier. The only way that these distribution patterns alter is because growth and death rates are different in each of the niche-spaces.

## Separating Cells from Solutes

It may be interesting therefore to separate cells from substrates using filter membranes (Figure 19c). Two systems can be devised: in the first, neighboring vessels are separated by a membrane which allows solutes to diffuse through but keeps cells apart. This is the basis of the Herbert diffusion system discussed in detail in Chapter 5 in this volume. Another alternative is that cells would be separated from the medium in their own chambers, while the solutes would be transferred as usual from vessel to vessel by pumping. The cell chamber could be closed or open. If the latter, it would itself be equipped with inputs and outputs at right angles to the main solute gradient flow and perhaps consisting of a basal medium lacking major solutes. Open systems are intrinsically more satisfying than the closed variety, since true steady states become possible.

In conclusion, I see the gradostat as a powerful tool in future investigations of microbial behavior where differences in habitat physical-chemistry play an important part in the expression of the system. While recognizing the complexity at least of the original tube coupled device, more recent developments mean that simple effective and convenient devices can be built, and this should encourage microbiologists to apply such techniques to their own problems.

## ACKNOWLEDGMENTS

I would like to express my gratitude to Robert Lovitt whose contributions to gradostat work were seminal, to Steve Jaffe who wrote the gradostat simulation program, to Richard Earnshaw some of whose unpublished results are quoted here, and finally to Howard Gest and his colleagues in Bloomington, Indiana, for their interest and active participation in developing the DC-gradostat.

# REFERENCES

1. **Van Gemerden, H.,** Coexistence of organisms competing for the same substrate: an example among the purple sulfur bacteria, *Microb. Ecol.,* 1, 104, 1974.
2. **Beeftink, H. H. and Van Gemerden, H.,** Actual and potential rates of substrate oxidation and product formation in continuous cultures of *Chromatium vinosum, Arch. Microbiol.,* 121, 161, 1979.
3. **Herbert, D.,** A theoretical analysis of continuous culture systems, *Soc. Chem. Ind. Monogr.,* 12, 21, 1960.
4. **Abson, J. W. and Todhunter, J. H.,** Plant for continuous biological treatment of carbonisation effluents, *Soc. Chem. Ind. Monogr.,* 12, 147, 1960.
5. **Pritchard, P. M., Ventullo, R. M., and Suflita, J. M.,** The microbial degradation of diesel oil in a multistage continuous culture system, in *Proc. 3rd Int. Biodegrad. Symp.,* Miles-Sharply, J. and Kaplan, A. M., Eds., Applied Sciences Publishers, London, 1976.
6. **Hawkes, H. A.,** Eutrophication of rivers, effects, causes and control, in *Treatment of Industrial Effluents,* Callely, A. G., Forster, C. F., and Stafford, D. A., Eds., Hodder and Stoughton, London, 1977, 159.
7. **Thompson, L. A., Nedwell, D. B., Balba, M. T., Banat, I. M., and Senior, E.,** The use of multiple-vessel open flow systems to investigate carbon flow in anaerobic microbial communities, *Microb. Ecol.,* 9, 189, 1983.
8. **Thompson, L. A. and Nedwell, D. B.,** Existence of different pools of fatty acids in anaerobic model ecosystems and their availability to microbial metabolism, *Fed. Eur. Microbiol. Soc. Microb. Ecol.,* 31, 141, 1985.
9. **Parkes, R. J. and Taylor, J.,** The relationship between fatty acid distribution and bacterial types in contemporary marine sediments, *Estuarine Coastal Mar. Sci.,* 16, 173, 1983.
10. **Margalef, R.,** Laboratory analogues of estuarine plankton systems, in *Estuaries: Ecology and Populations,* Lauff, G. M., Ed., Hornshafer, Baltimore, 1967, 515.
11. **Slezak, J. and Sikyta, B.,** Continuous cultivation of microorganisms: a new approach and possibilities for its use, *J. Biochem. Microbiol. Technol. Eng.,* 3, 357, 1961.
12. **Stephanopoulos, G. and Fredrickson, A. G.,** Effects of spatial inhomogeneities on the coexistence of competing microbial populations, *Biotechnol. Bioeng.,* 21, 1491, 1979.
13. **Cooper, D. C. and Copeland, B. J.,** Responses of continuous-series estuarine micro-ecosystems to point-source input variations, *Ecol. Monogr.,* 14, 213, 1973.
14. **Lovitt, R. W. and Wimpenny, J. W. T.,** The gradostat: a tool for investigating microbial growth and interactions in solute gradients, *Soc. Gen. Microbiol. Q.,* 6, 80, 1979.
15. **Lovitt, R. W. and Wimpenny, J. W. T.,** The gradostat: a bidirectional compound chemostat and its applications in microbiological research, *J. Gen. Microbiol.,* 127, 261, 1981.
16. **Lovitt, R. W. and Wimpenny, J. W. T.,** Physiological behaviour of *Escherichia coli* grown in opposing gradients of oxidant and reductant in the gradostat, *J. Gen. Microbiol.,* 127, 269, 1981.
17. **Monod, J.,** The growth of bacterial cultures, *Ann. Rev. Microbiol.,* 1, 371, 1949.
18. **Monod, J.,** La technique de culture continue: theorie et applications, *Ann. Inst. Pasteur,* 79, 390, 1950.
19. **Tang, B.,** Mathematical investigations of growth of microorganisms in the gradostat, Preprint No. 303, Sonderforschungsbereich 123, University of Heidelberg, Heidelberg, West Germany, 1984.
20. **Kuenen, J. G.,** Discussion to paper by J. W. T. Wimpenny, *Philos. Trans. R. Soc. London,* 297, 515, 1983.
21. **Jager, W.,** Personal communication.
22. **Abdollahi, H. and Wimpenny, J. W. T.,** Unpublished observations.
23. **Wimpenny, J. W. T. and Lovitt, R. W.,** The investigation and analysis of heterogeneous environments using the gradostat, in *Microbiological Methods for Environmental Biotechnology,* Grainger, J. M. and Lynch, J. M., Eds., Academic Press, London, 1984, 293.

Chapter 5

# BIDIRECTIONAL COMPOUND CHEMOSTATS: APPLICATIONS OF COMPOUND DIFFUSION-LINKED CHEMOSTATS IN MICROBIAL ECOLOGY

**R. A. Herbert**

## INTRODUCTION

Natural environments are inhabited by a great diversity of microorganisms which encompass a complete spectrum of physiological and nutritional types, ranging from obligate aerobes to exacting anaerobes, and from autotrophs to heterotrophs. One of the fundamental questions of microbial ecology remaining to be answered is how such physiologically distinct populations coexist and develop in natural ecosystems. The majority of natural microbial ecosystems are heterogeneous, and microorganisms generally develop in spatially organized physico-chemical gradients. Physico-chemical gradients can vary in size by several orders of magnitude, from the smallest, which may develop over a distance of only a few micrometers, e.g., microbial films[1,2] and sediment particles,[3] to larger aggregates of particles, e.g., flocs, soil crumbs, microbial mats,[4-6] while in exceptional circumstances they may extend over several tens of meters, e.g., hot springs, stratified lakes, and sewer outfalls.[7,8] In all these habitats, the microorganisms are spatially organized, either vertically or horizontally in relation to prevailing physico-chemical gradients and function as highly complex microbial communities. Surfaces can also provide spatial arrangements which may play an important role in the development of microbial communities through the effects of the metabolic activity of adjacent microorganisms. The physico-chemical heterogeneity of microbial ecosystems not only encourages a wide diversity of nutritional and physiological types of microorganisms to develop, but also facilitates interactions between organisms of dissimilar physiology, e.g., between aerobes and anaerobes. A further advantage of heterogeneous microbial populations is that they are better able to withstand environmental changes and consequently are more stable than comparable ecosystems which have a low species diversity. In addition to spatial organization, one must also consider temporal changes in microbial ecosystems, because though over an extended timescale many microbial communities appear to be in a steady state, they nonetheless undergo numerous short-term successional changes. For example, microbial communities are markedly influenced by fluctuations, either regular or irregular, in such environmental parameters as pH, temperature, salinity, dissolved oxygen concentration, and availability of organic carbon. These environmental factors clearly exert a significant selective pressure on microbial ecosystems and may play a major role in controlling the activity of microbial communities.

Having established that a characteristic feature of most microbial communities is that they are both spatially and temporally heterogeneous, it is important to consider the mechanisms whereby metabolites are transferred from one physiological group of microorganisms to another. For example, in anoxic estuarine sediments or waterlogged soils, the mineralization of organic detritus is a complex process which is due primarily to the restricted metabolic potential of the individual microorganisms in such habitats. Complete mineralization does not take place by one species of microorganism alone.[9] Each physiological group exploits only a fraction of the detrital energy available and supplies the remainder in the form of excreted metabolic endproducts (lower fatty acids, alcohols, etc.) to the next member of the detrital food chain. In anoxic environments such as sediments or waterlogged soils, then, the microbial processes which take place are related to the quantity and quality of inorganic carbon and concentrations of inorganic electron acceptors, such as $NO_3^-$, $SO_4^{2-}$, $Fe^{3+}$, $Mn^{4+}$ and $CO_2$. The movement of solute molecules, both organic and inorganic, is therefore essential for the successful functioning of these microbial communities, and molecular diffusion is the principal solute transfer mechanism involved.[10,11] In addition, mechanical forces such as capillary action and hydraulic pumping assist diffusion processes. For solute transport, mediated by molecular diffusion, to be effective, it must have direction to facilitate transfer from producer to utilizer, and this may be a characteristic feature of all spatially organized microbial communities especially those in which the microorganisms are attached to surfaces. In ecosystems such as the water column of stratified lakes, the majority of microorganisms are motile, or like some of the photosynthetic bacteria, contain gas vacuoles and are able to position themselves at optimum points along concentration gradients. Sorokin[12] reported that the purple-red layer of *Chromatium okenii* showed diurnal vertical migrations over distances of up to 2 m in the meromitic Lake Belovod, U.S.S.R.

From what has been said, it is clear that the complex structural and dynamic nature of microbial ecosystems makes *in situ* studies extremely difficult, and this, coupled with the constantly changing conditions within the environment during the period of investigation, makes valid interpretation of field data a major problem. The experimental methodology currently available to study microbial ecosystems *in situ* is either insufficiently sensitive or discrete to probe these microhabitats without destroying their structural integrity, although considerable advances have been made recently in the development and application of needle microelectrodes for measuring physico-chemical gradients.[6,13] The *in situ* studies carried out by Jorgensen and his coworkers[13] have elegantly demonstrated the effects of the diurnal cycling of $O_2/H_2S$ gradients and their influence on the daily vertical migrations of populations of *Oscillatoria* spp., *Beggiatoa* spp., and *Chromatium* spp. in the surface layers of a marine lake.

In order to determine how microbial communities function in natural environments, it is frequently necessary to develop simple laboratory model systems to elucidate the key processes involved. The principal advantage of laboratory-based models is that environmental variables such as temperature, $O_2$ concentration, pH, redox potential, nutrient concentrations, and growth rate can be precisely controlled for the duration of the experiment. As a consequence of this degree of control, individual parameters can be varied in an ordered and predictable manner so that the experimental data obtained are easier to interpret. However, if model systems are to advance our understanding of microbial communities *in situ*, it is vitally important that their design is relevant to the ecosystem under investigation. Problems usually only arise when data from simple model systems are extrapolated uncritically to interpret complex processes occurring in natural environments. Laboratory models should not be considered as substitutes for *in situ* investigations, since their primary function is to greatly simplify the interactions between microbial populations and the environment in order to identify the key processes involved.

It is now well recognized that microbial activity in natural environments results from the actions of mixed populations of microorganisms, yet until recently, the whole philosophy of microbial research has been to study single, pure cultures.[14,15] In devising laboratory model systems, we must take into account that in natural environments monocultures are the exception rather than the rule. Equally, we must recognize the fact that within mixed microbial populations, a whole range of complex interactions and relationships exist between individual groups of microorganisms. Bull and Slater[16] have reviewed the types of microbial interactions which occur in natural environments. These range, for example, from commensalism, where only one population benefits, to competition, where microorganisms compete for limiting nutrients. In designing laboratory models, all these features need to be incorporated if the model is to accurately reflect real microbial ecosystems.

# MODEL SYSTEMS FOR STUDYING MICROBIAL INTERACTIONS

A primary requirement of laboratory systems is to incorporate features which bear a close resemblance to the properties of the natural habitat. Batch culture systems are the usual traditional method, with the advantage of technical simplicity and the potential for establishing replicates. Batch culture liquid systems are characterized by continuously changing growth conditions and the necessity to start with artificially high concentrations of all the growth substrates. Since the majority of natural environments are oligotrophic, it is self-evident that the physiological and biochemical properties of microorganisms grown under such conditions are likely to be quantitatively irrelevant compared with growth under natural conditions.

With the introduction of the continuous flow systems, and in particular, chemostats by Monod[17] and Novick and Szilard,[18] a major advance in experimental methodology was available which enabled useful ecological parameters to be modeled. The basic principles of chemostat growth will not be discussed in this chapter, and the reader should consult reviews by Bull,[19] Pirt,[20] Veldkamp and Jannasch,[21] Jannasch and Mateles,[22] and Bull and Brown[23] for further details. The open nature of continuous flow systems reproduces in gross terms the dynamic nature of natural environments, and since medium input is balanced by outflow of metabolic products and cells, the system can be maintained under steady-state conditions. Similarly, a variety of parameters such as Eh, pH, dissolved oxygen, and temperature can be automatically controlled, thereby leading to stable and reproducible experimental conditions. A further advantage of continuous-culture systems is that dilution rate (flow rate of fresh medium/culture volume) determines the growth rate of the microbial populations. In most natural environments, it is probable that for much of the time microorganisms are growing at submaximal rates. This property of chemostat systems enables a systematic analysis of the influence of a growth of a microorganism on various physiological parameters. Chemostats can be used at low dilution rates and low nutrient concentrations to enrich microorganisms which never become dominant in nutrient-rich batch-culture experiments. Jannasch[24,25] has successfully exploited this feature of continuous flow systems to successfully isolate autochthonous microorganisms from seawater. Subsequent kinetic analysis of the growth characteristics of autochthonous bacteria confirmed the specialized nature of their scavenging role (i.e., low maximum specific growth rate, high substrate affinity), reflecting the nutrient-poor environment from which they were isolated. While open-flow systems such as chemostats offer a number of features which are ecologically relevant, they suffer from a number of disadvantages. For example, they are homogenous experimental systems, whereas most natural environments are highly heterogeneous. Equally they are designed to maintain constant conditions, whereas natural ecosystems are subject to numerous transient regimes of such parameters as temperature, nutrient concentrations, salinity, and day length. A few attempts have been made to incorporate temporal heterogeneity, particularily light/dark regimes, in continuous culture systems, but much still remains to be done to design systems which simulate both regular and irregular transient changes.[26,27] In recent years, continuous-culture systems have been used to isolate and examine stable mixed cultures or microbial communities and their ability to degrade xenobiotic compounds.[28-30] Furthermore, they are useful for looking at many aspects of microbial interactions such as competition and predator-prey relationships. Thus, despite their obvious limitations, single-stage chemostats have provided a valuable and powerful research tool to probe the physiology of microorganisms and the function of microbial processes.

An alternative approach to single-stage chemostat systems is to use a series of separate continuous-flow systems, which may be linked in a number of different configurations. Multistage systems result in the transfer of cells, metabolic end products, and substrates from one vessel to another. A variety of systems have been developed, of which the simplest is a series of homogenously stirred, linked vessels supplied with nutrients from one end only. Wimpenny[31] has used the term unidirectional to describe these systems. Numerous modifications of the basic system are possible, and, for example, Jannasch[24] successfully used a system involving three vessels and varied the dilution rate and therefore the growth rate by using vessels of different volumes. Staged systems based on these principles have been used to study sequential microbial processes, e.g., nitrification and denitrification[32] or to follow simple food chains and trophic structures.[33] Unidirectional continuous-flow systems therefore have all the advantages of single-stage systems, but still lack the facility to model spatially heterogeneous ecosystems.

With the development of bidirectional flow-systems, in which there is a simultaneous supply and counterflow of different nutrients from each end, spatially heterogeneous model systems became available.[11,31,34] The reader is referred to the preceding chapter by Wimpenny for a detailed discussion of bidirectional flow systems and their application in microbial ecology. In my laboratory, we have developed an alternative strategy to investigate microbial interations involving the development of multistage continuous-flow diffusion chemostats.

A number of attempts have been made to grow defined or natural populations of microorganisms within biologically isolated environments while still maintaining free access to solutes. The microbial populations under investigation are retained within semipermeable membranes, e.g., dialysis tubing, and are open to the dissolved nutrients of the surrounding environment.[35,36] The simplest experimental system is to enclose the microbial population in dialysis tubing, which is then suspended in the natural environment under investigation.[35] To eliminate the formation of solute gradients across the dialysis sac, the more sophisticated systems incorporate a motor and are driven at a constant speed to ensure good mixing. Diffusion chamber chemostats have been successfully used to grow microorganisms such as the colorless sulfur bacterium *Thiovulum*, which has exacting requirements for hydrogen sulfide and oxygen. Jannasch et al.[37] used a two-chamber diffusion system in which hydrogen sulfide generated in one chamber by a sulfate-reducing bacterium *Desulfovibrio aestuarii* diffused into the second chamber, containing a population of *Thiovulum*, which was constantly circulated with oxgenated seawater. The oxygen/hydrogen sulfide gradient which developed in this experimental system enabled populations of *Thiovulum* to be maintained for up to a year.

The compound continuous-flow diffusion chemostat system we have developed was designed to follow the path of carbon flow in defined mixed cultures of anaerobic bacteria isolated from estuarine sediments and the effect of carbon availability on the microbial interactions involved. The system has also been used to study the interactions between obligately aerobic nitrifying bacteria belonging to the genera *Nitrosomonas* and *Nitrobacter* and facultatively anaerobic bacteria.

## CONSTRUCTION OF COMPOUND CONTINUOUS FLOW DIFFUSION CHEMOSTATS

The compound bidirectional diffusion chemostat system developed in my laboratory by my graduate student Stephen Keith and myself was constructed from standard 1-$\ell$ Quickfit® reaction vessels linked together, as shown in Figure 1. The overflow weirs in each chamber were so arranged that the working volume of each vessel was approximately 1 $\ell$. Standard Quick-fit® socket joints (B19/26) were fitted to the upper surface of each vessel to facilitate the input of growth media, acid or alkali, sterile gas addition (air- or oxygenfree nitrogen), and location of sensing electrodes (pH and dissolved oxygen electrodes). The contents of each chamber were physically separated by 142 mm diameter 0.2 μm pore size Durapore® (polyvinylidene difluoride) tortuous path membrane filters (Millipore Corporation Inc.). Durapore® membranes were selected since they are chemically inert, tough, flexible,

Gas inlet

(air or N$_2$)

Mineral salts

medium input

Acid/Alkali

input

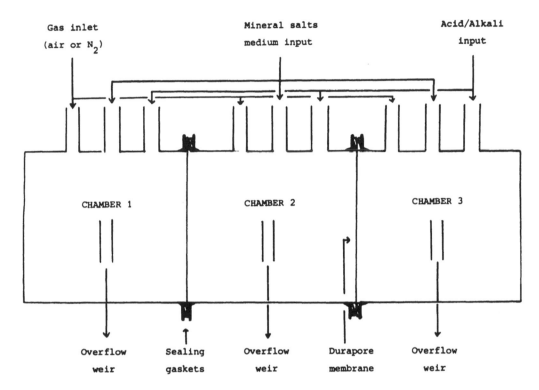

FIGURE 1. Schematic diagram of a three stage bidirectional diffusion chemostat. (Reproduced with permission from Keith, S. M. and Herbert, R. A., *FEMS Microbiol. Ecol.*, 31, 241, 1985.)

and steam sterilizable. Rubber gaskets were used to ensure a leak-proof seal between the membranes and the ground-glass flange joints. For anaerobic cultures, these were oxygen-impermeable butyl rubber gaskets; however, when growing autotrophic microorganisms, such as nitrifying bacteria, silicone rubber gaskets were selected since they do not contain growth inhibitory components. The diffusion boundary between each chamber is only the thickness of the Durapore® membrane, which is in the order of 100 to 125 μm thick, and since each chamber is continuously stirred by magnetic followers, solute diffusion gradients are greatly reduced. When required, each chamber can be operated independently at different dilution rates, and therefore different growth rates. The bidirectional compound system we have developed is similar in concept to the commercially produced "Ecologen®" (New Brunswick Scientific Co., New Jersey) in that both systems employ semipermeable membranes to separate the growth chambers. The Ecologen® system comprises a 500-mℓ capacity central diffusion reservoir surrounded by four 100- to 300-mℓ capacity satellite growth chambers (Figure 2a). Membrane filters are installed between the central diffusion reservoir and each growth chamber, thereby allowing metabolites but not cells to diffuse, via the central reservoir, into individual growth chambers. Crump and Richardson[38] reported that 0.1 μm polycarbonate membranes were required as *E. coli* was able to penetrate membranes with larger pores within 48 hr. The contents of the Ecologen® growth chambers are agitated by means of a reciprocating shaker (Figure 2b), although continuous shaking results in rupture of the membranes.[38] A major advantage of our system is that it is applicable to a greater number of ecological situations than the Ecologen®, and since the spatial arrangement of the growth chambers is linear, sequential microbial processes can be followed. This linear arrangement of the chambers confers a further advantage in that the individual vessels are only separated by the thickness of the semipermeable membrane, whereas in the Ecologen® system, the growth chambers are radially located around a central reservoir and no direct

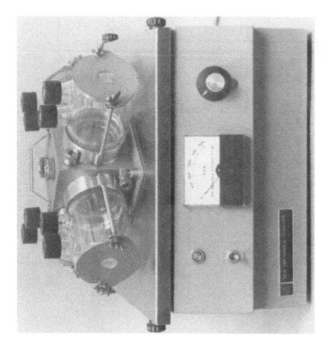

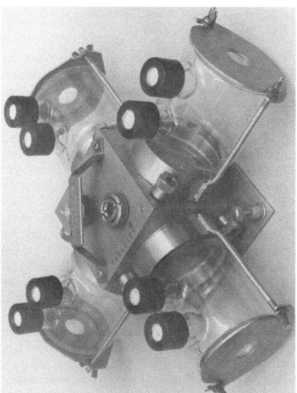

FIGURE 2. (a) Spatial layout of the Ecologen® central diffusion reservoir and growth chambers. (b) Location of Ecologen® system on reciprocating shaker.

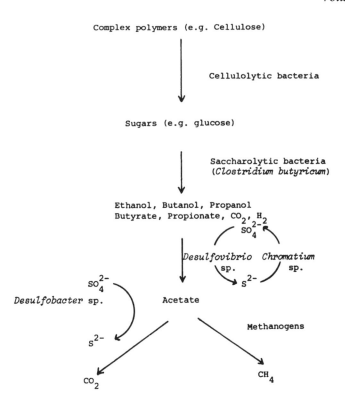

FIGURE 3.  Schematic representation of the role of saccharolytic clostridia, sulfate reducing bacteria, and phototrophic bacteria in carbon/electron flow in anaerobic estuarine sediments. (Reproduced with permission from Keith, S. M. and Herbert, R. A., *FEMS Microbiol. Ecol.*, 31, 241, 1985.)

exchange is possible. In our system, the individual microbial populations can be grown under precisely controlled conditions, and if required, at different growth rates. The bidirectional system we have developed can be used to study microbial interactions such as ammensalism, commensalism, mutualism, and competition. In the following section, we illustrate the application of this model system to the study of carbon flow in anaerobic estuarine sediments and to the study of interactions between autotrophic nitrifying bacteria and facultatively anaerobic nitrate-respiring bacteria.

## APPLICATIONS OF COMPOUND BIDIRECTIONAL DIFFUSION CHEMOSTATS TO ECOLOGICAL PROBLEMS

### Carbon Flow in Anaerobic Estuarine Sediments

Strict anaerobes are commonly isolated from sediments in the Tay estuary,[39,40] and for this study we have used a saccharolytic clostridium, identified as a strain of *Clostridium butyricum*, a sulfate reducing bacterium belonging to the genus *Desulfovibrio desulfuricans* and a purple sulfur bacterium identified as a strain of *Chromatium vinosum*. While these three bacteria form only a simple experimental system, they nonetheless enable the flow of carbon to be followed according to the simplified scheme presented in Figure 3. The objectives of this study were to investigate the effects of carbon and inorganic nitrogen source availability on the populations of these three bacteria when grown in a compound-diffusion chemostat. The layout of the experimental system is shown in Figure 4, with the population of *Clostridium butyricum* in Chamber 1, *D. desulfuricans* in Chamber 2, and *Chromatium*

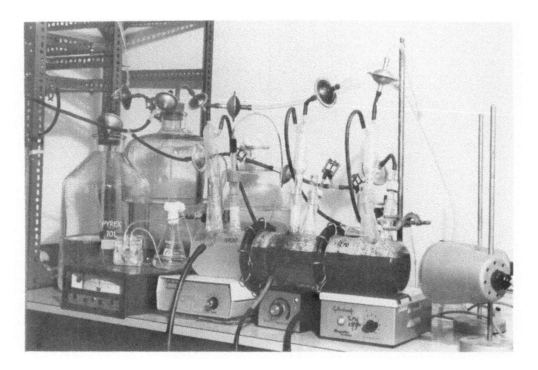

FIGURE 4.    Experimental layout for the investigation of carbon/electron flow in a bidirectional flow compound diffusion chemostat. Chamber 1 (left) contains *Clostridium butyricum*, Chamber 2 (center) *Desulfovibrio desulfuricans*, and Chamber 3 (right) *Chromatium vinosum*. (Reproduced with permission from Keith, S. M. and Herbert, R. A., *FEMS Microbiol. Ecol.*, 31, 242, 1985.)

*vinosum* in Chamber 3. The basal mineral salts growth medium was that described by Keith.[41] In addition, the medium reservoir supplying Chamber 1 was supplemented with either 50 m$M$ glucose and an inorganic nitrogen source of either 3.5 m$M$ $NH_4Cl$ or $KNO_3$ (N-limitation) or 10 m$M$ glucose and either 7 m$M$ $NH_4Cl$ or 7 m$M$ $KNO_3$ (C-limitation). The medium vessel supplying Chamber 2 was also supplemented with 10 m$M$ $Na_2SO_4$ as electron acceptor for the sulfate-reducing bacterium, while medium Reservoir 2 contained 5 m$M$ $NaHCO_3$ as carbon source for the phototroph. The cultures were maintained at pH 7.5 using automatic pH controllers and maintained at 25°C by means of a thermocirculator. Anaerobic conditions were maintained in the medium reservoirs and culture vessels by sparging with high purity nitrogen. All cultures were grown at a dilution rate of 0.03 per hour.

Data presented in Figure 5 and 6 show typical results obtained from these experiments. When the cultures were grown on $NH_4^+$ under C-limitation, a series of interactions were observed. Growth Chamber 1 containing the *Clostridium butyricum* was the only vessel provided with an organic carbon source (glucose) which can be fermented to produce lower fatty acids (principally butyrate and acetate) and ethanol[42] which were used as carbon and energy sources by the sulfate-reducing bacterium in Chamber 2. There is thus a commensal relationship between the clostridium and the sulfate reducer based upon the generation of carbon substrates (Figures 5 a and b). Similarly, the sulfate-reducing bacterium was involved in a mutualistic relationship with the purple sulfur bacterium *Chromatium vinosum*. The coupling between the two populations was mediated by the cyclical reduction and oxidation of $SO_4^{2-}$ and $S^{2-}$, the electron acceptor and donor, respectively (Figures 5 b and c). It is also apparent from the data in Figures 5 a, b, and c that the three microbial populations underwent cyclical oscillations, and when there was a population maximum for the clostridium, there were similar maxima for the sulfate reducer and the phototroph.

Under $NH_4^+$ limitation (carbon excess), the populations of the clostridium and the sulfate

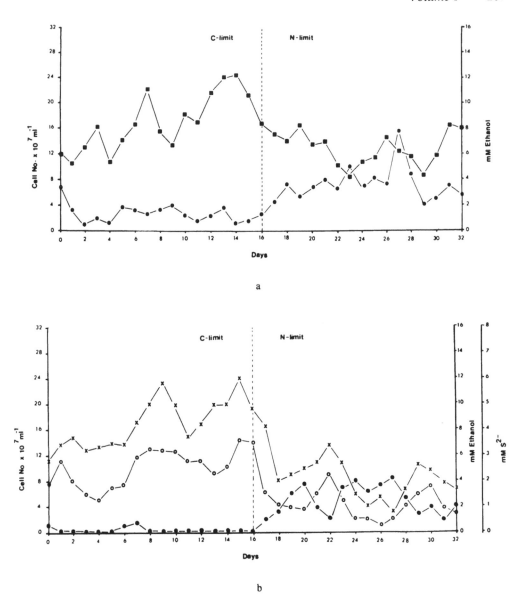

FIGURE 5.    (a) Changes in population densities of *Clostridium butyricum* (■) when grown on NH₄Cl as nitrogen source under either C or N limitation at a dilution rate of 0.03 per hour. Closed circles (●) are concentrations of ethanol. (b) Changes in population densities of *Desulfovibrio desulfuricans* (×) when grown on NH₄Cl. Closed circles (●) are ethanol concentration and open circles (○) are concentrations of free $S^{2-}$. (c) Changes in population densities of *Chromatium vinosum* (▲) when grown on NH₄Cl. Symbols as in b.(Reproduced with permission from Keith, S. M. and Herbert, R. A., *FEMS Microbiol. Ecol.*, 31, 242, 1985.)

reducer declined, and from the data in Figures 5 a and b, there is evidence that the two populations were competing for the limiting substrate, $NH_4^+$. While the sulfate-reducing bacterium remained involved in a mutualistic relationship with the clostridium, the *D. desulfuricans* population was not carbon limited, since excess carbon (ethanol) was available (Figure 5b). One possibility from these data is that the sulfate-reducing bacterium was less successful in competing for the limiting N-source than the clostridium. While the *D. desulfuricans* population declined under $NH_4$ limitation, there was no concurrent decrease in the phototroph population. This was not altogether unexpected, since the energy requirements for $CO_2$ fixation was provided by tungsten lamps and sufficient $S^{2-}$ was available as reductant

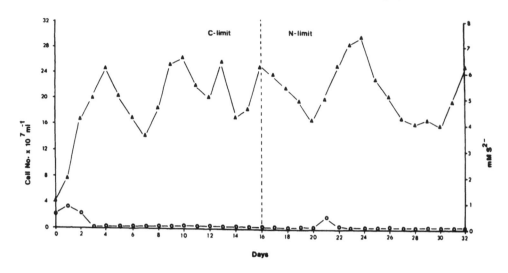

FIGURE 5c

for photosynthesis. The *C. vinosum* was therefore neither electron-donor nor energy limited, and since this strain is able to fix nitrogen, it is interesting to speculate that it overcame the problem of N-limitation by fixing nitrogen.[53]

In analogous experiments, the three populations were grown on nitrate as the inorganic nitrogen source, and data in Figures 6 a, b and c show typical results. Nitrate-grown cultures, with the exception of the phototroph, were more stable and showed smaller oscillations in cell populations than those grown on $NH_4^+$. The commensalism between the clostridium and sulfate-reducing bacterium was again observed with $NO_3^-$-grown cultures irrespective of nutrient limitation. Cell populations of the sulfate-reducing bacterium (Figure 6b) were constantly higher than those recorded for the equivalent $NH_4^+$-grown populations (Figure 5b), but paradoxically little sulfate reduction occurred. More detailed investigation of chemostat-grown cultures of the sulfate-reducing bacterium showed that it was able to utilize $NO_3^-$ as well as $SO^{4-}$ as a terminal electron acceptor which accounted for the observed reduction in the free $S^{2-}$ pool under $NO_3^-$ excess conditions.[39] The ability of the sulfate reducer to use $NO_3^-$ as an electron acceptor led to disruption of the mutualistic relationship between this bacterium and the phototroph, and in the absence of sufficient $S^{2-}$, the *C. vinosum* population decreased sharply (Figure 6c). On switching to $NO_3^-$ limitation, the population of the sulfate-reducing bacteria remained high but since $NO_3^-$ was the limiting substrate, there was an increased flow of electrons to $SO_4^{2-}$, as terminal electron acceptor leading to a rapid increase in free $S^{2-}$. *C. vinosum* was unable to oxidize the free $S^{2-}$ rapidly enough, and $S^{2-}$ concentrations reached levels which were inhibitory to the phototroph whose numbers declined sharply (Figure 6c). Under these conditions, the relationship between the sulfate reducer and the phototroph changed from one of mutualism when grown on $NH_4^+$, irrespective of limitation, to one of ammensalism. While the *C. vinosum* population was sensitive to elevated $S^{2-}$ levels, the *C. butyricum* population remained unaffected and was able to tolerate high $S^{2-}$ concentrations. These data are consistent with results from studies on the $S^{2-}$ tolerance of the *C. butyricum*, which showed that this bacterium would tolerate up to 10 m$M$ $S^{2-}$ without significant effect on growth rate or cell yield.[53] From these simple experiments, we have been able to identify and establish a number of relationships and interactions among this physiologically distinct group of anaerobic bacteria, and it is interesting to record that we first established that sulfate-reducing bacteria could respire using nitrate in this experimental system. In this example of the application of compound-diffusion chemostats, all the bacteria were obligate anaerobes. However, the

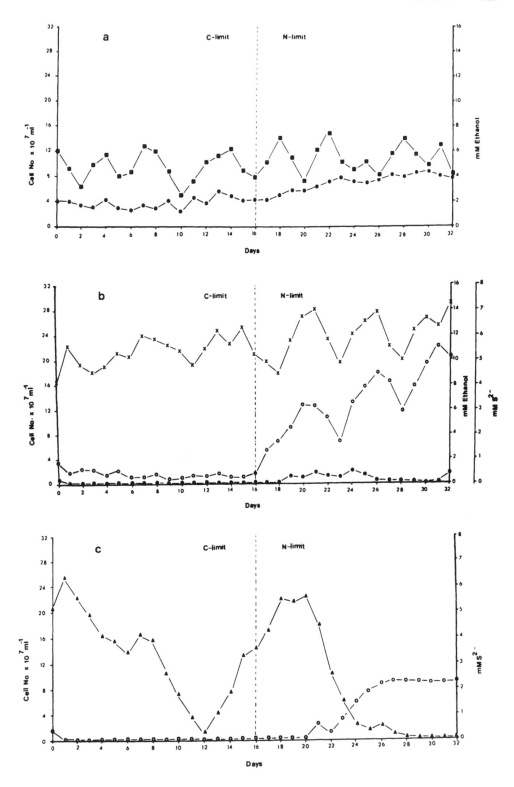

FIGURE 6. (a) Changes in population densities of *Clostridium butyricum* when grown on KNO₃ under either C or N limitation at a dilution rate of 0.03 per hour. Symbols as in Figure 5. (b) Changes in population densities of *Desulfovibrio desulfuricans* when grown on KNO₃. Symbols as in Figure 5. (c) Changes in population densities of *Chromatium vinosum* when grown on KNO₃. Symbols as in Figure 5. (Reproduced with permission from Keith, S. M. and Herbert, R. A., *FEMS Microbiol. Ecol.*, 31, 242, 1985.)

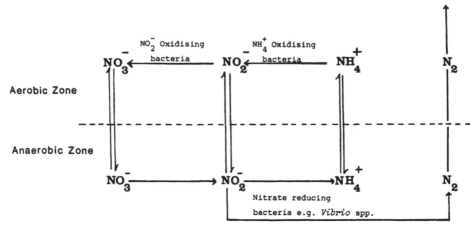

FIGURE 7.    Relationship between nitrifying bacteria and nitrate-respiring bacteria in estuarine and inshore marine sediments. (Reproduced with permission from Keith, S. M. and Herbert, R. A., *FEMS Microbiol. Ecol.*, 31, 242, 1985.)

experimental system can equally well be applied to studies of interactions between aerobic and facultatively anaerobic bacteria.

### Analysis of Interactions between Nitrifying Bacteria and Nitrate-Respiring Bacteria

A number of field studies have unequivocally demonstrated the simultaneous processes of nitrification and nitrate respiration (denitrification and $NO_3^-$ reduction to $NH_4^+$) in estuarine and inshore marine sediments.[43-45] The vertical distribution of bacteria within sediments is largely dependent upon the presence or absence of oxygen, and as a broad generalization, the distribution of the different physiological types exhibits a zonation similar to the physico-chemical gradients which develop.[46,47] Using the compound-diffusion chemostat system, we have investigated the relationship between a nitrate respiring *Vibrio* spp. and autotrophic-nitrifying bacteria belonging to the genera *Nitrobacter* and *Nitrosomonas*. The series of nitrogen intermediates involved in the anaerobic dissimilation of $NO_3^-$ to $NH_4^+$ are the reverse of those involved in nitrification, the aerobic oxidation of $NH_4^+$ to $NO_3^-$ by nitrifying bacteria (Figure 7). Nitrifying and denitrifying bacteria used in this study were isolated from estuarine sediments as described by MacFarlane and Herbert.[48,49] The *Vibrio* spp. was grown in Chamber 1, a $NH_4^+$-oxidizing bacterium identified as a *Nitrosomonas* spp. in Chamber 2, and a $NO_2^-$ oxidizing bacterium identified as a *Nitrobacter* spp. in Chamber 3. The basal mineral salts used were those described by MacFarlane.[50] In addition, the medium reservoir supplying Chamber 1 was supplemented with glycerol as carbon source and $NO_3^-$ as nitrogen source (see Figure 8 for details). Chamber 1 containing the *Vibrio* spp. was sparged with oxygen-free nitrogen to maintain anaerobic conditions while Chambers 2 and 3 containing the nitrifying bacteria were aerated to ensure aerobic conditions. The cultures were automatically maintained at pH 7.6 using pH controllers and a temperature of 25°C. All cultures were grown at a dilution rate of 0.025/hr.

Data presented in Figure 8 a, b, and c show a typical series of results from these experiments. Previous studies using single-stage chemostats have demonstrated that when grown under C-limitation with nitrate as nitrogen source and terminal electron acceptor, the *Vibrio* spp. produced $NO_2^-$ as the principal end product of respiration, whereas under N-limitation $NH_4^+$ was the predominant product of nitrate reduction.[48,51]

The *Vibrio* spp. was therefore grown initially under $NO_3^-$ limitation in order to generate $NH_4^+$ as an energy source for the *nitrosomonas* spp. Data in Figure 8b show that the

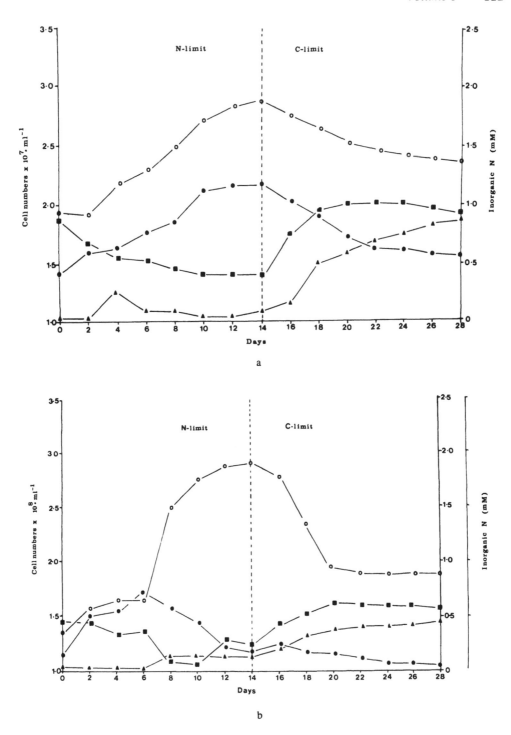

FIGURE 8. (a) Change in population density of *Vibrio* spp. (○) when grown on glycerol and KNO₃ under either C-limitation (30 m*M* glycerol, 10 m*M* KNO₃) or N-limitation (50 m*M* glycerol, 5 m*M* KNO₃) at a dilution rate of 0.025 per hour at 25°C. Nitrate concentration ● ———— ●, NO₂⁻ concentration ▲ ———— ▲, and NH₄⁺ concentration ■ ———— ■. (b) Change in population density of *Nitrosomonas spp.* (○) when grown under either C or N limitation at a dilution rate of 0.025 per hour at 25°C. Symbols as in a. (c) Change in population density of *Nitrobacter* spp. (○) when grown under either C or N limitation at a dilution rate of 0.025 per hour at 25°C. Symbols as in a. (Reproduced with permission from Keith, S. M. and Herbert, R. A., *FEMS Microbiol. Ecol.*, 31, 242, 1985.)

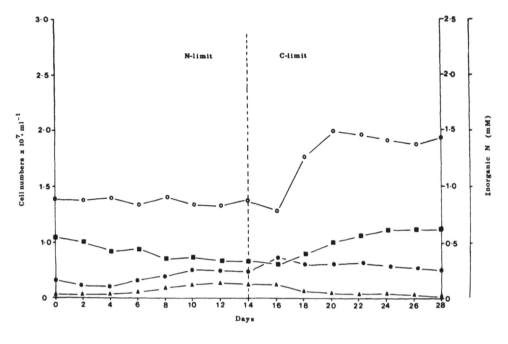

FIGURE 8c

*Nitrosomonas* population increased under these growth conditions and there was a concomitant reduction in the $NH_4^+$ concentration. Nitrite, the end product of $NH_4^+$ oxidation, produced by the *Nitrosomonas* population, was in turn used as an energy source by the *Nitrobacter* spp. (Figure 8c). Under these growth conditions ($NO_3^-$ limitation) the concentrations of $NO_2^-$ were always low but were sufficient to maintain a *Nitrobacter* population of *circa* $1.6 \times 10^7$ viable cells/m$\ell$. When the experimental system was changed to $NO_3^-$ excess, the *Vibrio* spp. produced greater quantities of $NO_2^-$ and the population of the *Nitrosomonas* spp. declined sharply as the available energy source ($NH_4^+$) decreased. In contrast, since more $NO_2^-$ was generated under these conditions, the *Nitrobacter* population increased significantly (Figure 8c). Upon returning the experimental system to $NO_3^-$ limitation, the cycle of events could be repeated. These data demonstrate that not only could populations of aerobic and facultatively anaerobic bacteria coexist in this model system, but that the availability of the carbon and nitrogen sources significantly influenced the development of individual populations. The principal interaction involved in this defined microbial community is commensalism, but it is interesting to speculate from the results obtained that these bacteria may operate an internal nitrogen cycle which is energetically advantageous to both groups of microorganisms. Under these circumstances, the relationship between these bacteria would be of a mutualistic nature and analagous to those observed with microorganisms of the sulfur cycle.[9,52] In this example of nitrogen-cycling bacteria, the diffusion of solutes couples the matabolism of the facultatively anaerobic *Vibrio* spp. to those of the obligately aerobic nitrifying bacteria. In natural environments, the spatial heterogeneity of the system enables such physilogocally dissimilar groups of microorganisms to coexist by diffusion of solutes across the oxic-anoxic interface.

## SUMMARY AND CONCLUSIONS

A fundamental objective of microbial ecology is to understand how microbial communities function within natural environments. A necessary prerequisite if we are to achieve this objective is to understand the nature of the microbial interactions involved. It is self-evident that no single experimental approach is universally applicable to study the complex interactions involved in microbial communities. The methods used to analyze interactions are important since the interpretation of the way in which microorganisms interact is frequently dependent upon the experimental approach used. As has been pointed out earlier in this chapter, monocultures of microorganisms are rare in natural ecosystems, yet for almost a century, microbial methodology has been dominated by the philosophy of pure cultures. Equally most natural environments are spatially and temporally heterogeneous, and yet until the last decade, the emphasis has been on homogenous laboratory methods of culture. While our present understanding of microbial interactions is still superficial, there is an acute awareness of the limitation of existing experimental methodology. Fortunately, experimental models which incorporate spatial and temporal heterogeneity are now becoming available, and there is good reason to be optimistic that microbial ecologists will now make as rapid advances in our understanding of microbial communities as did microbial physiologists in the study of microbial growth with the advent of the chemostat. In the final analysis, however, there is no substitute for field investigations, which must still form the basis of all laboratory studies.

## ACKNOWLEGDMENTS

The author gratefully acknowledges financial support from United Kingdom Natural Environment Research Council (NERC), which has supported many aspects of the work described in this chapter. My thanks are also due to S. M. Keith, G. T. MacFarlane, and Moya Russ for stimulating discussions.

## REFERENCES

1. **Atkinson, B. and Fowler, H. W.,** The significance of microbial films in fermenters, *Adv. Biol. Eng.,* 3, 221, 1974.
2. **Atkinson, B. and Knight, A. J.,** Microbial film fermenters, their present and future applications, *Biotechnol. Bioeng.,* 17, 1245, 1975.
3. **Stanley, S. O. and Malcolm, S.,** The sediment environment, in *Sediment Microbiology,* Nedwell, D. B. and Brown, C. M., Eds., Special Publications, Society for General Microbiology, Academic Press, Orlando, Fla., 1982, 3.
4. **Atkinson, B. and Daoud, I. S.,** Microbial flocs and flocculation in fermentation process engineering, *Adv. Biol. Eng.,* 4, 41, 1975.
5. **Armstrong, W.,** Waterlogged soils, in *Environmental and Plant Ecology,* Etherington, J. R., Eds., Wiley, London, 1975, 181.
6. **Jorgensen, B. B., Kuenen, J. G., and Cohen, Y.,** Microbial transformations of sulfur compounds in a stratified lake (Solar Lake), Sinai, *Limnol. Oceanogr.,* 24, 799, 1979.
7. **Brock, T. D.,** *Thermophilic Microorganisms and Life at High Temperatures,* Springer-Verlag, New York, 1978.
8. **Sorokin, Y. I.,** The bacterial population and the process of hydrogen sulphide oxidation in the Black Sea, *J. Cons. Int. Explor. Mer.,* 34, 423, 1972.
9. **Jorgensen, B. B.,** The sulphur cycle of a coastal marine sediment (Limfjorden, Denmark), *Limnol. Oceanogr.,* 22, 814, 1977.
10. **Bazin, M. J., Saunders, P. T., and Prosser, J. I.,** Models of microbial interactions in the soil, *Crit. Rev. Microbiol.,* 4, 463, 1976.
11. **Wimpenny, J. W. T., Lovitt, R. W., and Coombs, J. P.,** Laboratory model systems for the investigation of spatially and temporally organised microbial ecosystems, in *Microbes in Their Natural Environments, Symposium of the Society for General Microbiology,* 34, Slater, J. H., Whittenbury, R., and Wimpenny, J. W. T., Eds., Cambridge University Press, Cambridge, U.K., 1983, 67.

12. **Sorokin, Y. I.,** Interrelations between sulphur and carbon turnover in meromicitic lakes, *Arch. Hydrobiol.,* 66, 391, 1970.
13. **Jorgensen, B. B.,** Ecology of the bacteria of the sulphur cycle with special reference to anoxic-oxic interface environments, in *Sulphur Bacteria,* Postgate, J. R. and Kelly, D. P., Eds., The Royal Society, London, 1982, 113.
14. **Parkes, R. J.,** Methods for enriching, isolating and analysing microbial communities in laboratory systems, in *Microbial Interactions and Communities,* Vol. 1, Bull, A. T. and Slater, H. J., Eds., Academic Press, Orlando, Fla., 1982, 45.
15. **Bungay, H. R. and Bungay, M. L.,** Microbial interactions in continuous culture. *Adv. Appl.Microbiol.,* 10, 269, 1968.
16. **Bull, A. T. and Slater, H. J.,** *Microbial interactions and Communities,* Vol. 1, Academic Press, Orlando, Fla., 1982, chap. 1.
17. **Monod, J.,** La technique de culture continue, theorie et applications, *Ann. de l'Inst. Pasteur (Paris),* 79, 390, 1950.
18. **Novick, A. and Szillard, L.,** Description of the chemostat, *Science,* 112, 714, 1950.
19. **Bull, A. T.,** Microbial growth, in *Companion to Biochemistry,* Bull, A. T., Lagnado, J. R., Thomas, J. D., and Tipton, K. F., Eds., Longman, London, 1974, 415.
20. **Pirt, S. J.,** *The Principles of Microbe and Cell Cultivation,* Blackwell Scientific Publications, Oxford, U.K., 1975.
21. **Veldkamp, H. and Jannasch, H. W.,** Mixed culture studies with the chemostat, *J. Appl. Chem. and Biotechnol.,* 22, 105, 1072.
22. **Jannasch, H. W. and Mateles, R. I.,** Experimental microbial ecology studied in continuous culture, *Adv. Microb. Physiol.,* 11, 165, 1974.
23. **Bull, A. T. and Brown, C. M.,** Continuous culture applications to microbial biochemistry, in *Microbial Biochemistry, International Review of Biochemistry,* Vol. 21, Quayle J. R., Ed., University Park Press, Baltimore, 1979, 177.
24. **Jannasch, H. W.,** Growth of marine bacteria at limiting concentrations of organic carbon in seawater, *Limnol Oceanogr.,* 19, 716, 1967.
25. **Jannasch, H. W.,** Enrichments of aquatic bacteria in continuous culture, *Arch. Mikrobiol.,* 59, 165, 1967.
26. **Van Gemerden, H.,** Co-existence of organisms competing for the same substrate: an example among the purple sulphur bacteria, *Microb. Ecol.,* 1, 104, 1974.
27. **Chisholm, S. N. and Nobbs, A. A.,** Simulation of algal growth and competition in a phosphate limited cyclostat, in *Modelling Biochemical Processes in Aquatic Ecosystems,* Canale, R. P., Ed., Ann Arbor Science, Michigan, 1976, 357.
28. **Senior, E., Bull, A. T., and Slater, J. H.,** Enzyme evolution in a microbial community growing on the herbicide dalapon, *Nature, (London),* 263, 476, 1976.
29. **Daughton, C. G. and Hsieh, D. P. H.,** Parathion utilization by bacterial symbionts in a chemostat, *Appl. Environ. Microbiol.,* 34, 175, 1977.
30. **Boethling, R. S. and Alexander, M.,** Effect of concentration of organic chemicals on their biodegradation by natural microbial communities, *Appl. Environ. Microbiol.,* 37, 1211, 1979.
31. **Wimpenny, J. W. T.,** Spatial order in microbial ecosystems, *Biol. Rev.,* 56, 295, 1981.
32. **Hawkes, H. A.,** Eutrophication of rivers, effects, causes and control, in *Treatment of Industrial Effluents,* Cailely, A. G., Forster, C. F., and Stafford, D. A., Eds., Hodder and Stoughton, London, 1977, 159.
33. **Lampert, W.,** A directly coupled, artificial two step food chain for long term experiments with filter feeders at consistent food concentrations, *Mar. Biol.,* 37, 349, 1976.
34. **Cooper, D. C. and Copeland, B. J.,** Responses of continuous series estuarine microecosystems to point source input variations, *Ecol. Monogr.,* 43, 213, 1973.
35. **Skipnes, O., Eide, I., and Jensen, A.,** Cage culture turbidostat: a device for rapid determination of algal growth rate, *Appl. Environ. Microbiol.,* 40, 318, 1980.
36. **Jensen, A.,** The use of phytoplankton cage cultures for *in situ* monitoring of marine pollution, *Cons. Int. Explor. Mer,* 179, 322, 1980.
37. **Jannasch, H. W., Wirsen, C. O., and Eimjhellen, K.,** Continuous culture methods for the isolation of *Thiovulum, Bact. Proc.,* (Abstr. G179), 42, 1970.
38. **Crump, J. E. and Richardson, G.,** The suitability of a membrane diffusion growth chamber for studying bacterial interaction, *J. Appl. Bact.,* 58, 215, 1985.
39. **Keith, S. M. and Herbert, R. A.,** Dissimilatory nitrate reduction by a strain of *Desulfovibrio desulfuricans, Fed. Eur. Microbiol. Soc. Microb. Lett.,* 18, 55, 1983.
40. **Keith, S. M., Herbert, R. A., and Harfoot, C. G.,** Isolation of new types of sulphate reducing bacteria from estuarine sediments, *J. Appl. Bact.,* 53, 29, 1982.
41. **Keith, S. M.,** *Microbial Interactions in Anaerobic Estuarine Sediments,* Ph.D thesis, University of Dundee, Scotland, 1985.

42. **Keith, S. M., MacFarlane, G. T., and Herbert, R. A.,** Dissimilatory nitrate reduction by a strain of *Clostridium butyricum* isolated from estuarine sediments, *Arch. Microbiol.*, 132, 62, 1982.
43. **Koike, I. and Hattori, A.,** Simultaneous determination of nitrification and nitrate reduction in coastal sediments by a $^{15}$N-dilution technique, *Appl. Environ. Microbiol.*, 35, 853, 1978.
44. **Sorensen, J.,** Capacity for denitrification and reduction of $NO^-$ to $NH^+$ in a coastal marine sediment, *Appl. Environ. Microbiol.*, 35, 301, 1978.
45. **MacFarlane, G. T. and Herbert, R. A.,** Dissimilatory nitrate reduction and nitrification in estuarine sediments, *J. Gen. Microbiol.*, 130, 2301, 1984.
46. **Mechalas, B. J.,** Pathways and environmental requirements for biogenic gas production in the oceans, in *Natural Gases in Marine Sediments*, Kaplan, I. R., Ed., Plenum Press, New York, 1974, 12.
47. **Billen, G.,** Modelling processes of organic matter degradation and nutrient recycling in sedimentary systems, in *Sediment Microbiology*, Nedwell, D. B. and Brown, C. M., Eds., Special Publications, Society for General Microbiology, Academic Press, London, 1982, 15.
48. **MacFarlane, G. T. and Herbert, R. A.,** Comparative study of enrichment methods for the isolation of autotrophic bacteria from soil, estaurine and marine sediments, *FEMS Microb. Lett.*, 22, 127, 1984.
49. **MacFarlane, G. T. and Herbert, R. A.,** Nitrate dissimilation by *Vibrio* spp. isolated from estaurine sediments, *J. Gen. Microbiol.*, 128, 2463, 1982.
50. **MacFarlane, G. T.,** Physiology of Dissimilatory Nitrate Reduction and Nitrification in Estuarine Sediments, Ph.D. thesis, University of Dundee, Scotland, 1984.
51. **Dunn, G. M., Herbert, R. A., and Brown, C. M.,** Influence of oxygen tension on nitrate respiration by a *Klebsiella* spp. growing in continuous culture, *J. Gen. Microb.*, 112, 379, 1979.
52. **Jorgensen, B. B.,** The mineralisation and cycling of carbon, nitrogen, and sulphur in marine sediments, in *Contemporary Microbial Ecology*, Ellwood, D. C., Hedger, J. N., Latham, M. J., Lynch, J. M., and Slater, J. H., Eds., Academic Press, London, 1981, 239.
53. **Herbert, R. A. and Keith, S. M.,** Unpublished data.

Chapter 6

# STUDY OF ATTACHED CELLS IN CONTINUOUS-FLOW SLIDE CULTURE

**Douglas E. Caldwell and John R. Lawrence**

## CONTINUOUS-FLOW SLIDE CULTURES VS. CHEMOSTAT CULTURES

Continuous-flow slide culture provides more precisely defined growth conditions than chemostat culture. Although the term "chemostat" implies complete chemical control of the cell environment, absolute control is not possible. Substrate concentration is set in the supply reservoir but is unknown in the culture vessel and is constant only under steady-state conditions. Because steady-state is approached asymptotically, the time required to obtain it is poorly defined. Errors may consequently be introduced in the determination of growth constants. Metabolic wastes and secondary metabolites accumulate to unknown levels. These difficulties are compounded by inherent plumbing problems which make it difficult to meet the assumptions used in deriving chemostat kinetics. Among these are wall growth,[1] backgrowth, plugging of the outlet, and difficulties in obtaining instantaneous mixing of substrate.

Continuous-flow slide culture offers more precise physico-chemical control of the cell environment. The concentration of substrate is set by the irrigation solution. New medium is continuously supplied, thus preventing the depletion of substrate and the accumulation of metabolic wastes. The concentration or type of substrate can be changed instantaneously.[2] Similarly, an inhibitor can be instantaneously added or removed. By observing growth within the hydrodynamic boundary layer of surface microenvironments, ecological studies may be conducted. The growth kinetics of individual cells can be studied in detail within a single cell cycle.[3] Cells which tend to clump, aggregate, or adhere can be studied, as opposed to cell suspensions. The microenvironment can be equilibrated with the macroenvironment by using high laminar flow velocities and low cell densities. Alternatively, the microenvironment can be maintained in disequilibrium with the macroenvironment by using low laminar flow velocities or high cell densities.

## CONSTRUCTION OF CONTINUOUS-FLOW SLIDE-CULTURE CHAMBERS

One of the earliest continuous-flow slide chambers, for study of attached cells, was developed for tissue cells.[4] This chamber, which is similar to those currently in use for studies of bacterial chemotaxis,[2] consisted of two coverslips separated by a stainless steel support containing inlet and outlet. A sterile irrigation medium was continuously passed through the culture chamber. The inoculum was normally added as a pulse of suspended cells, some of which attached to the wall of the chamber. The attached cells were then studied microscopically. Alternative designs for continuous-flow slide-culture chambers include the shaped capillaries developed by Perfil'ev and Gabe.[5] These were formed by assembling and annealing large plates of glass, which were subsequently stretched in vertical furnaces to form capillaries with flat surfaces. However, their small size can cause difficulties in aligning the microscope condensor (light scattering caused by capillary walls), and they may contain stretch marks which reduce resolution. Other historical developments are discussed in detail elsewhere.[5]

An inexpensive alternative to earlier systems is the use of silicone adhesive (RTV 118, General Electric) to assemble slide-culture chambers, using two microscope slides and two 25- × 50-mm cover slips as shown in Figure 1. The interior dimensions are 1 × 3 × 50 mm and the volume of the cell is 0.1 m$\ell$. Medium enters and exits the chamber through

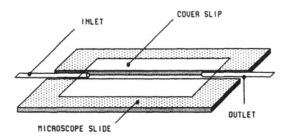

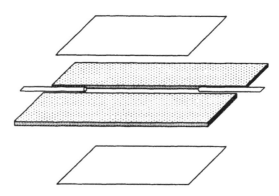

FIGURE 1. Diagram of continuous-flow slide-culture chamber. The chamber is used for microscopic investigation of attached cells. It consists of two microscope slides between two 25 × 50 mm cover slips (shown in lower portion of figure). The coverslips were bonded to the slides using silicone sealant. Medium entered and exited the chamber via silicone tubing with a 1 mm inner diameter. The size of the flow cell was 1 × 3 × 50 mm. The chamber is inexpensive and disposable.

silicone tubing. The chamber is autoclavable and disposable. An earlier version of this chamber was described by Caldwell and Lawrence.[3] In the original, medium entered the chamber through a 22 gauge needle and the length of the chamber was 25 mm. The needles were replaced with 1 mm (inner diameter) silicone tubing to prevent clogging and reduce hydrostatic pressure. The length of the flow cell was increased to ensure laminar flow. A peristaltic pump was used to provide a laminar flow velocity of up to 10 cm/sec. The chamber was taped to a rotating mechanical microscope stage (Zeiss) with the normal slide retainer removed as shown in Figure 2. Despite the removal of the slide retainer (and consequent loss of movement in the X dimension), the stage could still be rotated and adjusted in the X-Y plane by using the stage-centering screws. This aided in aligning the flow cell over the microscope condensor.

## OPERATION OF CONTINUOUS-FLOW SLIDE CULTURES

### Circulation System

Study of attached microorganisms under defined hydrodynamic conditions requires that the laminar flow velocity be controlled. This velocity is equal to the flow rate (cm³/sec) of the pump divided by the cross-sectional area (cm²) of the chamber. However, the circulation of sterile solutions requires the use of peristaltic pumps which produce periodic surges in flow. Consequently, laminar flow rates are usually mean values. Surge filters, consisting

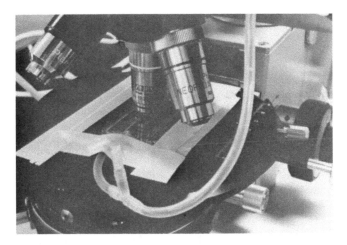

FIGURE 2.    Photograph of continuous-flow slide chamber mounted on the microscope stage. The silicone tubing shown at the front of the chamber supplies fresh medium at a laminar-flow velocity of approximately 5 cm/ sec. The T-connector in the supply line serves as an injection port for the inoculum. The septum consists of a short section of silicone tubing filled with cured silicone sealant.

of 3-m$\ell$ syringe bodies containing an air space, can be used to suppress the peristaltic effect. The syringes are "tuned" by adjusting the volume of the air space.[6] The use of a pulse-free syringe pump eliminates this problem but is more expensive.

During the circulation of medium, small air bubbles may spontaneously form within pump tubing. When these bubbles pass through the culture chamber, the movement of the meniscus along the glass surface may dislodge attached cells or alter the conformation of microcolonies and their associated exopolymers. This can be avoided by degassing (autoclaving) the supply and/or dilution reservoirs, by subsequently saturating these solutions with an 80% helium/ 20% oxygen mixture,[6] and by ensuring that the culture inlet tube supplies medium in a downward direction to prevent the descent of bubbles to the culture chamber.

Chemical conditions within the surface microenvironment depend, in part, upon the length of time the medium is exposed to the biofilm. This is due to the nature of laminar flow. The liquid moves in parallel layers along the chamber wall. Layers near the wall move very slowly, and layers near the center of the chamber (bulk phase) move more rapidly. Because very little mixing between these layers occurs, the biofilm rapidly depletes substrate from the layers adjacent to the chamber walls. The amount of depletion depends primarily on the metabolic activity of the biofilm and the length of time fluid is exposed to the biofilm as it passes through the chamber. Thus, it is important to define the upstream laminar-flow distance. This is the distance liquid has moved over the biofilm before it reaches the microorganisms being studied microscopically. If this distance is divided by the laminar-flow velocity, the exposure time of the liquid to the microbial biofilm can be calculated. This is most important when the macroenvironment and microenvironment are in disequilibrium due to high cell densities and/or low laminar-flow velocities.

### Preparation of Culture Chamber Surfaces Prior to Inoculation

Surfaces are cleaned using acid-alcohol (3 m$\ell$ concentrated HCl in 100 m$\ell$ 95% ethanol) prior to assembly and sterilization of slide culture chambers. The acid-alcohol is applied with a cotton swab and immediately absorbed by rubbing with tissue. The chamber is then oxidized and sterilized by baking 12 hr at 205°C and then purged with the irrigation medium for 5 min to allow adsorption of solutes to chamber surfaces prior to inoculation. This is necessary to define the chemistry of the surface before adsorption or attachment of the microbial population.

## Irrigation Solution and Substrate Concentration

The irrigation solution may consist of a synthetic minimal medium with a single limiting substrate, sterile-filtered stream water, stream water containing *in situ* communities, etc. If a rate-limiting substrate is used, it is important to perform a control experiment. A temporary increase in substrate concentration should result in a corresponding increase in growth rate within the surface microenvironment. This test is necessary to ensure that oxygen (or possibly other substrates) does not become rate-limiting within the microenvironment, although the primary substrate may be rate-limiting in the macroenvironment.

## Inoculation

The slide culture may be inoculated with a pure culture, a series of pure cultures, or with a natural community. When inoculating with a single pure culture, a constant laminar-flow velocity is maintained. A suspension of cells is then injected at the culture chamber inlet. This pulse of cells leads to some cell attachment and this is confirmed by microscopic observation. Additional pulses are added until the desired cell density is obtained. The final cell density achieved is critical, owing to the effect of cell density on the rate of substrate depletion within the hydrodynamic boundary layer.

If a mixed population is to be studied, the first population is allowed to attach and the image of the microscope field is recorded photographically. A second population can then be added and the image of the field again recorded. The identity of the second set of cells can be obtained later by using difference imagery.[7] This is done by subtracting the first image from the second. This shows only the differences between images and quickly pinpoints members of the second population. Difference images of additional populations can also be produced to record the position of third, fourth, and fifth populations. In this way, the growth kinetics of identified, individual populations can be studied within mixed communities. This is particularly important in investigating species which fail to colonize surfaces unless other primary colonizers are already established. One example is the case of *Veillonella* spp. colonizing tooth surfaces.[8]

Natural communities can also be used as an inoculum, although it is not possible to identify individual populations unless their morphology is distinct or fluorescein-conjugated antibodies are used. However, it is possible to retain the identity of a pure culture while studying its behavior as part of a natural community. This is done by sequentially adsorbing the natural community and the pure culture as described above for mixed cultures.

## Microscope Resolution and Photomicroscopy

To ensure that accurate measurements of bacterial growth can be obtained, images must be of the highest quality. This necessitates phase microscopy and the use of an interference filter to provide monochromatic light and eliminate chromatic aberration. A Zeiss® VG-9 46-78-05 filter with a band pass of 546 nm ± 10 nm or a comparable filter is recommended. Online image processing is normally used to quantitate microbial growth kinetics. However, computer errors in the midst of an experiment can result in the loss of data. In some cases, the growth form of the organism may change during the experiment or debris may accumulate in the background and interfere. This requires modification of the measuring program and that the images be reprocessed. Thus, in some cases, images are recorded as a series of black and white photographic negatives. This provides an inexpensive hard copy of the data and allows the investigator to address image-processing difficulties and experimental problems independently. Videotape can also be used, however, the quality of images is insufficient for analytical analysis unless high resolution systems are utilized (Super VHS).

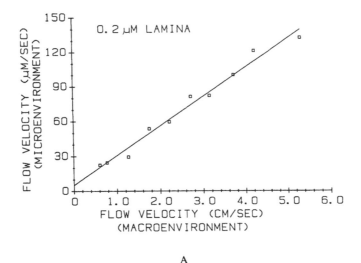

A

FIGURE 3. Laminar-flow velocity within the hydrodynamic boundary layer plotted vs. laminar-flow velocity in the bulk phase of the continuous-flow slide culture shown in Figures 1 and 2. The flow velocities for (A) 0.2 μm, (B) 4.0 μm, (C) 6.0 μm, and (D) 10.0 μm laminae are shown. Note that within the surface microenvironment, 0.2 μm from the surface, the laminar-flow velocity is more than two orders of magnitude less than that of the macroenvironment. (From Lawrence, J. R., Delaquis, P. J., Korber, D. R., and Caldwell, D. E., *Microb. Ecol.*, 14, 8, 1987. With permission.)

### Irradiation between Observations

Due to the effects of intense microscope illumination on cell temperature, growth rate, and viability,[9] the culture chamber should normally be shielded from irradiation between observations unless a low light television system is used.

## HYDRODYNAMICS OF CONTINUOUS-FLOW SLIDE-CULTURE CHAMBERS

Due to the small size of slide-culture chambers, flow is laminar rather than turbulent.[2] Liquid movement is thus seen as a series of water layers flowing parallel to the walls of the chamber. Liquid moving near the center of continuous-flow slide-culture chambers moves more rapidly than liquid adjacent to the chamber wall due to drag caused by viscosity.[10-12] This is demonstrated in continuous-flow slide culture by determining the velocity of 0.2 μm latex spheres at various distances from the wall of the chamber (i.e., within various laminae as shown in Figure 3).

The relatively quiescent layers of water near surfaces have been referred to by a variety of terms, including mass transfer resistance layer, hydrodynamic boundary layer, and viscous sublayer. Within these layers, inertial forces become relatively unimportant as compared to viscous forces.[12] This is due to the small size scale of surface microenvironments. Viscosity tends to decrease with the square of size (i.e., surface area) while inertia tends to decrease with the cube of size (i.e., volume). Consequently, surface microenvironments become diffusion limited. Motile bacteria, sequestered within the hydrodynamic boundary layer, can move upstream against a bulk flow of 10 cm/sec or greater. In addition, microorganisms attached to the wall of continuous-flow culture chambers may easily deplete the surface microenvironment of substrate.[3] Consequently, at high laminar-flow velocities and low cell densities, the surface microenvironment may be nearly at equilibrium with the macroenvi-

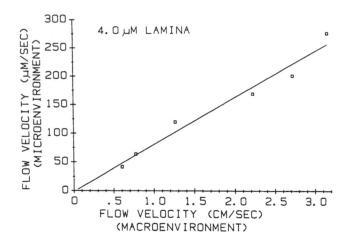

FIGURE 3B

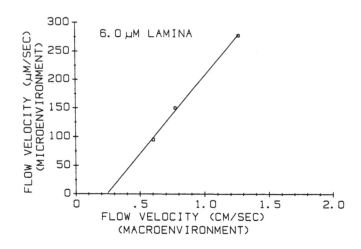

FIGURE 3C

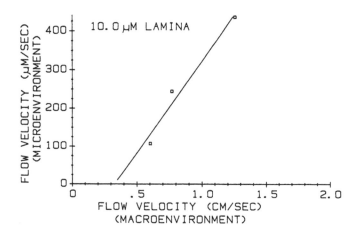

FIGURE 3D

ronment and the conditions within each are similar. However, at low laminar flow velocities and/or high cell densities, the macro- and microenvironments are in disequilibrium, due primarily to the hydrodynamic characteristics of surfaces. To quantitate the mass transfer resistance of the hydrodynamic boundary layer in continuous-flow slide culture, colored dyes have been added and washed out while monitoring concentration within the hydrodynamic boundary layer by microphotometry.[2] Dye in the center of the flow cell was displaced instantaneously, while dye within the hydrodynamic boundary layer was displaced more slowly. The time required for 95% displacement of both blue dextran (m.w. 2,000,000) and bromophenol blue (mol wt 600) was approximately 3 sec.

## QUANTITATION OF MICROBIAL GROWTH IN CONTINUOUS-FLOW SLIDE CULTURE

Although it is possible to quantitate the growth rate of attached cells by determining the increase in cell number as a function of time, this procedure is time-consuming and inaccurate. It also fails to provide sufficient detail to quantitate events which occur within the period of a single cell cycle.[3] As stated by Marr et al.,[13] "Microscopic measurement of bacteria growing in microculture is less satisfactory as a basis for determining the kinetics of growth than for measurement of interdivision times. The random error of measurement by microscopy obscures the detailed kinetics of growth . . . " However, due to the development of automated computer image enhancement and analysis, a new level of precision is possible and the resolution of the light microscope has been effectively increased by an order of magnitude.[3,7,13,15]

Growth in continuous-flow slide culture can be measured in terms of cell area.[3,14] Cell area is the space ($\mu m^2$) which each cell occupies within computer images. Like optical density and cell counts, cell area is proportional (but not equal) to biomass (grams dry weight of cell material) for those organisms which grow as a monolayer of cells within the surface microenvironment. The growth rate is thus equal to the slope of a plot showing the natural logarithm of cell area vs. time.

Cell area may be determined manually or preferably by automated image processing (Figure 4). Due to the limit of microscope resolution (0.2 $\mu m$) relative to the size of bacterial cells (0.5 $\mu m$), cell contours are poorly defined. The image of cells must thus be sharpened through image enhancement. The effectiveness of the procedure is indicated by correlation coefficients of 0.998 for natural log plots of cell area vs. time (Figure 5).[3] Processing normally involves input averaging, shading correction, electronic contrast enhancement, normalization, and discrimination.[3] These processing steps are necessary for computer recognition and enhancement of the cells to be analyzed. Cell area may also be determined using difference imagery.[7] Using the latter procedure, the initial image of cells in slide culture is subtracted from subsequent images to reveal the growing portion of cells and to remove debris, dead cells, etc. (Figure 6).

Computer enhancement results in a well-defined cell boundary which is drawn very precisely and reproducibly. This boundary may not necessarily represent the actual position of the cell boundary. However, because the processing error is constant, it cancels when natural logarithms are subtracted to obtain the specific growth rate. All that is required is that the measurement of the "apparent" cell area be proportional to the actual biomass. This has made it possible to determine that *Pseudomonas fluorescens* grows exponentially at a constant rate even within a single cell cycle and that microbial growth is unbalanced during the formation of surface microcolonies.[3] This information has been useful in testing assumptions used to derive the surface colonization equation and the surface growth rate equation.[3,15,22,23] It has also aided in determining the effect of laminar-flow velocity on specific growth rate[3] as shown in Figure 7.

FIGURE 4. Photo of Interactive Biological Analysis System (IBAS, Kontron, West Germany) image processor. The host computer is shown in the lower left. It utilizes a CPM operating system, controls the menu, and performs routine calculations. The array processor is shown in lower right. It simultaneously processes multiple image elements and controls automated cell recognition, enhancement, and analysis.

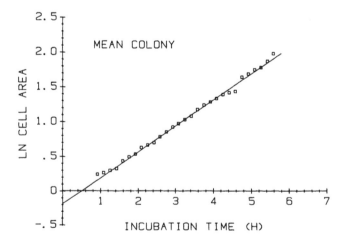

FIGURE 5. Growth rate of *Pseudomonas fluorescens* microcolonies at a laminar-flow velocity of 10 cm/sec. Natural logarithm of cell area vs. time for the mean of four colonies. Cell area was determined by automated image analysis of cells attached to the wall of the continuous-flow slide culture and is proportional to cell biomass for cells growing as a surface monolayer. The slope equals the specific growth rate (0.37/hr). The doubling time was 1.9 hr and the correlation coefficient was 0.998. (From Caldwell, D. E. and Lawrence, J. R., *Microb. Ecol.*, 12, 303, 1986. With permission.)

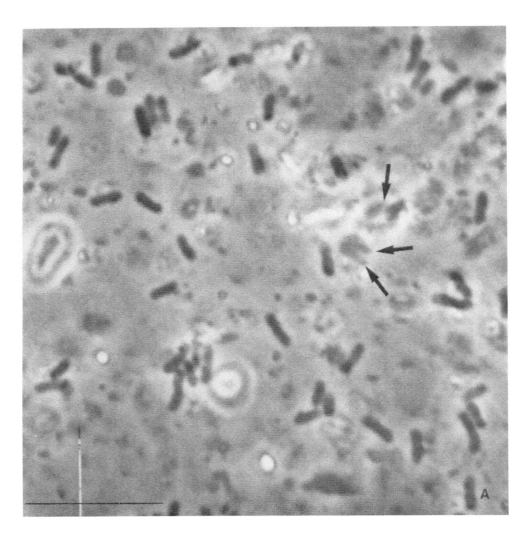

A

FIGURE 6. Original image (A) and difference image (B) of *Ensifer adhearens* growing in slide culture. Note that the growing cells are discriminated from clay particles (arrows) in the difference image. Phase micrograph. Bar equals 10 μm (From Caldwell, D. E. and Germida, J. J., *Can. J. Microbiol.*, 31, 39, 1985. With permission.)

## GROWTH KINETICS IN GRADIENT SLIDE CULTURE

### Determination of $K_s$ in Batch Culture, Chemostat Culture, and Continuous-Flow Slide Culture

In continuous-flow slide culture, the macroenvironment and microenvironment approach equilibrium (equality) when the laminar-flow velocity is high and the density of cells on the surface is low. Under these conditions, the maximum specific growth rate ($\mu_{max}$) and the half-saturation constant ($K_s$) can be determined from Equation 1[16] by measuring growth rate as a function of substrate concentration.

$$\mu = \mu_{max} \{S/(K_s + S)\} \tag{1}$$

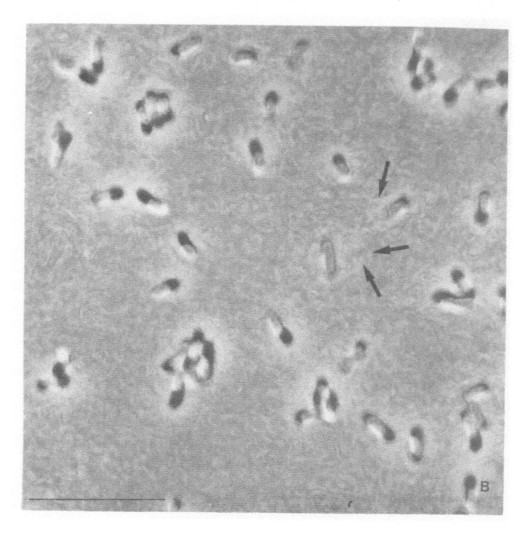

FIGURE 6B

where $\mu$ = specific growth rate, $\mu_{max}$ = maximum specific growth rate, S = substrate concentration, and $K_s$ = half-saturation constant.

A linearized form of Equation 1 is Equation 2.

$$1/\mu = (K_s/\mu_{max})(1/S) + 1/\mu_{max} \qquad (2)$$

As can be seen from Equation 2, the slope of a plot of $1/\mu$ vs. $1/S$ is equal to $K_s/\mu_{max}$. The Y intercept is equal to $1/\mu_{max}$ and the X intercept is equal to $-1/K_s$.

It is difficult to determine $K_s$ in batch culture due to the depletion of substrate as $\mu$ is measured. When cells are suspended in a defined minimal medium with a known substrate concentration, the concentration of substrate changes as the rate of growth is determined. Thus, growth rate continually varies as it is measured. Consequently, the rate of growth must be measured as a function of time and extrapolated to time zero to obtain the initial growth rate at the initial substrate concentration. Monod realized that this procedure was inaccurate and that $\mu$ equals D in steady-state chemostat cultures. Thus, he substituted D for $\mu$ in Equation 1 to obtain Equation 3, which allows determination of $K_s$ from S, $\mu_{max}$, and D.

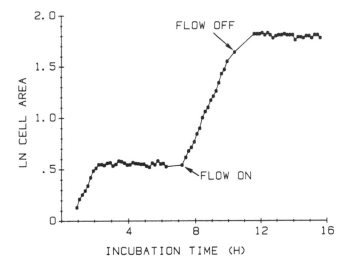

FIGURE 7.   Effect of laminar flow (10 cm/sec) on the growth kinetics of a *Pseudomonas fluorescens* microcolony within the hydrodynamic boundary layer of a surface microenvironment (continuous-flow slide culture at a glucose concentration of 100 mg/ℓ). The microcolony began as a single cell growing in the absence of laminar flow. Growth continued exponentially until the microenvironment became glucose-depleted and growth decreased to zero. Initiation of laminar flow (flow on) resulted in the resumption of exponential growth until flow was stopped (flow off). In control experiments with 1 g/ℓ glucose, exponential growth was independent of laminar flow. (From Caldwell, D. E. and Lawrence, J. R., *Microb. Ecol.*, 12, 308, 1986. With permission.)

$$K_s = \{S(\mu_{max} - D)\}/D \qquad (3)$$

where D = dilution rate.

Although the chemostat approach is an improvement over batch culture, there are several potential sources of error. To utilize Equation 3, the substrate concentration must be determined experimentally. However, the substrate concentration at half-maximal growth rate is often only a few mg/ℓ. For many substrates, this is too low a concentration for routine chemical determination. At high-input substrate concentrations ($S_r$ values), cell density is high and it is difficult to obtain instantaneous mixing of substrate as it enters the culture vessel. If mixing is not instantaneous, substrate is consumed before it can be uniformly dispersed in the culture. Thus, the concentration of substrate in the culture reservoir varies depending upon position with respect to the inlet and upon the frequency with which droplets of medium periodically fall into the culture vessel. At low values of $S_r$, cell density is low and the consumption of substrate by adsorbed cells becomes an increasingly important source of error. In some cases, steady-state conditions in continuous culture may be approached very slowly, and it is then difficult to determine the time required to achieve a specified degree of accuracy. For these and other reasons, results obtained in chemostat culture vary, depending upon the construction of culture apparatus and the operating conditions.

In continuous-flow slide culture, the chemical environment is well defined and under direct control. The concentration of rate-limiting substrate can be set at a desired value and maintained. Cell metabolism does not result in substrate depletion, the accumulation of metabolic wastes, or the accumulation of secondary metabolites. Consequently, $K_s$ can be determined directly and is not dependent upon assumptions concerning the mechanics of the culture system.

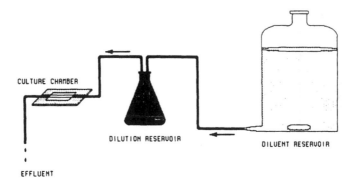

CULTURE CHAMBER

DILUTION RESERVOIR

DILUENT RESERVOIR

EFFLUENT

FIGURE 8.    Diagram of the apparatus used to produce substrate concentration gradients in continuous-flow slide culture. The diluent reservoir supplies basal salts. The dilution reservoir contains basal salts and the limiting substrate. The dilution reservoir inlet is positioned to the right of the stirrer vortex and the outlet is positioned to the left of the vortex to ensure adequate mixing. The inlet is positioned above the outlet to ensure that mixing of the diluent solution occurs before it leaves the dilution reservoir.

### Kinetics of Negative Substrate Gradients

Although it is possible to determine $\mu_{max}$ and $K_s$ by performing a series of continuous-flow slide-culture experiments, it is often desirable to run a preliminary gradient experiment to determine which substrate concentrations should be used. A gradient of substrate concentration is thus supplied in the irrigation solution, and the growth rate is determined as a function of substrate concentration.

To produce a substrate concentration gradient in the irrigation solution, a dilution reservoir is placed in series between the diluent reservoir and the continuous-flow slide culture (Figure 8). The initial concentration of substrate is gradually diluted by the flow of diluent which contains a basal salts solution which lacks the limiting substrate. The rate of decrease in substrate concentration within the dilution reservoir is directly proportional to the concentration of the substrate present. The dilution rate (D) is the proportionality constant (Equation 4).

$$dS/dt = -DS \tag{4}$$

Integration of Equation 4 gives Equation 5.

$$S = (e^{-Dt})(S_0) \tag{5}$$

where S = concentration of substrate at time t, t = time diluted, D = dilution rate, and $S_0$ = initial substrate concentration.
Rearranging Equation 5 to solve for the dilution rate gives Equation 6.

$$D = (\ln S_0 - \ln S)/t \tag{6}$$

As can be seen from Equation 6, the slope of a plot of ln S vs. t is equal to the dilution rate. The dilution rate is also equal to the flow rate (F) divided by the volume (V) of the dilution reservoir. Substituting F/V for D in Equation 6 gives equation 7.

$$F/V = (\ln S_0 - \ln S)/t \tag{7}$$

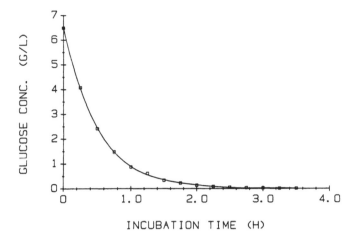

A

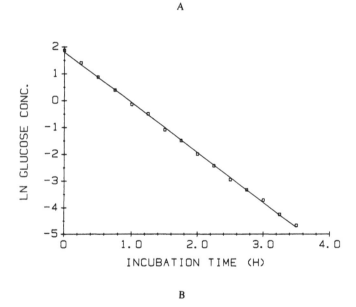

B

FIGURE 9.    Negative gradient of glucose concentration produced using
the apparatus shown in Figure 8. (A) Glucose concentration (g/$\ell$) vs. time,
(B) natural log of glucose concentration vs. time (correlation coefficient
equals 0.9996).

This equation is used to compute either the flow rate (F) or dilution reservoir volume (V)
which is required to obtain a concentration gradient between the desired initial and final
concentrations during any desired time period.

To avoid the need for calibration of the dilution reservoir volume and medium flow rate,
an internal concentration standard is added to the dilution reservoir (0.1 mg/$\ell$). It is diluted
at the same rate as the substrate and thus provides an accurate measure of both the dilution
rate and the substrate concentration in the irrigation solution. A typical gradient of glucose
concentration vs. time is shown in Figure 9. The effect of a glucose concentration gradient
on the growth rate of *P. fluorescens* is shown in Figure 10.

## Kinetics of Positive Substrate Gradients

It is also possible to obtain positive substrate concentration gradients using the system

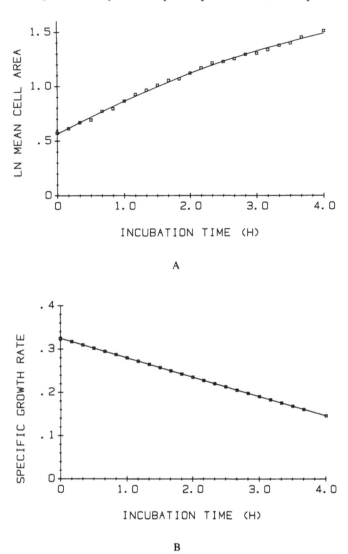

FIGURE 10.    Growth rate of attached *Pseudomonas fluorescens* cells growing in continuous-flow slide culture. A negative glucose concentration gradient was established in the irrigation solution. (A) Natural logarithm of cell area ($\mu m^2$) vs. incubation time, (B) specific growth rate (per hour) vs. incubation time, (C) specific growth rate (per hour) vs. glucose concentration.

described above. However, the final concentration of substrate desired is added to the diluent reservoir and a lower concentration of substrate is added to the dilution reservoir. In this case, the rate of change in substrate concentration (S) within the dilution reservoir is equal to the input rate (I) from the diluent reservoir minus the output (DS) which is due to dilution (Equation 8).

$$dS/dt = I - DS \tag{8}$$

Integration of Equation 8 gives Equation 9.

$$S = \{S_0 - (I/D)\}(e^{-Dt}) + I/D \tag{9}$$

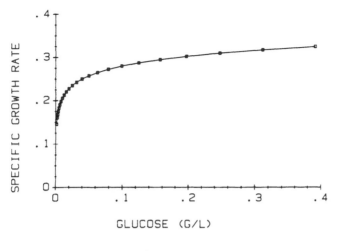

The input rate (I) equals the product of the flow rate (F) and the substrate concentration in the diluent reservoir ($S_r$) divided by the volume of the dilution reservoir (V) (Equation 10).

$$I = (F)(S_r)/V \qquad (10)$$

where I = input rate of substrate into dilution reservoir and $S_r$ = substrate concentration in diluent reservoir.

Substituting D for F/V in Equation 10 gives Equation 11.

$$S_r = I/D \qquad (11)$$

Substituting $S_r$ for I/D in Equation 9 gives Equation 12.

$$S = (S_0 - S_r)(e^{-Dt}) + S_r \qquad (12)$$

If $S_0$ equals zero, then Equation 12 can be rearranged to give Equation 13.

$$S = S_r (1 - e^{-Dt}) \qquad (13)$$

Substituting F/V for D in Equation 13 gives Equation 14.

$$S = S_r (1 - e^{-Ft/v}) \qquad (14)$$

Equation 14 can be rearranged to give Equation 15.

$$F/V = -\ln \{1 - (S/S_r)\}/t \qquad (15)$$

Equation 15 determines the flow rate or dilution reservoir volume required to obtain a concentration gradient from zero to S in any desired time, t.

## DETERMINATION OF SUBSTRATE CONCENTRATION WITHIN SURFACE MICROENVIRONMENTS

At high laminar-flow velocity and low cell density, the microenvironment approaches equilibrium with the macroenvironment. Thus, for a defined minimal medium in which the substrate concentration is rate-limiting, the substrate concentration in the microenvironment ($S_{mic}$) approaches the substrate concentration in the macroenvironment ($S_{mac}$). However, when the laminar flow velocity approaches zero or if the density of cells is high, the biofilm may deplete the hydrodynamic boundary layer of substrate. In this case, the growth rate of the cells becomes diffusion-limited rather than concentration-limited. This disequilibrium between macro- and microenvironment is the normal condition of mature biofilms even at high-flow velocities. In these situations, the Monod equation (Equation 1) cannot be used to obtain $K_s$ because the concentration of substrate in the ambient environment of the cells is unknown. Under these conditions if $1/\mu$ is plotted vs. $1/S_{mac}$ the plot will not necessarily be linear. The concentration of substrate at half maximum growth rate being only an "apparent $K_s$".

In situations where the macroenvironment and microenvironment are in disequilibrium (unequal) the microenvironment substrate concentration can be determined indirectly if cells with known $K_s$ and $\mu_{max}$ are used as biological probes of microenvironment conditions.[3] By substituting $S_{mic}$ for S and $\mu_{mic}$ for $\mu$ in Equation 1 and then rearranging to solve for $S_{mic}$, Equation 16 is obtained.

$$S_{mic} = K_s \{(\mu_{mic})/(\mu_{max} - \mu_{mic})\} \qquad (16)$$

where $S_{mic}$ = substrate concentration in the microenvironment, $\mu_{mic}$ = specific growth rate in the microenvironment, and $K_s$ = half-saturation constant.

This equation allows calculation of the substrate concentration within the microenvironment from the specific growth rate ($\mu$) observed in the microenvironment, half-saturation constant ($K_s$), and maximum specific growth rate ($\mu_{max}$). If the cell yield (Y) is known, the rate of substrate diffusion in surface microenvironments can also be determined from the specific growth rate and initial biomass.

## STUDY OF MICROBIAL BEHAVIOR IN CONTINUOUS-FLOW SLIDE CULTURE

Traditional microculture methods, using semi-solid agar, agar blocks, or static liquid media[17-20] have several disadvantages in the study of microbial behavior. These include oxygen supply, nutrient supply, waste accumulation, and substrate-accelerated death. These techniques also fail to provide a suitable surface microenvironment for observing chemotaxis, chemoadherence, immigration, emmigration, and colonization maneuvers. Continuous-flow slide culture overcomes these problems by providing an open system, similar to the solid-liquid interfaces which occur within many *in situ* environments. It has been used successfully to study the effect of selective inhibitors on the motility of *Flexibacter* spp.[21] In these studies, inhibitors were pulsed through the culture chamber and their effect on the motility of individual filaments was recorded. Similar studies have been conducted to investigate chemotaxis in *Escherichia coli*. Cells were tethered to the wall of the culture chamber by their flagella. The effects of attractants and repellants were observed as they affected cell rotation.[6]

Lawrence et al.[22] observed the colonization maneuvers associated with positioning, attachment, and growth of *P. fluorescens* on surfaces in continuous-flow slide culture. Studies of natural stream communities were also conducted and a variety of colonization maneuvers were identified[23] (Figure 11). Each type of maneuver resulted in unique growth kinetics[23] and represented adaptation to the environmental stresses associated with surface colonization.

Observation of predator-prey interactions and microbial competition is critical in understanding the development, structure, and function of any ecosystem. Continuous-flow slide culture provides an opportunity to directly observe and quantify some of these dynamic relationships during the development of biofilm communities. This is possible due to the application of difference imagery[7] to retain the identity of cells in mixed cultures or when pure cultures are added to natural communities. The location of the first population of cells to be attached is recorded as digital images. After a second population is added, the image of the mixed population is also recorded. The first image is subtracted from the second to reveal only the second population. This procedure can be continued indefinitely to retain the identity of cells and study cell interactions within complex mixed cultures.

## STUDY OF ENVIRONMENTAL SELECTION IN CONTINUOUS-FLOW SLIDE CULTURE

A central problem in microbial ecology is the understanding of microbial growth and behavior under *in situ* conditions. By conducting relevant pure-culture experiments and relating these to carefully controlled field studies, the gap between lab and field research has narrowed. However, microorganisms are often genetically altered during the process of purification and consequently the activity of lab cultures sometimes misrepresents the activity of natural populations. To study microbial colonization kinetics in surface microenvironments, it has thus been necessary to preserve microbial adaptation to surfaces[24,25] by using continuous-flow culture systems which eliminate nonadherent cells.[3,26] This maintains a controlled selection pressure in favor of microbial adherence, as opposed to "nonselective" culture conditions which result in nonadherence. The development of this and other model culture systems is thus necessary to control environmental selection under defined conditions in the laboratory.

The development of microbial ecology has been slowed somewhat by the tacit assumption that microorganisms, like elements and molecules, have fixed properties. Attention remains focused on improved biochemical and genetic methods of examining microorganisms rather than on the development of model systems for the study of selection theory. The organism is isolated, characterized, and placed in a collection for later reference. However, microorganisms adapt genetically and thus change their intrinsic properties as they are isolated and studied. Precautions must thus be taken to ensure that native characteristics are conserved and that new characteristics do not arise spontaneously.

The genetic adaptation of bacteria can occur through the movement and amplification of plasmids,[27] and through both the cryptification and decryptification of genetic information.[28] The latter phenomena make it possible for a single mutation to result in the apparent "instant evolution" of an entire metabolic pathway. Due to the rapid rate of genetic adaptation which results, the continuous application of environmental stress is necessary to retain many of the original characteristics of microbial populations. If stress is relieved or altered during isolation, the properties of the population change. This process might be referred to as "deselection" due to the tendency of native populations to lose behavioral characteristics and metabolic capability upon isolation. Deselection is thus the reverse of diversification and occurs due to the relief of environmental stress (selective pressure) when natural populations are suspended in nonselective media. Although this is the most common alteration of native populations, there is also the possibility of misselection. In this case, the process of culture purification imposes new stresses which were not present in the environment. Consequently, the native population acquires new behavioral and metabolic mechanisms which were not necessarily expressed *in situ*. Thus, when isolation imposes stresses unlike those in the environment, microbial cultures may develop features which are unrelated to those of *in situ* populations.

Continuous-flow slide culture provides a model system in which the hydrodynamic and other stresses associated with surface colonization can be defined, controlled and varied. Unless these stresses are maintained, surface adaptations are lost through deselection. This

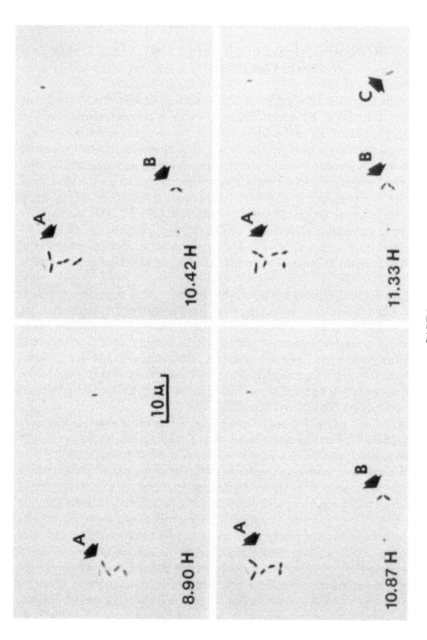

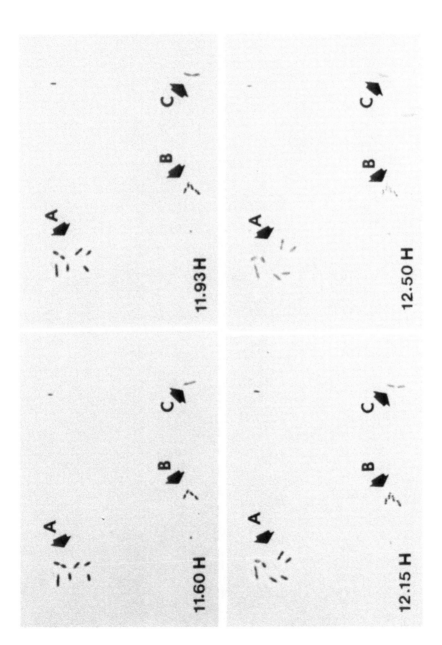

PART II

FIGURE 11.   Colonization maneuvers exhibited by members of a natural stream community incubated in continuous-flow slide culture (sterile-filtered stream water was used as the irrigation solution). The time course shown demonstrates the (A) spreading maneuver, (B) packing maneuver, and (C) shedding manuever. (From Lawrence, J. R. and Caldwell, D. E., *Microb. Ecol.*, 14, 21, 1987. With permission.)

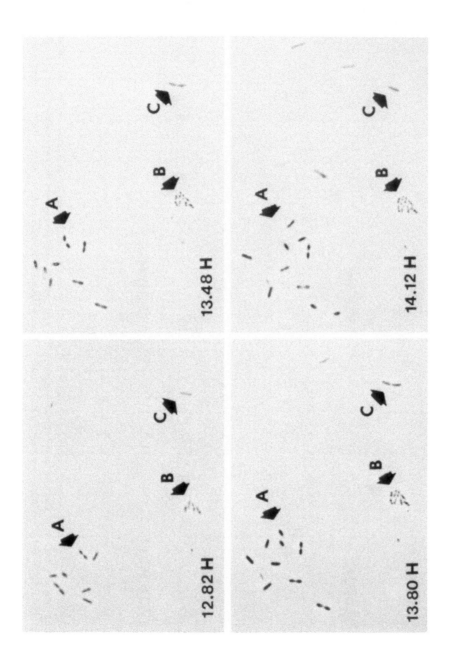

FIGURE 11, PART III

leads to populations which are unable to swarm on surfaces due to the loss of lateral flagella,[24] to the loss of proteinaceous adhesins,[25] and to the loss of adhesive exopolysaccharide.[26] By varying the environmental stresses associated with surfaces (laminar-flow velocity, cell-surface density, surface tension, etc.) the relationship between environmental selection and microbial adaptation (emigration rate, immigration rate, colonization maneuvers, dispersement of mature colonies, flagellation, cell surface tension, cell surface chemistry, etc.) may be quantitatively defined and new mechanisms of adaptation discovered.

## IMPORTANCE OF CONTINUOUS-FLOW SLIDE CULTURE IN THE STUDY OF SURFACE MICROENVIRONMENTS

Microorganisms form surface biofilms in bioreactors, biological contactors, trickling filters, soil, on minerals, on particulate substrates, within the rhizosphere, in the phyllosphere, within the intestines, on the skin, in the oral cavity, and elsewhere. It is difficult to imagine a plant, animal, mineral, or other exposed surface which does not possess attached microorganisms. Surface attachment increases the resistance of microorganisms to hypochlorite and other antimicrobial agents.[29] In addition, the maximum specific growth rate and half-saturation constant vary depending on attachment[30,31] as well as on surface hydrodynamics, cell-surface density,[3] and cell-surface distribution. However, most knowledge concerning microorganisms on surfaces is still extrapolated from studies of cell suspensions and more direct information is needed. Unfortunately, surface microbiology, like surface chemistry, has developed slowly due largely to lack of methodology. It is hoped that the application of image processing to continuous-flow slide culture will provide the opportunity to study surface microbiology more directly and to investigate cells under more highly controlled conditions.

## ACKNOWLEDGMENTS

Jim Malone, Tom Trumper, V. V. S. R. Gupta, Tom Pfeifer, and Fiona Crocker (AP MC 831, Microbial Ecology Laboratory) are acknowledged for the data shown in Figure 10. Darren Korber, Michael Whiting, Alison Jones, and Scott Gilbert are acknowledged for technical assistance. Zhang Lu is acknowledged for review of mathematical derivations. The Natural Sciences and Engineering Research Council of Canada (NSERC) and the U.S. Office of Naval Research are acknowledged for research funding.

## REFERENCES

1. **Baltzis, B. C. and Fredrickson, A. G.,** Competition of two microbial populations for a single resource in a chemostat when one of them exhibits wall attachment, *Biotechnol. Bioeng.*, 25, 2419, 1983.
2. **Berg, H. E. and Block, S. M.,** A miniature flow cell designed for rapid exchange of media under high-power microscope objectives, *J. Gen. Microbiol.*, 130, 2915, 1984.
3. **Caldwell, D. E. and Lawrence, J. R.,** Growth kinetics of *Pseudomonas fluorescens* within the hydrodynamic boundary layer of solid-liquid interfaces, *Microb. Ecol.*, 12, 299, 1985.
4. **Schwobel, W.,** Eine einfache Durchstromungsapparatur zur Gewebezuchtung aus nichtrostendem Stahl, *Exp. Cell Res.*, 6, 79, 1954.
5. **Perfil'ev, B. V. and Gave, D. R.,** *Capillary Methods of Investigating Microorganisms*, Shewan, J. M., Trans., University of Toronto Press, 1969.
6. **Block, S. M., Segall, J. E., and Berg, H. C.,** Adaptation kinetics in bacterial chemotaxis, *J. Bacteriol.*, 154, 312, 1983.
7. **Caldwell, D. E. and Germida, J. J.,** Evaluation of difference imagery for visualizing and quantitating microbial growth, *Can. J. Microbiol.*, 31, 35, 1985.

8. **Wittenberger, C. L., Beaman, A. J., Lee, L. N., McCabe, R. N., and Doonkersloot, J. A.,** Possible role of *Streptococcus salivarious glucosyltransferase* in adherence of *Veillonella* to smooth surfaces, in *Microbiology 1977*, Schlessinger, D., Ed., American Society of Microbiology, Washington, D. C., 1977, 417.

9. **Kuhn, D. A. and Starr, M. P.,** Effects of microscope illumination on bacterial development, *Arch. Microbiol.* 74, 292, 1970.

10. **Goldstein, S.,** *Modern Developments in Fluid Mechanics*, Vol. 1, Dover, New York, 1965.

11. **Landau, L. D. and Lifshitz, E. M.,** *Fluid Mechanics*, Addison-Wesley, Reading, Mass., 1959.

12. **Langlois, W. E.,** *Slow Viscous Flow*, Macmillan, New York, 1964.

13. **Marr, A. G., Painter, P. R., and Nilson, E. H.,** Growth and division of individual bacteria, in *Microbial Growth*, 19th Symp. Soc. Gen. Microbiol., Meadows, P. M. and Pirt, S. J., Eds., Cambridge University Press, London, 1969.

14. **Caldwell, D. E.,** Computer enhanced microscopy (CEM), *J. Microbiol. Methods*, 4, 117, 1985.

15. **Caldwell, D. E.,** Surface colonization parameters from cell density and distribution, in *Microbial Adhesion and Aggregation*, Marshall, K. C., Ed., Springer-Verlag, New York, 1984, 125.

16. **Monod, J.,** La technique de culture continui: theorie et applications, *Ann. Inst. Past. (Paris)*, 79, 390, 1950.

17. **Casida, L. E., Jr.,** Interval scanning photomicrography of microbial cell populations, *Appl. Environ. Microbiol.*, 23, 190, 1972.

18. **Marshall, K. C., Stout, R., and Mitchell, R.,** Mechanism of the initial events in the sorptions of marine bacteria to surfaces, *J. Gen. Microbiol.*, 68, 337, 1971.

19. **Meadows, P. S.,** The attachment of bacteria to solid surfaces, *Arch. Mikrobiol.*, 75, 374, 1971.

20. **Torella, F. and Morita, R. Y.,** Microcultural study of bacterial size changes in microcolony and ultra-microcolony formation by heterotrophic bacteria in seawater, *Appl. Environ. Microbiol.*, 41, 518, 1981.

21. **Duxbury, T., Humphrey, B. A., and Marshall, K. C.,** Continuous observation of bacterial gliding motility in a dialysis microchamber: the effects of inhibitors, *Arch. Microbiol.*, 124, 169, 1980.

22. **Lawrence, J. R., Delaquis, P. J., Korber, D. R., and Caldwell, D. E.,** Behavior of *Pseudomonas fluorescens* within the hydrodynamic boundary layer of solid-liquid interfaces, *Microb. Ecol.*, 14, 1, 1987.

23. **Lawrence, J. R. and Caldwell, D. E.,** Behavior of lotic bacterial communities within the hydrodynamic boundary layer of solid-liquid interfaces, *Microb. Ecol.*, 14, 15, 1987.

24. **Hall, P. G. and Krieg, N. R.,** Swarming of *Azospirillum brasilense* on solid media, *Can. J. Microbiol.*, 29, 1592, 1983.

25. **Jones, G. W. and Isaacson, R. E.,** Proteinaceous bacterial adhesins and their receptors, *Crit. Rev. Microbiol.* 10, 229, 1983.

26. **Pringle, J. H., Fletcher, M., and Ellwood, D. C.,** Selection of attachment mutants during the continuous culture of *Pseudomonas fluorescens* and relationship between attachment ability and surface composition, *J. Gen. Microbiol.*, 129, 2557, 1983.

27. **Reanney, E.,** Extrachromosomal elements as possible agents of adaptation and development, *Bacteriol. Rev.*, 40, 552, 1976.

28. **Hall, B. G.,** Adaptation by acquisition of novel enzyme activities in the laboratory, in *Current Perspectives in Microbial Ecology*, Klug, M. J. and Reddy, C. A., Eds., American Society for Microbiology, Washington, D.C., 1984, 79.

29. **LeChevallier, M. W., Hassenauer, T. S., Camper, A. K., and McFeters, G. A.,** Disinfection of bacteria attached to granular activated carbon, *Appl. Environ. Microbiol.*, 48, 918, 1984.

30. **Bright, J. J. and Fletcher, M.,** Amino acid assimilation and respiration by attached and free-living populations of a marine *Pseudomonas* sp., *Microb. Ecol.*, 9, 215, 1983.

31. **Bright, J. J. and Fletcher, M.,** Amino acid assimilation and electron transport system activity in attached and free-living marine bacteria, *Appl. Environ. Microbiol.*, 45, 227, 1983.

Chapter 7

# MICROBIAL ADHESION TO SURFACES

**Harold W. Fowler**

## SIGNIFICANCE OF CELL/SURFACE INTERACTIONS

Although interest in ecology commonly relates to the "natural" environment, the urban and industrial situation is attracting increasing attention in terms of ecological response to human intervention. In addition, from the biotechnological viewpoint, a piece of process equipment, such as a biochemical reactor, an associated heat exchanger, or even a simple pipeline may be regarded as an ecosystem, just as the mouth and teeth together form an ecological entity (Chapter 9, this volume).

In all of these habitats, whether natural, man-modified, or manmade, interactions between cells and substrata (surfaces) may have an important effect on the behavior of the system as a whole.

Therefore, an understanding of the effects of cell attachment, of the physical principles involved, of the forces which may operate, together with a meaningful evaluation of these forces are important factors in investigations of any ecosystem.

This chapter gives an indication of the relevance of cell adhesion to laboratory and to "real-world" situations; it outlines the principles of cell adhesion and discusses techniques for the evaluation of the forces between the cells and the substrata.

### Relevance to Laboratory Practice

From other chapters in this book, it is clear that the chemostat plays an important role in the laboratory simulation of ecosystems.

The theory of the chemostat depends on the assumption that the microorganisms are in suspension and that there is a point at which "washout" will occur. Thus, a chemostat with a dilution rate greater than the maximum growth rate should become sterile. However, experience indicates the improbability of such an event in practice. For example, Larsen and Dimmick[1] increased the dilution rate of a culture of *Serratia marcescens* beyond washout to reduce the number of organisms in the vessel. The flow was then stopped and batch culture conditions allowed to operate when it was observed that the growth rate was greatly in excess of the predicted value, based on the number of cells which had been left in suspension. It was concluded that the increased growth resulted from surface growth, and that 98% of the cells present in the culture during the period of batch growth must have originated from the walls. It should be noted that the vessel used as the chemostat had a volume of only 18 mℓ so that the surface/volume ratio was high in their experiments.

Clearly, wall growth can be the cause of anomolous results in laboratory chemoststat studies and will lead to complications if laboratory data are used to scale-up to larger volumes where the surface to volume ratio will be smaller.[2]

The basic equations describing chemostat growth have been extended by Topiwala and Hamer[3] by the introduction of a second term describing the amount of biomass attached to the wall and the specific growth rate of the organisms in the film. While this provides a theoretical correction, it is not easy to obtain suitable practical values to define the relevant parameters. In addition, attachment to surfaces is rarely uniform, depending on the hydrodynamic shear forces in the system. It is also clear that biomass can accumulate on the wall above the liquid level where the conditions can be favorable to growth. Gas transfer is simple, as the film is in contact with the atmosphere, and nutrients are provided freely by splashing of the medium or by foaming, but there are no liquid shear forces to remove the attached cells.

The practical effect of wall growth in laboratory chemostat systems, therefore, is that a population will remain, even when the dilution rate exceeds the specific growth rate. In addition, in mixed cultures, a slow-growing organism can persist in the protected environment of a surface film in competition with faster growing species in suspension.

Another problem is that cells may attach to probes or sensors. Microbial film, possibly invisible to the naked eye, constitutes a diffusional resistance leading to concentration gradients, which are seen as spurious signals by the control system. As a result, the value of pH or dissolved oxygen, for example, at the film/probe interface, may differ considerably from the value in the bulk of the system. The practical effect of this can be to negate the benefits of the sophisticated control systems that are available for bioreactors or to lead to misleading conclusions from experiments. Film thicknesses capable of causing such effects will be considered later.

### Relevance to the Environment

The properties of natural ecosystems have been discussed elsewhere in this book, but it should be emphasised that microbial adhesion play an important role. Microorganisms attached to surfaces in an aquatic environment commonly have the advantage of enhanced nutrient supplies from materials adsorbed to the substratum,[4] but they enjoy the tranquility of laminar flow or the stationary sublayers rather than the turbulence of the mainstream.

The effect of surfaces in the natural environment has long been recognized, and Russell pointed out as early as 1891[5] that the bacterial distribution in the Bay of Naples showed higher counts in the bottom deposits than near the surface of the water.

The importance of such attachment to sediments was demonstrated by Heukelekian and Dondero[6] who showed that less than 0.04% of the bacteria that were found in the waters of the river Nile were in free suspension.

The effect of the nutrient adsorption to added surface was illustrated quantitatively by Heukelekian and Heller,[7] who studied the effect of glucose concentration, in the presence and absence of glass beads, on the growth of *Escherichia coli*. At 0.5 mg/$\ell$ in the absence of beads, the count did not rise above the inoculum level, whereas in the presence of beads, the count increased by more than two orders of magnitude. The effect of the beads was significant up to a nutrient concentration of 25 mg/$\ell$.

The significance of bacterial adhesion in the natural environment has been discussed in more detail by Marshall.[8] Specific accounts in relation to roots and plant surfaces, to soils and sediments, and to surfaces in freshwater systems, for example, have also been given.[9]

### Relevance to Manmade Systems

Cooling water systems provide an ideal environment for the development of microbial film and the effects of such fouling have been reviewed in detail by Kent and Duddridge.[10]

The effects of biofouling fall into three main categories:

1. Fluid flow effects are due to either channel blockage if there is a heavy deposition or biomass detachment, or an increase in frictional losses if the surfaces become roughened by the film. In fact, considerable losses can result from deceptively thin layers of microbial film, and Characklis[11] quotes a number of case histories relating to water supply. In one, the flow capacity of a concrete pipe approximately 1 m in diameter fell by 12% over a 2 year period as the result of a slime layer about 1.5 mm thick. In another case, a steel pipe 60 cm in diameter lost as much as 55% of its design capacity due to as little as 0.6 mm of microbial slime. Studies by Characklis[12] show that pipe frictional resistance is approximately proportional to biofilm thickness, up

to the typical values seen before hydrodynamic shear limits the thickness of the layer.

2.  The presence of a biofilm on a heat transfer surface provides another resistance which has to be taken into account in calculating the overall heat transfer coefficient. Since a biofilm consists of about 90% water as well as other components with even lower coefficients of thermal conductivity, it is not surprising that there is a considerable resistance to heat transfer due, again, to very thin layers. Characklis[12] presents data which show that heat transfer resistance is proportional to film thickness up to about 120 $\mu$m. Even this amount of film is seen to have a significant effect on the overall heat transfer coefficient, reducing it by as much as 30 to 40%.

3.  A third effect of biofilm on a substratum is that corrosion of the surface may occur as a result of the accumulation of metabolic products beneath the film, and while this may not be unexpected when the surface is metal, other materials, such as concrete, may be affected also. Dental caries due to accumulation of the plaque "biofilm" provides an interesting medical analogy to these industrial problems.

## FLOCCULATION

"Microbial film" and "microbial floc" are commonly regarded as distinctly different forms of microbial aggregation. However, this is to ignore the fact that it is only a monolayer of cells which directly interacts with the substratum. Subsequent layers of cells depend on cell/cell interactions and solute diffusion gradients, and so are governed by the same factors which control cell/cell aggregation to form flocs. A further complication is that, outside carefully controlled laboratory experiments, a particle of debris may provide a nucleus for the formation of a floc. Clearly, to a 1 $\mu$m cell, a particle perhaps 100 $\mu$m in diameter is just as much a surface as a rock, a pipeline, or the wall of a bioreactor.

The principal practical difference is that films usually remain static within an ecosystem, while flocs are mobile and may be washed-out of the system. It follows, therefore, that the presence of microbial film has the effect of converting a chemostat from an open system to a partially closed or even a closed system.

The aggregation of cells to flocs has been discussed, both in general and some areas more specifically, by a working group.[13] In addition, Atkinson and Daoud[14] have comprehensively reviewed the whole subject of flocculation.

## PRINCIPLES OF MICROBIAL ADHESION

The variation in attachment mechanisms involving macromolecules, cells, and substrata, in any combination, seems to be infinite. However, as Gingell and Vince[15] point out, evolution seems to have produced highly specialized adhesion-recognition mechanisms which complement a basic process with little specificity.

There are a number of useful reviews which may be consulted for details. Baddiley[16] summarizes biochemical aspects in a report for the Science and Engineering Research Council, while the proceedings of a conference on microbial adhesion[17] provides a comprehensive coverage of the whole field. Of particular value are the conclusions of a workshop on microbial adhesion and aggregation[13] which presents working papers by a number of specialists and reports of group discussions on various related topics. Both this report and another series of papers edited by Bitton and Marshall[9] include considerations likely to be of special relevance to ecology, for example, the adhesion of microorganisms to animate surfaces, both animal and plant.

In general, two main stages of cell/substratum adhesion occur: a primary attraction which is usually physical or physico-chemical, followed by a secondary stabilization of a biological nature.

In microbial systems, the strength of the initial physical attraction is of particular interest. Usually, cells are small and their density is similar to that of the medium, so that gravitational effects on individual cells are insignificant, although flocs of aggregated cells may be affected. Hence, apart from the possibility of chemotaxis by motile bacteria, the only mechanisms which will cause bacteria to approach surfaces are diffusion and bulk fluid transport and, by definition, these are as likely to take cells away from the surface as towards it. As a result, such approaches are random and transitory and are likely to be insufficient on their own to permit secondary biological stabilization, which is usually a slow process involving morphological changes or biochemical secretions. Hence, the necessity for the temporary attachment provided by short-term, reversible physical attraction between the cell and the surface of the substratum.

This would suggest that a knowledge of the charges on the cell and the surface could be used to predict and quantify the adhesion. In practice, this is not the only property to be considered and must be related to the overall biological adhesion potential of the cell, which depends on genetically determined cell factors and the physiology of the cell. In some cases, a strong physical attraction energy may provide adhesion without apparent biological stabilization. On the other hand, a much lower physical attraction energy may suffice to hold the cell to the surface if the biological adhesion is rapid.

In addition to cell properties, account must be taken of the substratum character. Apart from the obvious significance of surface charges in primary attraction, surface roughness can be important. Grooves or protuberances on natural surfaces, or machining marks on man-made surfaces, may be almost invisible to the naked eye, but may be enormous in relation to the scale of a single microorganism.

The link between the cell and the surface is the environment, and this may influence the interactions. Factors such as the pH, the presence or absence of particular ions (calcium, for example), and the ionic concentration commonly affect cell/substratum binding.

Additional complications arise if the surface, as it almost always does, acquires an organic conditioning layer of protein or polysaccharide. At present, not enough is known about such layers to quantify the effects, although these can be demonstrated.

The physico-chemical effects involved in the primary attraction can be described in various ways, such as the DLVO theory of colloid stabilization or the critical surface tensions of the system. These phenomena have received a good deal of attention[17] with equations formulated and the relevant parameters determined experimentally, but the values are usually obtained under controlled laboratory conditions, which are "clean" and "pure" and are likely to bear little relationship to heterogeneous systems in the natural environment or found in a commercial bioreaction.

Thus, while these theoretical phenomena have a value in understanding and explaining the adhesive characteristics of the ideal system, commonly they are of little help in attempting to predict and evaluate the overall behavior of "real-world" situations. Techniques which can yield data in this context are considered later.

Secondary biological stabilization mechanisms in microbial systems fall into two main categories. Filamentous appendages, known as pili, are found in certain families of bacteria and may be involved in the adhesion process. More commonly, extracellular slimes of polymeric materials are formed, which act as a "cement" securely attaching cells to the surface. Usually, such materials are high molecular weight polysaccharides, although proteins may be involved in some cases.

Clearly, the presence of two separate adhesion mechanisms, each of which may be modulated to a differing degree by environmental factors, complicates the definition of the adhesive behavior of a cell. This makes it harder to provide a straightforward means of evaluating the adhesive potential of a system which will predict or explain the attachment of particular cells to a particular surface, when in a specified environment under known hydrodynamic conditions.

# THE MEASUREMENT OF MICROBIAL ADHESION

From this outline of the principles of cell adhesion, it is clear that attempts to study adhesion in the context of microbial ecosystems demands a method for quantifying the forces involved, under at least four headings.

First, one must be able to determine the physio-chemical forces which take the cells to the surface in the first place. Second, one must be able to measure the strength of attachment following secondary biological stabilization, which is likely to be related to the forces required to remove the cells from the surface. Third, a range of test forces must be provided simultaneously; adhesion is a time-dependent process since conditions in a layer of microbial film will not remain static, but are changing continuously. Finally, the technique must not depend on "ideal" conditions but must be able to reproduce the natural conditions under investigation.

Many different methods have been proposed for the measurement of adhesion, and these have been reviewed by Fowler and McKay.[18] Almost all involve "distraction" (that is physical detachment, usually by shear forces) of established layers that have been formed under static conditions.

The methods may be divided into two main groups:[18]

1. The "adhesion number" type of test applies a fixed detachment force to adhering cells and expresses the adhesive strength in terms of the proportion of cells which can withstand this force.
2. The other group determines a "critical force" by applying a steadily increasing shear until detachment occurs, the minimum shear which results in detachment being regarded as equal and opposite to the adhesive forces holding the cells to the surface. This critical force is then defined as the shear force causing a predetermined proportion of cell distraction (not necessarily total) within a specified time.

While all of these tests may serve the particular purpose for which they were designed, in general, the shear conditions are not well defined, no account is taken of time dependence, and they do not measure the forces of adhesion between the cell and the surface, but the strength of the fully stabilized biological "joint" which may be quite different. It should be appreciated that distraction often does not occur at the original interface between cells and substratum, but results from cleavage of the cell layer, of the conditioning layers on the surface, or of the substratum itself. In the same way, a glued wooden joint will often be broken by splitting of the wood adjacent to the joint, rather than by failure of the adhesive itself.

There are occasions, of course, when all that needs to be determined is the resistance of a microbial film to detachment, an explanation of the film separation being quite irrelevant. Whatever the system, the force required to distract the stabilized layer usually bears little or no relationship to the original attractive forces, and the latter are often of considerable importance.

In order to overcome these limitations, the concept of the "Dynamic Assessment of Cell Adhesion" has been developed.

## THE DYNAMIC ASSESSMENT OF CELL ADHESION

The dynamic assessment of cell adhesion is based on subjecting the cell/substratum system to defined hydrodynamic conditions, by providing a continuous range of shear forces across a surface. In the case of cell attachment, a critical shear stress can be determined which is just overcome by the attractive forces between the cell and the surface. For detachment, the critical shear stress needed to remove cells from the surface is observed.

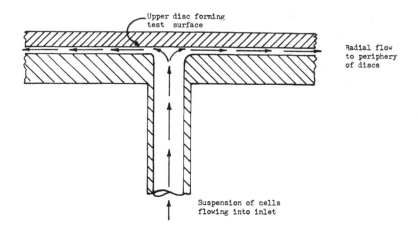

FIGURE 1.   Diagram of radial flow system.

The required range of shear stresses, with appropriate hydrodynamic control, can be obtained in a flow chamber with radial-flow conditions.

## THE RADIAL-FLOW CHAMBER

The radial-flow chamber comprises two flat disks, mounted precisely parallel to one another, one of which is be made from the material or substratum under test. A fluid inlet is provided at the center of one of the disks, so that flow occurs radially between the disks, from the center to the periphery. Since the flow area increases across the disk, the fluid velocity decreases with a consequent reduction in the shear stress at the fluid/surface interface (Figure 1).

When attachment is under investigation, a suspension of cells is pumped between the disks, and attachment takes place when the attractive forces overcome the surface shear stress. This leads to a clear central zone and an accumulation of cells in the outer zone. A typical result from such a test is illustrated in Figure 2.

To measure detachment, cells are allowed to attach to the disk under static conditions, for example, in a large petri dish. When a suitable layer of cells has been formed, depending on the objectives of the experiment, the disk is placed in the flow chamber and subjected to flow with the desired test medium, again forming a clear central zone. In both cases, the radius of the clear zone, described as the critical radius, is an indirect measure of the forces involved in the adhesion. Alternatively, the critical radius may be used to calculate the critical surface shear stress, which is a more direct assessment of the forces involved.

## THE LH-FOWLER CELL ADHESION MEASUREMENT MODULE (CAMM)

The original version of the radial flow chamber has been described by Fowler and McKay.[18] The technique has proved to be of value wherever there are cell/substratum/environment interactions and is appropriate to the study of fundamental aspects of cell adhesion, as well as to the solution of applied problems. As a result, the equipment is now available commercially as the *L-H FOWLER CELL ADHESION MEASUREMENT MODULE (CAMM)**.

The CAMM consists of two units (Figure 3). The radial flow unit, (Figure 4) has its three components (radial flow chamber, pump body, and flowmeter) combined in a single unit which may be sterilized by autoclaving and requires only to be connected aseptically to a

*   L. H. Fermentation Limited, Bells Hill, Stoke Poges, Bucks, SL2 4EG, U.K.

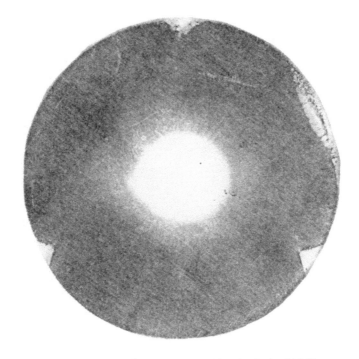

FIGURE 2.    Pyrex® glass test disk with stained microbial film.

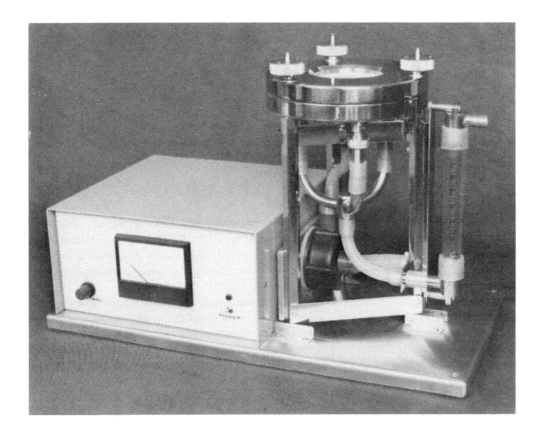

FIGURE 3.    The "LH-Fowler Cell Adhesion Measurement Module (CAMM)".

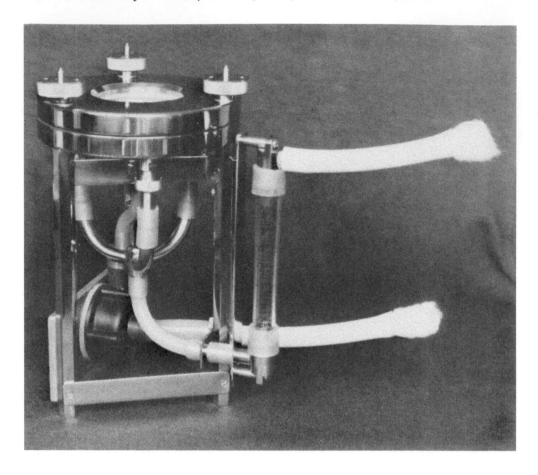

FIGURE 4. The radial flow unit of the LH-Fowler CAMM.

chemostat or to another cell source. The radial flow unit is placed on the base of the control unit, when the pump body fits into the magnetic drive, making the system ready for immediate operation (Figure 5).

## INTERPRETATION OF THE RESULTS

Where the objective of the work is a simple ranking procedure (for example, comparing the adhesive characteristics of different organisms or the effect of changing an environmental variable, such as pH or the nutrient concentration), critical radius values may be used directly. Care must be taken, of course, to keep other parameters (particularly the volumetric flowrate through the chamber) constant.

A better measure of the energy of adhesion between cell and substratum is obtained by calculating the critical surface shear stress. A full analysis of the hydrodynamics of radial flow for all conditions can be complicated, and it has been suggested that the use of complex mathematical models is desirable.[19] However, experience indicates that with proper design of the flow chamber and by careful observation of the recommended operating procedures, a very simple equation provides results which have a precision appropriate to normal laboratory experimental practice.

The surface shear stress in radial flow, with laminar flow conditions, may be calculated from:

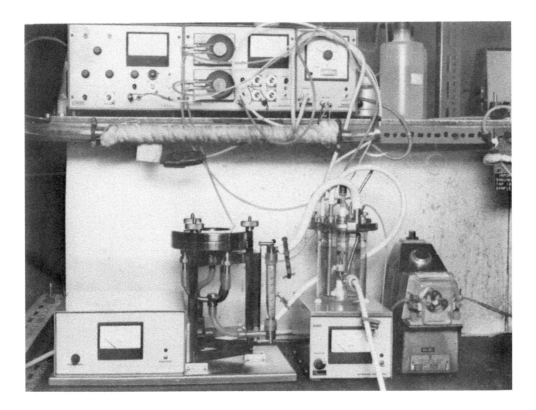

FIGURE 5.   The LH-Fowler CAMM in use with a chemostat.

$$\tau = \frac{3\,Q\,\mu}{\pi\,r\,h^2} \tag{1}$$

where Q = volumetric flowrate through the radial flow chamber, $\mu$ = viscosity of the fluid, h = disk separation distance, and $\tau$ = surface shear stress at the radius r.

## TYPICAL RESULTS

The photograph of a stained microbial film in Figure 2 shows the form of attachment of cells to a 10 cm Pyrex® glass disk. The direct use of critical radii (Figure 6) demonstrates the different attachment behavior of microorganisms to Pyrex® glass and to stainless steel and that there is a clear optimum pH for attachment.

It can be more informative to carry out microscopic counts in a predetermined pattern on the test disk when the numbers of microorganisms adhering at a specified radius may be plotted against the critical surface shear stress at that point (Figure 7).

This method has a number of advantages. First, quicker results may be obtained, since it has been found that cell attachment can give countable numbers in a short time — as little as 5 min in some cases — so that it is not necessary to wait for a visible layer of cells to accumulate. Furthermore, in this context, it should be remembered that the formation of a visible microbial film involves more than one adhesion mechanism. The first monolayer actually depends on the cell/substratum interactions, whereas additional layers arise from cell/cell attachment or from cell divisions.

A second advantage of the cell count vs. shear stress plot is that it gives a clearer picture of the sharpness of the critical shear stress. In addition, the effects of other variables can

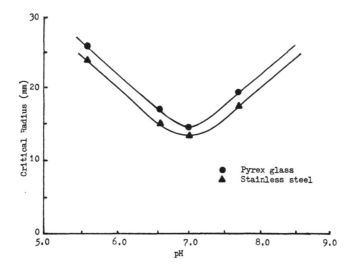

FIGURE 6.   Use of the critical (clear zone) radius to show the effect of pH and of substratum on microbial adhesion.

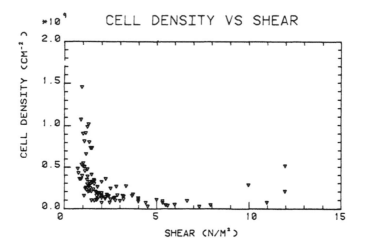

FIGURE 7.   Count of attached cells at various critical shear stresses across a test disk.

be seen, for example, by the attachment of some cells at higher shear stresses, which may be due to surface inhomogeneities or to more adhesive cells, either from a mixed culture or as mutants in a monoculture.

The dynamic assessment of cell adhesion has proved to be capable of yielding reproducible data in a wide range of situations, such as the study of the forces of attachment and of detachment,[20] which may differ by quite substantial ratios.

## THE SIGNIFICANCE OF MICROBIAL FILM ON SURFACES

The implications of microbial film growth on surface structures have been considered in some detail by Atkinson and Fowler.[21] Although their discussion emphasized the effects in bioreactors, the same principles must apply in most ecological studies. After all, a bioreactor is no more than a specialized ecosystem, while those used for wastewater treatment are only controlled enhancements of natural ecosystems. Furthermore, chemostats and other forms of bioreactor are used as models for laboratory studies of natural ecological systems.

Some effects of microbial films have been described earlier, but it should be pointed out that, while corrosion of the walls of a bioreactor or the structure of an offshore oil rig are problems in an industrial context, the same phenomena are an essential part of the recycling process of the natural environment.

Studies of laboratory or industrial situations involving microbial film commonly do not report the film thickness, although this can be a matter of great importance. However, from investigations where values were quoted or could be deduced, Atkinson and Fowler[21] concluded that microbial films could be divided into four main groups if film thickness is the criterion.

1.   Where the film is able to develop freely in the presence of abundant nutrients, as occurs in trickling filters for wastewater treatment, for example, the film is unlikely to be less than 200 μm and can be millimeters or even centimeters in thickness.
2.   If the film thickness is subjected to some form of control, however, either by hydrodynamic shear forces or by the use of mechanical devices such as scrapers, the values are usually from 70 to 200 μm.
3.   In the case of wall growth in bioreactors with pure cultures, the film is commonly little more than a monolayer, so that the film thickness is in the range 1 up to, perhaps, 10 μm.
4.   At the lowest level of attachment is the casual deposition of microorganisms on substrata where the film thickness in numerical terms is less than 1 μm, since there is not even a complete monolayer.

A few workers have made comparisons between film thickness and parameters such as oxygen tension or substrate removal, and it is apparent that there is a limiting distance over which diffusion can occur in the film, described by Atkinson and Fowler[21] as the penetration depth.

Therefore, dependent on the variety of factors which affect growth, there will be an optimum thickness for an active microbial film, and such data as are available suggest that this will normally be in the second group, that is between 70 and 200 μm, most commonly in the lower part of that range.

In a very thin film, there is insufficient biomass to provide the maximum activity for the available surface, and it has been shown that substrate removal or product formation is proportional to the film thickness up to the penetration depth, when diffusional limitation occurs. On the other hand, in a thicker film beyond the oxygen-penetration depth, anaerobic conditions will develop, even in a well-aerated culture system. Similarly, substrate limitation beyond the nutrient penetration depth will give rise to endogenous respiration, while the accumulation of waste products can lead to metabolic changes or to complete inhibition. In very thick layers, this leads to the sloughing characteristic of trickling filter operation. This is not a problem in wastewater treatment; in fact, the sloughing provides a natural control of film thickness. However, the cells exposed after sloughing may differ in behavior, and in an immobilized cell bioreactor, this could result in different products, or in laboratory experimentation, give rise to anomalous results.

It must be emphasized that microbial film up to about 100 μm in thickness may be invisible to the naked eye, but can have significant effects. Some of these have been described earlier, but, from the point of view of laboratory experiments, probably the most important problem is the possibility of the coating of probes and sensors. Modern laboratory equipment is becoming increasingly automated, with sophisticated instrumentation and greater use of computers, thereby reducing human involvement in the operation. While this has a number of advantages, there are some circumstances where the human eye observes changes which are not detected by, and may mislead, instruments. It cannot be overstated that this microbial film, only a few μm in thickness and probably visible only on very careful inspection, can negate the advantages associated with the developments that have occurred in the equipment.

## THE CONTROL OF MICROBIAL FILM

Clearly, it is necessary to be able to control microbial film development in laboratory systems. In some cases, this may mean the complete elimination of attached cells, while in others the presence of the film may be important, but control of its thickness is required. Again, this may mean simply the prevention of the accumulation of excess film and the maintenance of an active film thickness not exceeding the penetration depth, but without specifying a precise thickness. On other occasions, the film thickness is the variable of interest and a series of layers of defined thicknesses must be provided.

It has been indicated that microbial film will be, to some extent, self-regulating when diffusional limitation causes excess biomass to slough off and, if operating under batch conditions (which are inherently in unsteady state), this may be acceptable. It will be appreciated, however, that the cells may be subjected to significant shock changes. When a portion of film detaches, cells which were under anaerobic conditions with low, or even zero, substrate concentration are transferred suddenly to higher concentrations of oxygen and of substrate. In addition, there may be effects, possibly deleterious, on the remainder of the culture due to biochemical products released from the depth of the film.

Clearly, this random removal of portions of film will be unacceptable in continuous systems, where the objective to achieve a steady state in which there will be an equilibrium film thickness and accumulation by attachment or growth is balanced by controlled removal of film. As might be anticipated, it is the latter which presents a problem, both in industrial equipment and in the laboratory. This may be carried out mechanically or physico-chemically.

Mechanical removal by scraping the surface has been used successfully for many years[22,23] and is simple and satisfactory for nonaseptic systems, so that it may be useful for some ecological studies.

If aseptic conditions have to be maintained, the procedure is more difficult, although rotating rubber wipers have been used[24] and a similar method used oscillatory movement.[25] While such devices are capable of removing excess film, there is the disadvantage that the thickness is not controlled at a definite value.

The control of microbial film thickness to a precisely defined value has been achieved on the small scale, however, and Peters and Wimpenny (see Chapter 9, this volume) have devised a system in which a rotating (poly) tetra-fluoroethylene (PTFE) disk is in contact with a thin PTFE scraper blade. A series of PTFE pans, 20 mm in diameter, are set flush within the disk. Each pan has a number of holes in which there are tight-fitting plugs, and these may be pushed down to produce a recess of the desired depth by means of a recessing tool. Film growth occurs on the surface of these plugs, excess growth being wiped away by the scraper blade, so that small areas of film of constant thickness are maintained.

In all cases where mechanical scraping is used, it should be noted that, although the desired film thickness is obtained, it is an artificial method which is not strictly analogous to natural ecological systems. Furthermore, the movement of the scraper blade across the film affects the fluid hydrodynamics adjacent to the film intermittently, thereby changing the boundary layer conditions and influencing mass transfer to, or from, the cells. Nevertheless, it would appear that it is a realistic compromise if a specified film thickness is required.

A closer analogue of some natural systems may be obtained, without resorting to mechanical scrapers, by employing the abrasive forces arising from physical contact between randomly moving surfaces. A convenient way to achieve this is to use small, inert particles as support surfaces for microbial growth and to maintain these in suspension when the film thickness is controlled by the attrition resulting from the frequent contact between the moving particles.

This may be achieved simply by the addition of the particles to a conventional stirred vessel, when the impact of the particles on the walls of the vessel and on surfaces such as baffles detaches thick film. Furthermore, the thin active film on the added surface provided by the particles increases the efficiency of the bioreactor,[21] so that there is a double benefit. There are limitations, however, both in the size and quantity of the particles which can be held in suspension by the agitator and in poor control of the particle movement.

A more efficient way to suspend the support particles is to use a fluidized bed, with the further advantage that agitation is provided by the recirculating medium so that mechanical stirring is not required. Atkinson and Fowler[21] describe a system based on this principle and employing glass beads to provide the support surface, although particles of other materials may be used. Since the behavior of a fluidized system depends largely on the particle size and density (which determine the fluid velocity required), it is possible to optimize the surface area (and thereby the hold-up of biomass) and the mass transfer to the film. The authors also give an example of the method of calculation to determine the particle size needed in the design of a fluidized bed bioreactor.

It might appear that attrition could remove too much of the attached film, but experience indicates that this is not the case. The fluidized bed bioreactors described by Atkinson and Fowler[21] and Atkinson and Knights[26] developed an attached layer of the order of 70$\mu$m, which is ideal for a biologically active film and within the typical values for the penetration depth to avoid diffusional limitation. It should be borne in mind that a surface which appears "smooth" to the touch or to the naked eye may be very rough to a microorganism. Such roughness can provide considerable protection and permit the formation of an equilibrium film which will allow the bioreactor to operate under steady-state conditions in a continuous process.

The alternative to mechanical removal of the film is to use physico-chemical control, and factors such as pH can be very effective in governing the formation of film.[27,28] However, there may be difficulties, for example, the conditions correct for the control of adhesion may be inappropriate for the desired bioreaction. Furthermore, there may not be a straight-forward relationship between the physico-chemical parameter that can be varied and its effect on the adhesion. This is illustrated clearly by a study of the attachment of *Chlorella* to glass [29] in which the charges on the algal cells and the glass were measured as the concentration of ferric chloride was increased.

As the molar concentration of the ferric chloride changed from $5 \times 10^{-6}M$ to $0.05M$, the charges changed through like, negative, strong $\rightarrow$ unlike, slight $\rightarrow$ like, positive, slight $\rightarrow$ like, positive, strong $\rightarrow$ unlike, very strong. During these changes the adhesion between the algae and the glass varied through slight $\rightarrow$ strong $\rightarrow$ slight $\rightarrow$ moderate $\rightarrow$ very strong. Obviously, prediction of the behavior to utilize the effects presents considerable problems.

The same study provided an excellent illustration of the effect of changing a physico-chemical parameter on the mechanical removal of the cells. The method used to evaluate the adhesion was to subject the attached cells to increasing velocities of liquid flow (and thereby increasing hydrodynamic shear) through a flow chamber and to observe the numbers of cells that were detached. When the adhesion, as referred to above, was "slight", in excess of 75% of the cells were removed at average liquid velocities of about 1 mm/sec, but when the adhesion was "very strong" almost none of the cells were removed at average liquid velocities of more than 3 m/sec.

Clearly, it is not always easy to predict the practical effects of the cell/surface interactions and experimentation, which is to some extent trial and error, is necessary. The LH-Fowler

Cell Adhesion Measurement Module (CAMM) has been found to be ideal for this purpose, since it provides the range of velocities instantaneously and gives a direct and immediate evaluation of the adhesion.

## CONCLUSIONS

1.  Cell/surface interactions play a significant role, for good or ill, in the environment — natural, man-modified, and manmade.
2.  Any study which aims at modeling an ecosystem must take account of such interactions; otherwise the work may yield anomalous results or will not be truly representative of that ecosystem.
3.  Care must be taken in the laboratory to anticipate the effects of adhesion. Even if the system which is the subject of the investigation does not involve adhesion to a significant extent, a laboratory study in a chemostat, for example, may become meaningless if wall-growth affects mixed cultures or if growth on probes has the effect of controlling the medium at the wrong pH or oxygen tension.

## REFERENCES

1.  **Larsen, D. H. and Dimmick, R. L.,** Attachment and growth of bacteria on surfaces of continuous culture vessels, *J. Bacteriol.,* 88, 1380, 1964.
2.  **Howell, J. A., Chi, C. T., and Pawlowski, U.,** Effect of wall growth on scale-up problems and dynamic operating characteristics of the biological reactor, *Biotech. Bioeng.,* 14, 253, 1972.
3.  **Topiwala, H. H. and Hamer, G.,** Effect of wall growth in steady state continuous culture, *Biotech. Bioeng.,* 13, 919, 1971.
4.  **Zobell, C. E.,** The effect of solid surfaces upon bacterial activity, *J. Bacteriol.,* 46(1), 39, 1943.
5.  **Russell, H. L.,** *Z. Hyg. Infektionskr.,* 11, 165, 1891.
6.  **Heukelekian, H. and Dondero, N. C.,** *Principles and Applications in Aquatic Microbiology,* Wiley & Sons, New York, 1964.
7.  **Heukelekian, H. and Heller, A.,** Relation between food concentration and surface for bacterial growth, *J. Bacteriol.,* 40, 547, 1940.
8.  **Marshall, K. C.,** Bacterial adhesion in natural environments, in *Microbial Adhesion to Surfaces,* Berkeley, R. C. W., Lynch, J. M., Melling, J., Rutter, P. R., and Vincent, B., Eds., Ellis Horwood, Chichester, U.K., 1980, chap. 9.
9.  **Bitton, G. and Marshall, K. C., Eds.,** *Adsorption of Micro-Organisms to Surfaces,* Wiley & Sons, New York, 1980.
10. **Kent, C. A. and Duddridge, J. E.,** Microbial Fouling of Heat Transfer Surfaces in Cooling Water Systems, AERE-R-10065, United Kingdom Atomic Energy Authority, Harwell, 1981.
11. **Characklis, W. G.,** Attached microbial growth. II. Frictional resistance due to microbial slimes, *Water Res.,* 7, 1249, 1973.
12. **Characklis, W. G.,** Int. Conf. on Fouling of Heat Transfer Equipment, Rensselaer Polytechnic, Troy, N.Y., 1979.
13. **Marshall, K. C., Ed.,** *Microbial Adhesion and Aggregation,* Dahlem Workshop Report, Life Sciences Research Report No. 31, Springer-Verlag, Berlin, 1984.
14. **Atkinson, B. and Daoud, I. S.,** Microbial flocs and flocculation in fermentation process engineering, *Adv. Biochem. Eng.,* 4, 41, 1976.
15. **Gingell, D. and Vince, S.,** Long-range forces and adhesion: an analysis of cell-substratum studies, in *Cell Adhesion and Motility,* Curtis, A. S. G. and Pitts, J. D., Eds., Cambridge University Press, Cambridge, 1980, 1.
16. **Baddiley, J.,** *Cell Adhesion,* Science and Engineering Research Council, Swindon, U.K., 1983.
17. **Berkeley, R. C. W., Lynch, J. M., Melling, J., Rutter, P. R., and Vincent, B., Eds.,** *Microbial Adhesion to Surfaces,* Ellis Horwood, Chichester, U.K., 1980.
18. **Fowler, H. W. and McKay, A. J.,** The measurement of microbial adhesion, in *Microbial Adhesion to Surfaces,* Berkeley, R. C. W., Lynch, J. M., Melling, J., Rutter, P. R., and Vincent, B., Eds., Ellis Horwood, Chichester, U.K., 1980, chap. 7.

19. **Fryer, P. J., Slater, N. K. H., and Duddridge, J. E.,** Suggestions for the operation of radial flow cells in cell adhesion and biofouling studies, *Biotech. Bioeng.,* 27, 434, 1985.
20. **Duddridge, J. E., Kent, C. A., and Laws, J. F.,** Effect of surface shear stress on the attachment of *Pseudomonas fluorescens* to stainless steel under defined flow conditions, *Biotech. Bioeng.,* 24, 153, 1982.
21. **Atkinson, B. and Fowler, H. W.,** The significance of microbial film in fermenters, *Adv. Biochem. Eng.,* 3, 221, 1974.
22. **Maier, W. J., Behn, V. C., and Gates, C. D.,** *J. Sanit. Eng. Div. Am. Soc. Civ. Eng.,* 93, 91, 1967.
23. **Atkinson, B., Daoud, I. S., and Williams, D. A.,** A theory for the Biological Film Reactor, *Trans. Inst. Chem. Eng.,* 46, 245, 1968.
24. **Anderson, P. A.,** Automatic recording of the growth rates of continuously cultured micro-organisms, *J. Gen. Physiol.,* 36, 733, 1953.
25. **Northrop, J. H.,** Apparatus for maintaining bacterial cultures in the steady state, *J. Gen. Physiol.,* 38, 105, 1954.
26. **Atkinson, B. and Knights, A. J.,** Microbial film fermenters: their present and future applications, *Biotech. Bioeng.,* 17, 1245, 1975.
27. **Daniels, S. L. and Kempe, L. L.,** The separation of bacteria by adsorption onto ion exchange resins, *Chem. Eng. Prog., Symp. Ser. 69,* 62, 142, 1966.
28. **Fowler, H. W.,** The adhesion of microbial film to submerged surfaces *J. Appl. Chem. Biotechnol.,* 26, 348, 1976.
29. **Nordin, J. S., Tsuchiya, H. M., and Fredrickson, A. G.,** Interfacial phenomena governing adhesion of *Chlorella* to glass surfaces, *Biotech. Bioeng.,* 9, 545, 1967.

Chapter 8

# MODEL BIOFILM REACTORS

### W. G. Characklis

## INTRODUCTION

Bacteria stick firmly, and often highly specifically, to almost any surface submerged in an aqueous environment. The bacteria attach by means of a matrix of extracellular polymeric substances (EPS), primarily polysaccharide, that extend from the cell surface and form a gel composed of tangled polymer fibers. The adhesion mediated by EPS may determine spatial distribution of bacteria in many aquatic environments. The cells grow and reproduce at the surface increasing the biomass and associated material. The entire deposit is termed the *biofilm*.

Biofilm processes may be beneficial, for example, in fixed-film wastewater treatment processes (e.g., trickling filters and rotating biological contactors). In addition, biofilms frequently play a major role in stream purification processes. It is clear that microbial activity in natural waters has been found predominantly at interfaces.[1,2] Biofilms can be quite troublesome in certain engineering systems. For example, biofilms in water conduits can cause energy losses resulting from increased fluid frictional resistance and increased heat transfer resistance. The role of biofilms in other natural and technological processes has been discussed elsewhere.[3]

The majority of the methods described in this paper were developed for research concerned primarily with biofilm development in relation to fouling. Fouling refers to the formation of inorganic and/or organic depositis on surfaces. In cooling systems, these deposits form on condenser tube walls, increasing fluid frictional resistance and heat transfer resistance. Fouling biofilm development, referred to as biofouling, is a direct result of the attachment to, and growth of, microorganisms on surfaces.

In our view, the nature of biofilms demands a fundamental approach to establish a systematic framework for analysis of biofilm structure and activity irrespective of the specific environments where they are found or their particular engineering application. This chapter reviews some laboratory systems and methods for measuring biofilm parameters in the laboratory and in the field.

## MODEL SYSTEMS

Model reactor systems have been used in the laboratory and in the field to investigate microbial film accumulation. Laboratory systems permit the observation of biofilm behavior with relatively strict environmental control of a number of variables including the following.

1. Chemical
   a. Substrate type
   b. Substrate concentration
   c. pH
   d. Inorganic composition
   e. Dissolved oxygen
   f. Microbial inhibitors
2. Physical
   a. Temperature
   b. Fluid shear stress

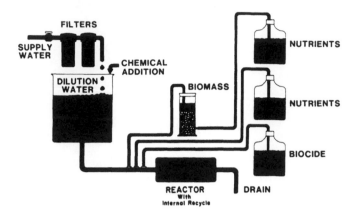

FIGURE 1. A schematic diagram of a typical biofilm reactor system including the reactor (sometimes with internal recycle), the biomass reactor for continuous inoculation of microorganisms, and the chemical (nutrients, biocide) feed system.

    c. Heat flux
    d. Surface composition
    e. Surface texture (roughness)
    f. Hydraulic residence time
3.    Biological
    a. Organism type (pure or mixed culture)
    b. Organism concentration

In laboratory systems, the effect of any one of the variables can be observed by maintaining the other variables constant. This is not usually possible in the field. Nevertheless, field experiments are still useful and necessary and have been discussed elsewhere.[4,5]

Three types of ideal reactor sytems are available for such experiments: *batch reactors*, with no inputs or outputs; *plug flow reactors*, in which reactants and products move as a "plug" from inlet to outlet; *continuous flow stirred tank reactor* (CFSTR), in which no concentration gradients exist within the liquid volume. All of the laboratory systems described in this chapter will be discussed in the context of CFSTR. The CFSTR provides significant advantages for observing, separating, and evaluating the kinetics and stoichiometry of each biofilm process:

1.    The liquid phase is uniform which makes sampling, chemical analysis, and mathematical modeling simple.
2.    The steady state condition is convenient and reproducible.
3.    Biofilms developed in CFSTR with constant shear stress at the walls are relatively uniform.

Any biofilm reactor system must contain certain components, whatever the application, if quantitative data regarding biofilm processes are desired (Figure 1):

1.    The *reactor feed* requires a continuous water supply. The water must contain all cellular growth requirements (energy, carbon, trace elements) at appropriate concentrations. pH and temperature should be maintained at desired levels. Typically, when large volumetric water flow rates are needed, tap water is treated continuously to remove residual organic carbon, chlorine, and suspended solids, and the necessary nutrients

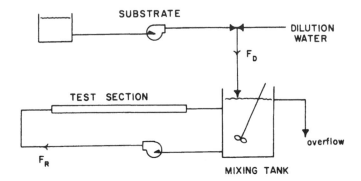

FIGURE 2. Schematic diagram of the tubular reactor with recycle.

are added as shown in Figure 1. A continuous flow of microorganisms is an option which may be required for certain experiments. A simple packed-bed reactor is suitable for providing a continuous, stable supply of microbial biomass (Figure 1).
2. A *pumping system* and method for controlling and measuring the input and recycle flow rate(s) are necessary.
3. The *reactor* where the substratum for biofilm accumulation is located. The reactors can take several forms which are described below.

These components are shown incorporated into our particular system in Figure 1. Two of these components, the reactor and the reactor feed, will be described in more detail below.

Reaction time in the CFSTR bulk water is dependent on the hydraulic residence ($\theta$) which equals reactor liquid volume divided by water supply flow rate ($\theta$ = volume/flow rate). Control of $\theta$ provides a number of benefits, compared to operating with once-through flow as is the case in a plug flow reactor.

1. Growth rate of bacteria in suspension, as in a chemostat, is equal to $\theta^{-1}$ (dilution rate, D). If wall growth only is needed, the CFSTR can be operated with D significantly greater than the maximum specific growth rate, and suspended bacteria will washout before reproducing. The activity in the reactor can then be attributed to the biofilm alone.
2. Operating with long hydraulic residence times requires less water and generally therefore less nutrient chemicals compared to a once-through system.
3. $\theta$ can be varied during an experiment if an approximation of a once-through system is desired. For example, an experiment may be operated at $\theta$ = 30 min during development of the biofilm and operated with a $\theta \ll 60$ sec while a biocide (microbial inhibitor) is applied.

**The Reactor**
   A number of different reactor geometries can be inserted in the recycle flow system described in Figure 2.

1. Tubular systems are used because circular tubes are the prevalent geometry in heat exchangers and also because a wide variety of alloys are available in this form. Fluid dynamics in this geometry are well defined.
2. Open channel systems are a variation of the tubular systems except that one side of the tube is open to the atmosphere. These systems have been used to simulate streambed conditions.

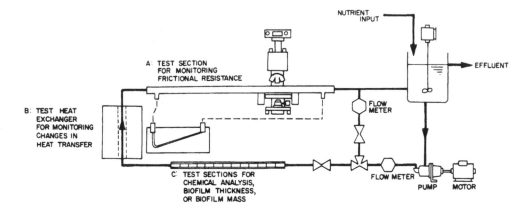

FIGURE 3.    Composite diagram of a tubular reactor incorporating the important features of three tubular reactors.

3.    Rotating annular reactors are used because of their compactness and ease of operation.
4.    Radial flow and rotating disc reactors provide a defined range of fluid shear stress conditions simultaneously at the substratum.
5.    Packed or fluidized bed reactors increase the surface-area-to-volume for biofilm formation through the addition of specific packing material to the reactor.

*Tubular Reactor*

Tubular reactors are CFSTRs with internal recycle (Figure 3). This type of system is generally used when modeling heat transfer tubing or water supply conduits. Often, the experimental system may use tubing of the same composition and diameter as that being simulated. Advantages of the tubular configuration include the following:

1.    At high recycle rates (recycle flow rate ≫ dilution water flow rate), the reactor contents are completely mixed and no longitudinal concentration gradient exist in the liquid phase. This simplifies mathematical descriptions and sampling. It also provides a relatively uniform biofilm in the recycle section while allowing simple control of pH and temperature. From a practical standpoint, this system minimizes the consumption of water, microbial nutrients, and other chemical additives.
2.    The tubular system can be operated as a plug flow reactor if flow is once-through and the tube is long enough to establish concentration gradients.
3.    If desirable, a short hydraulic residence time can be maintained which minimizes biomass activity in the bulk fluid and restricts microbial activity in the system to the reactor surfaces.
4.    Fluid shear stress at the wall in the recycle loop is independent of mean residence time in the reactor system.

Each tubular reactor system may incorporate one or more of the following types of tubular sections:

1.    A tubular test section in which pressure drop is monitored by a manometer or pressure transducer during biofilm development. Fluid frictional resistance can be calculated from flow rate and pressure drop measurements.
2.    A test heat exchanger section in which changes in heat transfer resistance are monitored as a function of biofilm development.
3.    A tubular section or sleeve which contains removable sections of tube. These sample tubes are removed periodically for determining biofilm thickness, *biofilm mass, or* biofilm chemical analysis.

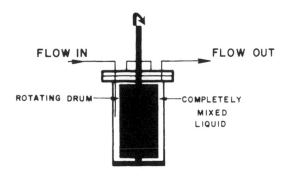

FIGURE 4.   The annular reactor.

4.   A glass tubular section for visual or microscopic observations. Periodic micrographs or video recordings of biofilm development may be obtained using such a transparent section.
5.   A rectangular duct section may be used when samples of biofilm developing on a flat-plate are needed. This is particularly desirable when periodic access to a deposit is required, for instance, using probes to obtain chemical profiles within a deposit.

*Open Channel*

The open channel geometry is ideal for simulating stream beds, aqueducts, or culverts. An important difference between this geometry and a closed tube geometry is that gas transfer between the atmosphere and water occurs throughout the length of the open-channel reactor, whereas the tubular reactor cannot be operated with atmospheric contact.

*Annular Reactor*

The annular reactor is potentially an excellent method for monitoring biofilm development because of its sensitivity, particularly to changes in fluid frictional resistance. Furthermore, changes in biofilm can be monitored continuously and nondestructively.

The annular reactor consists of two concentric cylinders, a stationary outer cylinder and a rotating inner cylinder (Figure 4). A torque transducer, mounted on the shaft between the cylinder and the motor drive, monitors the drag force on the surface of the inner cylinder (Figure 5). Fluid frictional resistance is calculated from rotational speed and torque measurements. Removable slides (4 to 12 of them), which form an integral part of the inside wall of the outer cylinder, are used to determine biofilm thickness, biofilm mass, and/or provide samples for determining biofilm chemical composition (Figure 6). The reactor is completely mixed either by an external recycle or by draft tubes in the inner cylinder which are designed to enhance fluid mixing. The fluid shear stress at the wall (function of rotational speed) can be varied independently of mean residence time. The annular reactor has a high surface-area-to-volume ratio and most of its surface area is exposed to a uniform fluid shear stress. The reactor also is convenient for material balances.[6,7]

The reactor liquid phase is completely mixed (i.e., it is a CFSTR) by virtue of draft tubes bored through the solid inner cylinder. The draft tubes are positioned at angles so that the rotation of the inner cylinder "pumps" the fluid through the tubes. By virtue of the complete mixing, effluent liquid samples represent the reactor liquid composition.

Fluid shear stress at the wall is a function of rotational speed. Mean liquid residence time depends on dilution flow rate through the reactor. Thus, fluid shear stress and residence time can be varied independently. Reactor residence time can be short (e.g., 10 min) so that planktonic growth is negligible and all reactor activity can then be attributed to the biofilm. The reactor has a high surface-area-to-volume ratio (300 to 350 m$^{-1}$), and most of its surface area is exposed to a uniform wall shear stress.

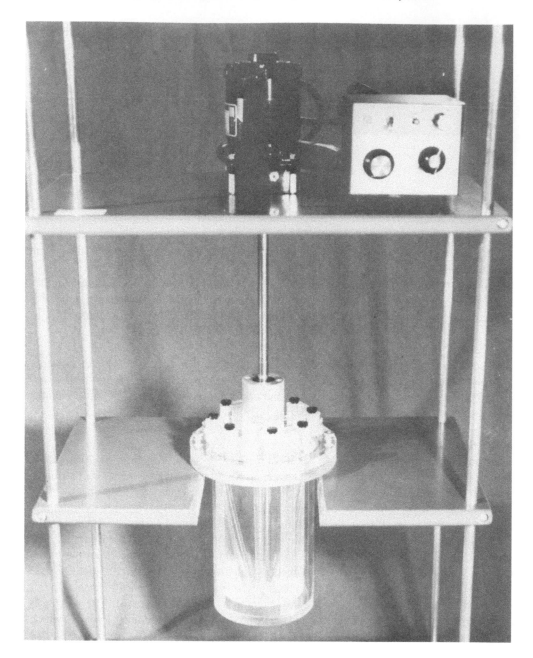

FIGURE 5.   The annular reactor system with variable-speed motor for turning the inner cylinder.

Since it is operated as a CFSTR, material balances are conveniently accomplished permitting useful kinetic analyses of biofilm rate processes including substrate removal, growth, production formation, and detachment. Trulear and Characklis[6] have reported such an analysis for a mixed culture biofilm while Bakke et al.[7] have similarly described *Pseudomonas aeruginosa* biofilms. Bakke[49] has used chemostat data and bacterial adsorption data to predict the progression of a biofilm experiment in the annular reactor. From material balance equations, Bakke et al.[7] have also shown that *P. aeruginosa* activity is not significantly different in a biofilm than in a chemostat (Figure 7). The reactor permits a nonintrusive,

FIGURE 6. The annular reactor with eccentrically drilled draft tubes and four removable slides. Tubing fittings in top and one at the bottom permit continuous feed and overflow for nutrients.

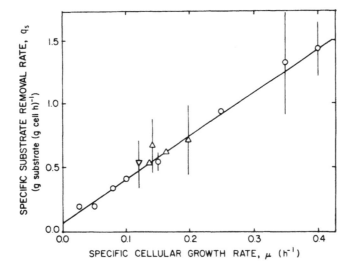

FIGURE 7. Steady-state biofilm-specific substrate-removal rate vs. specific cellular growth rate in the biofilm: ($\triangle$) D = 6 per hour; ($\nabla$) D = 3 per hour. The ($\bigcirc$) line is based on chemostat data. (Bakke, R., Trulear, M. G., Robinson, J. A., and Characklis, W. G., *Biotech Bioeng.*, 26, 1418, 1984. With permission.)

nondestructive measure of biofilm activity normalized over reactor surface area. Specific activities can be determined by removing a slide and determined cell mass or numbers per unit area. Oxygen balances are also simple since oxygen only enters the reactor with the feed water.

The torque transducer provides a continuous indicator of biofilm accumulation, thus

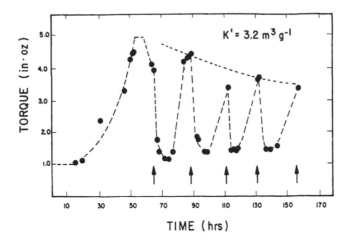

FIGURE 8. The influence of periodic chlorination on biofilm accumulation in an annular reactor can be observed indirectly by measuring torque. The arrows indicate chlorine additions (12.5 mg/ℓ for 0.5 hr). (From Characklis, W. G., Trulear, M. G., Stathopoulos, N. A., and Chang, L. C., *Water Chlorination*, Jolley, R. L., Ed., Ann Arbor Science, Ann Arbor, Mich., 349, 1980. With permission.)

permitting rather effortless, though indirect, observation of biofilm behavior. For example, torque has been used to estimate the effectiveness of chemical additives (e.g., chlorine) for biofilm removal (Figure 8).

The annular reactor has been operated with external recycle for various purposes:

1.  A gas diffuser in the recycle increases oxygen input or continuously deaerates the reactor for anaerobic experiments.
2.  A sidestream filter continuously removes suspended (planktonic) microorganisms from the reactor liquid.
3.  A sidestream packed column continuously leaches biocide into the reactor liquid.

The recycle flow rate can be operated independently of the dilution flow rate for more operating flexibility.

*Rotating Disk and Radial Flow Reactor*

These two reactor designs (Figures 9 and 10) are useful when a range of fluid shear stress values are desired simultaneously for investigating the effect of fluid shear stress on microbial attachment.

**Radial Flow Reactors** — The radial flow reactor, as reported by Fowler and McKay,[8] consists of two parallel disks separated by a narrow gap. This reactor is described in more detail in another chapter. Culture fluid is pumped into the center of one of the disks at a constant volumetric flow rate and flows out radially between the disks to a collection manifold. As the cross sectional area available for flow increases with increasing radius, the linear velocity and fluid shear stress decreases. Thus, high shear forces are present near the inlet and lower shear forces are present toward the outlet. This geometry is only useful for observing initial events of biofilm accumulation because of the narrow spacing between the disks.

**Rotating Disk Reactor** — The rotating disk reactor consists of a rotating disk placed in a solution (Figure 10). The rotating disk has been used to study *the effect of fluid* shear stress on biofilm development[9] because fluid shear stress varies with the radius of the disk,

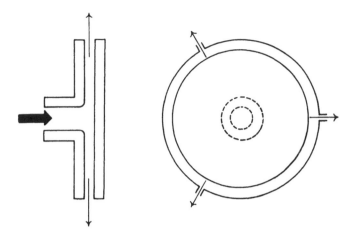

FIGURE 9.   The radial flow reactor.

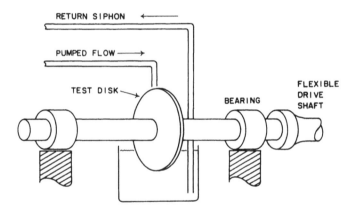

FIGURE 10.   The rotating disk reactor.

with the highest fluid shear stress at the outer edge of the disk and lower fluid shear stress values toward its center. Gulevich et al.[10] used this system to study substrate removal rate by biofilms because the mass transfer boundary layer is constant over the entire disk.

*Packed and Fluidized Beds*

By adding aggregates to a CFSTR, the surface area for biofilm formation can be greatly increased. In a packed bed, aggregates are packed or are sufficiently heavy that they are not fluidized under flow condition. The packed bed restricts mixing in the axial direction, which may result in the bed operating more as a plug flow reactor than as a CFSTR. A high recycle rate will minimize this effect. A fluidized bed consists of aggregates which are suspended and therefore well mixed, so long as sufficient turbulence is maintained. In both packed and fluidized beds, the aggregates are maintained within the reactor and cannot be washed out (Figure 11). The increased surface area may be desirable when high biofilm mass concentration is required.

**Reactor Feed**

The reactor feed consists of three components: dilution water, nutrient solution, and microbial inoculum.

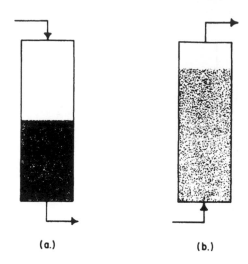

(a.)                    (b.)

FIGURE 11.   The packed bed and fluidized bed reactors.

*Dilution Water*

We have used distilled and reverse osmosis-treated water for dilution water. In some cases, tap water was also used. Tap water was treated before entering the experimental system in the following way:

1. Flow through an in-line 5 μm filter to remove particulates.
2. Flow through a downflow carbon adsorption column to remove residual chlorine.
3. Storage and aeration in a 200 ℓ expoxy-lined tank.
4. Flow through an in-line 0.2 μm filter to remove bacteria.

*Microbial Inoculum*

When undefined mixed cultures are selected, we prepare a standard inoculum to minimize the effects of population distribution differences in laboratory experiments. Mixed liquor from a domestic wastewater treatment plant is settled and the concentrated sludge mixed with glycerol to approximately 25% v/v glycerol. Aliquots (10 mℓ) of the resulting suspension are transferred to glass ampules (10 cm³) which are "quick frozen" in liquid nitrogen and stored at −20°C. Growth rate tests on the standard inocula are conducted periodically. No significant changes in overall growth rate have been observed over one year.

When pure cultures are used, aseptic technique is used. Regular contamination checks of the reactor effluent should be conducted.

*Nutrient Solution*

Nutrient solutions are generally prepared in a concentrated form to keep storage and sterilizing volumes small. We generally autoclave nutrient solutions while dilution water is sterilized by filtration. The nutrients are blended with the dilution water before entering the reactor.

## ANALYTICAL METHODS

**Direct Measurement of Biofilm Quantity**

The only two direct measures of film quantity are thickness (Th) and mass (M) (Table 1). The two quantities are related by the film density ($\rho_{Th}$), i.e., $\rho_{Th} = X_b/Th$ where $X_b$ is film mass per unit area, M/A.

## Table 1
## MEASUREMENT OF BIOFILM ACCUMULATION

| Classification | Analytical Method | Ref. |
|---|---|---|
| A. Direct measurement of biofilm quantity | Biofilm thickness | 6, 13, 14, 15, 45 |
| | Biofilm mass | 6, 11, 36 |
| B. Indirect measurement of biofilm quantity: specific biofilm constituent | Polysaccharide | 17, 38 |
| | Total organic carbon (TOC) | 7,28 |
| | Chemical oxygen demand (COD) | 38 |
| | Protein | 38, 44 |
| C. Indirect measurement of biofilm quantity: microbial activity within the biofilm | Viable cell count | 2, 21, 22 |
| | Epifluorescence microscopy | 7, 24 |
| | ATP | 27, 28, 29 |
| | Lipopolysaccharide | 40, 47 |
| | Substrate removal rate | 2, 6, 7 |
| D. Indirect measurement of biofilm quantity: effects of biofilm on transport properties | Frictional resistance | 6, 11, 14 |
| | Heat transfer resistance | 37, 41, 42 |

0.02m

FIGURE 12.    The tubular test section.

*Biofilm Thickness*

Biofilm thickness has been determined in tubular reactor systems by Picologlou et al.[11] Small sample tubes (1.27 cm inside diameter and 5 cm in length) were inserted as an integral part of the tubular reactors. The sample tubes were inserted end-to-end in an acrylic plastic test section (1.9 cm inside diameter and 76 cm in length) as shown in Figure 12. A modified design for the test section was used more recently to study the effects of roughness on biofouling.[12] The test section is connected to the recycle loop with pipe unions to provide easy access to the sample tubes. At designated intervals, a sample tube is removed from the reactor and a clean sample tube inserted in its place. The sample tube with biofilm is then drained to reduce excess water in the biofilm. Drainage time is generally 10 min, but, in some cases, is reduced to 2.5 min when the biofilm appears to be drying around the end of the sample tube.

The volumetric displacement of the biofilm is determined using a displacement cell (Figure 13). The latter is filled with a water-surfactant solution (0.3% v/v turgitol). Initial liquid level before immersing the sample tube is measured by lowering the conductive probe with a micromanipulator until contact is made with the water surface. Contact is indicated by visual observation or by deflection of an ammeter connected in series with the cell and a 1.5 V power source. A sample tube with biofilm is then immersed in the cell and the new liquid level (and, hence displacement) due to the sample tube and biofilm is determined. The sample tube is cleaned and its volumetric displacement is determined in the same way. The difference between the displacements of the fouled and clean sample tubes is the biofilm volume. The biofilm thickness is determined by dividing biofilm volume by the inner surface area of the sample tube. Overall sample standard deviation for this method of determining biofilm thickness is approximately $\pm$ 10 $\mu$m based on 5 replicate measurements on 73 samples varying in mean thickness from 50 to 300 $\mu$m.

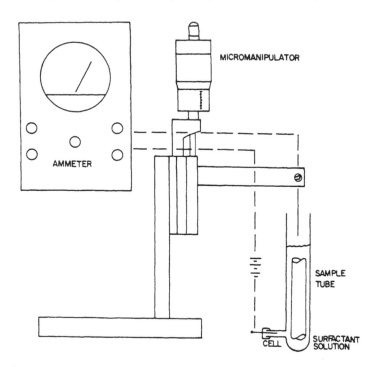

FIGURE 13.    Apparatus for measuring wet biofilm volume. (Reprinted with permission from *Water Res.*, 16, Characklis, W. G., Trulear, M. G., Bryers, J. D., and Zelver, N., Dynamics of biofilm processes: methods, Copyright 1982, Pergamon Journals, Ltd.)

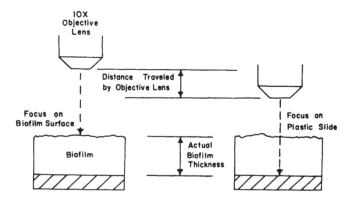

FIGURE 14.    Method for measuring wet biofilm thickness with an optical microscope.

Biofilm thickness has also been determined using various methods which locate the biofilm-fluid interface and the biofilm-substratum interface. Trulear and Characklis[6] used an optical microscope method to determine biofilm thickness in the annular reactor. The technique was adapted from Sanders[13] and requires biofilm growth on a thin acrylic plastic slide which forms an integral part of the annular reactor wall. The slide is withdrawn from the reactor and placed on a microscope stage. The 10 × objective (100 × magnification) is lowered until the biofilm surface is in focus and the fine adjustment dial setting is recorded. The objective is then lowered further until the inert plastic growth surface is in focus (*Figure* 14). The differences in fine adjustment settings is compared with a calibration curve which

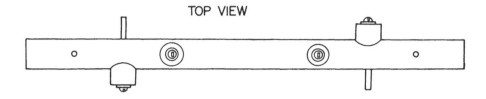

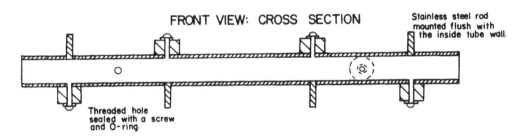

FIGURE 15. Tubular test section used in measurement of wet biofilm thickness by electrical conductance. (From Norrman, G., Characklis, W. G., and Bryers, J. D., *Dev. Ind. Microbiol.*, 18, 581, 1977. With permission.)

is generally linear and the thickness obtained. Sample standard deviation of the measurement is approximately 10 to 12 μm on biofilms ranging from 11 to 130 μm mean thickness. The variation includes any irregularities in the biofilm surface. The accuracy of the method is influenced by the refractive index of the wet biofilm.

Norrman et al.[14] located the biofilm-fluid and biofilm-substratum interfaces by means of electrical conductance, using a technique adapted from Hoehn and Ray[15] which uses an apparatus consisting of a steel needle mounted on a micromanipulator. Figure 15 shows details of the test section in which biofilm is grown and thickness measured. The test section, if not metal, consists of six measurement points, each one being a stainless steel rod (3 mm O.D.) mounted flush with the inside tube wall. Opposite each rod is a threaded hole which can be sealed with a screw and O-ring. The thickness of the biofilm on the stainless steel rod surface is measured using the electrical conducting probe depicted in Figure 16. The probe and one of the stainless steel rods are connected to an electrical circuit completed by an electrometer. To obtain a measurement, the test section is removed from the reactor, the screws are withdrawn, and the test section drained for 2 min. The probe is then lowered into the test section through the threaded hole until contact is made with the biofilm surface; a current of about $10^{-8}$ A is detected and the depth noted on the micromanipulator (Figure 16a). Next, the probe is lowered until contact is made with the stainless-steel rod surface when a greater current flow of around $3 \times 10^{-5}$ A is registered, and the depth is again noted (Figure 16b). The difference in depth is the biofilm thickness. The procedure is repeated at the other five locations. Precision is about ± 6% and accuracy, compared to Vernier micrometer measurements, is within 5%. If the test section is metal, the stainless steel studs are superfluous.

### Biofilm Mass

Wet biofilm is scraped from sample tubes (tubular reactors), acrylic slides (annular reactor), or a sample of the support medium (packed or fluidized beds) and dried to constant weight at an elevated temperature (100 to 105°C). For substrata unaffected by heat, the biofilm is left on the substratum and the entire section is placed in the oven. After drying, the combined weight of biofilm and substratum is obtained. The substratum is then cleaned, dried, and weighed again. The difference in the two measurements is the dry biofilm mass.

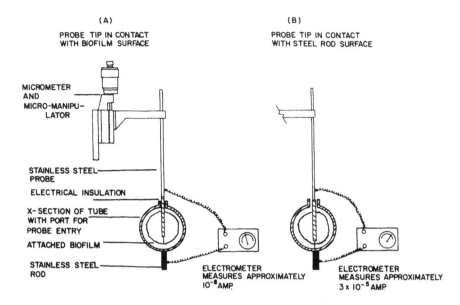

FIGURE 16.   Apparatus for measuring wet biofilm thickness by electrical conductance. (From Norrman, G., Characklis, W. G., and Bryers, J. D., *Dev. Ind. Microbiol.*, 18, 581, 1977. With permission.)

The surface area available for growth is generally known so an areal mass density can be determined. If biofilm thickness or biofilm volume has been measured, volumetric biofilm density ($\rho_{Th}$) can be determined. The volumetric biofilm density has units of dry mass per unit wet volume. The areal film density, $X_b$, has units of dry mass per unit of substratum area.

Biofilm density can increase with increasing fluid shear stress and nutrient loading *between* experiments.[36,45] There is even some indication that density increases with time during a given experiment of several days duration. Biofilm thickness and mass should be determined simultaneously in a given experimental system, so that a consistent relationship between the two can be established.

### Indirect Measurement of Biofilm Mass: The Determination of Specific Constituents

Various biofilm constituents have been used to monitor film development. Some of them are listed in Table 1. A calibration curve relating the specific constituent and biofilm mass is needed if material balances are desired. This is less important if a quantity like carbon is used for measuring biomass, substrate, and product formation. There is usually no substitute for measuring if this is possible. Mass measurements however, may not be as sensitive as some other analytical techniques. Table 2 presents the detection limits and standard errors for the various analysis used in our laboratory.

Two chemical procedures have been used by us as indirect measurements of early biofilm formation:

1.   Total bound polysaccharide
2.   Total organic carbon or chemical oxygen demand

*Total Polysaccharide*

Biofilms contain relatively large quantities of polysaccharide.[2,16] One method for measuring polysaccharide concentration is based on the reaction of *carbohydrate* "*reducing*" ends (ketones or aldehydes) with strong nonoxidizing acid to yield hydroxymethyl furfural

**Table 2**
**SENSITIVITY AND PRECISION OF VARIOUS BIOFILM ACCUMULATION**
**MEASUREMENT TECHNIQUES**

| Method | Sensitivity | | Precision ($\pm$) | | |
| | Measured quantity | Equivalent COD[a] | Measured quantity | Equivalent COD[1] | Ref. |
| --- | --- | --- | --- | --- | --- |
| Biofilm thickness | 9 $\mu$m | 25.6 mg cm$^{-2}$ | 9 $\mu$m | 25.6 mg cm$^{-2}$ | 36 |
| | 10 | 28.5 | 10 | 28.5 | 27 |
| | 10 | 28.5 | 9 | 25.6 | 45 |
| Biofilm mass | 0.11 mg cm$^{-2}$ | 0.13 mg cm$^{-2}$ | 0.01 mg cm$^{-2}$ | 0.01 mg cm$^{-2}$ | 36 |
| Biofilm COD | 0.006 mg cm$^{-2}$ | — | 0.0001 mg cm$^{-2}$ | — | 18, 38 |
| Biofilm TOC | 0.002 mg cm$^{-2}$ | 0.0057 | 0.00045 mg cm$^{-2}$ | 0.00128 mg cm$^{-2}$ | 46 |

[a]  Chemical oxygen demand (COD) calculated from biofilm thickness values assuming biofilm density = 25 mg/cm$^3$, COD = 1.14 mg COD per milligram and biofilm carbon content = 0.4 mg carbon per milligram biofilm.[18,45,46] TOC, total organic carbon.

plus other by-products. Condensation between these activated aldehydes and phenolic compounds like resorcinol, naphthol, anthrone, or phenol leads to the formation of colored compounds as a direct function of polysaccharide concentration.

The phenol-$H_2SO_4$ method of Dubois et al.[17] has been used as an indirect assay for biofilm, during the early stages of its development. Figure 17 demonstrates the increase in attached polysaccharide with time in a tubular reactor.[18] The problem with this method is that EPS contains many different sugar residues but the calibration method usually uses one sugar (e.g., glucose). This can lead to errors in determining actual amounts present. .

*Organic Carbon or Chemical Oxygen Demand*

Indirect measurements of biofilm are possible by measuring oxidizable organic biofilm material which is mostly organic carbon. Bryers[18] reports good sensitivity (6 $\mu$g COD/cm$^2$) and precision ($\pm$ 0.10 $\mu$g COD/cm$^2$) with a modified chemical oxygen demand (COD) analysis in the tubular reactor system. Modifications to the *Standard Methods*[19] COD procedure consisted of diluting the dichromate oxidant and ferrous ammonium sulfate titrant.

Bakke et al.[7] have monitored accumulation in the annular reactor using total organic carbon (TOC). Sensitivity and precision of the method are less than 2.0 $\mu$g TOC/cm$^2$ and $\pm$ 0.45 $\mu$g TOC/cm$^2$, respectively. Carbon, being an atomic element, is a conservative quantity useful for material balances and related kinetic analyses in microbial reactors.

**Indirect Measurements of Biofilm Quantity: Microbial Activity**

Table 3 enumerates several analytical methods — currently used or proposed — for the indirect monitoring of biofilm using metabolic activity as an indicator. White et al.[20] have reviewed these and many other spohisticated biochemical techniques for these measurements.

*Direct Enumeration*

Corpe[21] measured biofilm development on glass slides immersed in seawater using *total cell count* and *viable cell count* per unit area. Results indicated an increasing total count attached while numbers of viable cells fluctuated. Total biofilm quantity (i.e., mass or thickness) was not determined.

Gerchakov[22] used *viable cell counts* to monitor biofilm development on a variety of metal surfaces exposed to tropical seawater. Results indicate an exponential increase in viable cell counts with exposure time for all material. Stainless steel showed the highest rate of development, followed by brass, glass, and copper-nickel. No direct measure of biofilm quantity was made.

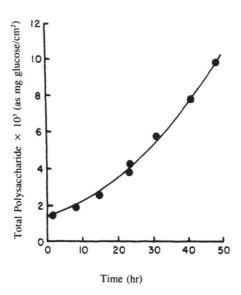

FIGURE 17.   Increase in biofilm polysaccharide content during a typical experiment in a tubular reactor. (Reprinted with permission from *Water Res.*, 16, Characklis, W. G., Trulear, M. G., Bryers, J. D., and Zelver, N., Dynamics of biofilm processes: methods, Copyright 1982, Pergamon Journals, Ltd.)

**Table 3**
## MEASURES OF MICROBIAL ACTIVITY WITHIN THE BIOFILM

| | |
|---|---|
| Direct enumeration | Total and viable cell count[2,21,22] |
| | Total cell count by epifluorescence microscopy[7,24] |
| Cellular component chemical analysis | Adenosine triphosphate (ATP)[27,29,28] |
| | Lipopolysaccharide (LPS)[31,40,47] |
| | Total proteins[20,38,44] |
| | Alkaline phosphates[20] |
| | HAEM catalyzed luminescence[32] |
| | Muramic acids[29,20] |
| | Poly-β-hydroxybutyrate[29,20] |
| Substrate removal | Direct measurement[6,7,2] |
| | Heterotrophic potential[20,24] |

Epifluorescence techniques for *total cell counts* involve the disruption of large aggregates of attached biofilm, recovery of bacterial cells by filtration, staining cells with acridine orange (a fluorescent stain), and direct counting by epifluorescence microscopy.[23] Bakke et al.,[7] using direct counts by epifluorescence microscopy, have counted up to $5.6 \times 10^7$ cells/$cm^2$ of biofilm in the annular reactor. Geesey et al.,[24] using direct counts by epifluorescence, demonstrated the importance of attached bacteria in alpine streams. Nelson et al.[25] have used epifluorescence techniques to observe initial adsorption of *P. aeruginosa* to glass surfaces. Measurement of the adenosine triphosphate (ATP) content of biofilms has been widely used as a chemical assay for biofilm activity. ATP is found only in living organisms: no residual ATP is detectable after cell death.[26] Activity appears limited to the upper layers of the biofilm. LaMotta[27] observed an increase in ATP content with biofilm thickness up to approximately 320 μm in an annular reactor. However, the ATP content remained constant as biofilm thickness increased beyond 320 μm, *suggesting the presence of an* "active" layer of biofilm. This may also suggest analytical limitations to the ATP extraction technique, though this has not been confirmed. Attached ATP per unit area has been measured by Little and Lavoie[28] and Bobbie et al.[29] in two separate field heat exchange units.

Lipopolysaccharides (LPS) are unique to Gram-negative bacterial cell walls. Dexter and coworkers[30] report that attached LPS is proportional to total bound cell population as determined by scanning electron microscopy, however, the precision is poor. LPS from Gram-negative bacteria is not confined to cell walls but may "leach" into surrounding fluids.[31]

Several other methods, previously used in quantifying free-floating bacterial populations, have been proposed for the indirect determination of biofilm development. Mason et al.[32] report an accurate, practical, and rapid method for the estimation of immobilized whole cell biomass based on the chemiluminescent oxidation of luminol in the presence of cellular hematin catalysts. Sensitivity of less than 1 mg biomass was reported.

White[20] proposed several measurements of biomass activity: (1) alkaline phosphatase, a bacterial cell wall component; (2) muramic acid, a unique prokaryotic storage material; (3) poly-β-hydroxybutyrate; and (4) phospholipids. At this time, there is little data available for determining the accuracy or precision of these methods applied to biofilms.

*Substrate Removal Rate:* Microbial activity within the biofilm can be determined by measuring the removal rate of a particular substrate, nutrient, or electron acceptor. Using the annular reactor, Trulear and Characklis[6] observed an increase in glucose removal rate with increasing biofilm accumulation up to some critical biofilm thickness termed the *active thickness*. Similar observations have been reported by many others.[27,33-35] The active thickness is the effective depth of penetration of substrate, nutrient, or electron acceptor before it is exhausted.

Heterotrophic potential is a similar test developed to measure microbial activity in natural waters.[2] The test essentially measures removal of glutamic acid (or other substrate) due to microbial activity in the sample. The heterotrophic potential test may provide spurious results since the sample is subjected to some rather drastic environmental changes during sampling and processing.

In biofilm research, substrate removal can be reported as substrate mass removed per unit biofilm area, per unit biofilm volume, per unit cell mass, or per unit biofilm mass. The quantities are related through the biofilm thickness, biofilm cell density, and overall biofilm density.

**Indirect Measurements of Biofilm Quantity: Effects on Transport**

The development of a biofilm in a flowing fluid system affects fluid frictional resistance (momentum transport) as described by Picologlou[11] and by Zelver[36] in the tubular reactor and by Trulear and Characklis[6] in the annular reactor. Biofilm accumulation also influences heat transfer resistance (heat transport) as described by Characklis et al.[37] in the tubular reactor. Frictional resistance (e.g., torque measurement in the annular reactor) and heat transfer resistance measurements can both be automated[4] which can be useful in many cases.

## SUMMARY

Various laboratory model systems are suitable for observing biofilm processes. This chapter has described some of the more useful of these. Not only the specific biofilm process under consideration but convenience of sampling and general handling will influence choice of a particular system.

Methods available for monitoring film are classified as follows:

* Direct measurement of biofilm quantity
* Indirect measurement of biofilm quantity using specific biofilm constituents

- Indirect measurement of biofilm quantity by determining microbial activity within the biofilm
- Indirect measurement of biofilm quantity: effects on biofilm transport properties

This chapter has discussed methods for accomplishing measurements under the first three headings.

Activity measurements are most useful in investigation of film from natural waters and wastewater treatment systems using reactors such as trickling filters or rotating biological contactors. In these, activity of the biofilm in terms of substrate removal is critically important.

Measurements of biofilm quantity become necessary when fouling processes are involved. For example, biocides used to ameliorate the detrimental effects of biofilms on fluid frictional resistance and heat transfer resistance must *remove* attached biomass to be effective. Merely "killing" or inactivating the microbial cells in the biofilm will not be sufficient since the "hydraulic roughness" and thermal conductive resistance caused by the biofilm will be unaffected. Consequently, activity determinations are not very useful for such a task.

The choice of experimental system and analytical method for monitoring biofilm processes will be dependent on the questions which prompt the investigation.

## ACKNOWLEDGMENTS

The author gratefully acknowledges financial support in part from the National Science Foundation, Office of Naval Research, and Phillips Petroleum Company. Jean Julian prepared the manuscript.

## REFERENCES

1. **Marshall, K. C.,** *Interfaces in Microbial Ecology,* Harvard University Press, Cambridge, Mass., 1976.
2. **Costerton, J. W. and Colwell, R. R., Eds.,** *Native Aquatic Bacteria: Enumeration, Activity and Ecology,* ASTM Press, Philadelphia, Pa., 1979.
3. **Characklis, W. G. and Cooksey, K. E.,** Biofilms and microbial fouling, *Adv. Appl. Microbiol.,* 29, 93, 1983.
4. **Zelver, N., Characklis, W. G., Robinson, J. A., Roe, F. L., Dicic, Z., Chapple, K., and Ribaudo, A.,** Tube Material, Fluid Velocity, Surface Temperature, and Fouling, CTI Paper TP-84-16, Cooling Tower Institute, Houston, Tex., 1984.
5. **Matson, J. V. and Characklis, W. G.,** Biofouling control in recycled cooling water with bromo chloro dimethylhydantoin, *J. Cooling Tower Inst.,* 4, 27, 1983.
6. **Trulear, M. G. and Characklis, W. G.,** Dynamics of biofilm processes, *J. Water Pollut. Control Fed.,* 54, 1288, 1982.
7. **Bakke, R., Trulear, M. G., Robinson, J. A., and Characklis, W. G.,** Activity of *Pseudomonas aeruginosa* in biofilms: steady state, *Biotech Bioeng.,* 26, 1418, 1984.
8. **Fowler, H. W. and McKay, A. J.,** The measurement of microbial adhesion, in *Microbial Adhesion to Surfaces,* Berkeley, R. W., Lynch, J. M., Melling, J., Rutter, P. R., and Vincent, B., Eds., Ellis Horwood, Chichester, U.K., 143, 1980.
9. **Loeb, G. I., Laster, D., and Gracik, T.,** The influence of microbial fouling films on hydrodynamic drag of rotating discs, in *Marine Biodeterioration: An Interdisciplinary Study,* Costlow, J. D. and Tipper, R. C., Eds., Naval Institute Press, Annapolis, Md., 88, 1984.
10. **Gulevich, W., Renn, C. E., and Liebman, J. C.,** Role of diffusion in biological waste treatment, *Environ. Sci. Technol.,* 2, 113, 1968.
11. **Picologlou, B. F., Zelver, N., and Characklis, W. G.,** Biofilm growth and hydraulic performance, *J. Hydraul. Eng.,* 106(HY5), 733, 1980.
12. **Turakhia, M. H. and Characklis, W. G.,** An observation of microbial cell accumulation in a finned tube, *Can. J. Chem. E.,* 61, 873, 1983.

13. **Sanders, W. M.,** 3rd, Oxygen utilization by slime organisms in continuous culture, *Air Water Pollut. Int. J.,* 10, 253, 1966.

14. **Norrman, G., Characklis, W. G., and Bryers, J. D.,** Control of microbial fouling in circular tubes with chlorine, *Dev. Ind. Microbiol.,* 18, 581, 1977.

15. **Hoehn, R. C. and Ray, A. D.,** Effects of thickness on bacterial film, *J. Water Pollut. Control Fed.,* 46(11), 2302, 1973.

16. **Jones, H. C., Roth, I. L., and Sanders, W. M.,** Electron microscopic study of a slime layer, *J. Bact.,* 99, 316, 1969.

17. **Dubois, M., Giles, D. A., Hamilton, J. K., Rebers, P. A., and Smith, I.,** Colorimetric method for determination of sugars and related substances, *Anal. Chem.,* 28(3), 350, 1956.

18. **Bryers, J. D.,** Dynamics of early biofilm formation in a turbulent flow system, Ph.D. thesis, Rice University, Houston, Tex., 1980.

19. American Public Health Association (APHA), *Standard Methods for the Examination of Water and Wastewater,* 14th ed., APHA, Washington, D.C., 1976.

20. **White, D. C., Bobbie, R. J., Herron, J. S., King, J. D., and Morrison, S. J.,** Biochemical measurements of microbial mass and activity from environmental samples, *Native Aquatic Bacteria: Enumeration, Activity and Ecology,* ASTM STP 695, Costerton, J. W. and Colwell, R. R., Eds., American Society for Testing and Materials, Philadelphia, Pa., 1979, 69.

21. **Corpe, W. A.,** Microfouling: the role of primary film-forming marine bacteria, in *Proc. Int. Congress on Marine Corrosion and Fouling,* Northwestern University Press, Evanston, Ill., 598, 1973.

22. **Gerchakov, S. M., Marszalek, D. S., Roth, F. J., Sallman, B., and Udey, L. R.,** Observations on microfouling applicable to OTEC systems, in Proc. OTEC Biofouling and Corrosion Symp., Seattle, Wash., 1977, U.S. Department of Energy, Washington, D.C., 63.

23. **Zimmerman, R. and Meyer-Riel, L. A.,** A new method for fluorescent staining of bacterial populations, *Kiel. Meeresforsch.,* 30, 24, 1974.

24. **Geesey, G. G., Mutch, R., Costerton, J. W., and Green R. B.,** Sessile bacteria: an important component of the microbial population in small mountain streams, *Limnol. Oceanogr.,* 23(6), 1214, 1978.

25. **Nelson, C. H., Robinson, J. A., and Characklis, W. G.,** Bacterial adsorption to smooth surfaces: rate, extent, and spatial pattern, *Biotech. Bioeng.,* 27, 1662, 1985.

26. **Hamilton, R. D. and Holm-Hanson, O.,** Adenosine triphosphate content of marine bacteria, *Limnol. Oceanogr.,* 12, 319, 1967.

27. **LaMotta, E. J.,** Evaluation of diffusional resistances in substrate utilization by biological films, Ph.D. thesis, University of North Carolina at Chapel Hill, 1974.

28. **Little, B. and Lavoie, D.,** Gulf of Mexico OTEC biofouling and corrosion experiment, in Proc. OTEC Biofouling, Corrosion and Materials Workshop, Rosslyn, Va., 1979, U.S. Department of Energy, Washington, D.C., 60.

29. **Bobbie, R. J., Nickels, J. S., and Davis, W. M.,** Measurement of microfouling mass and community structure during succession in OTEC simulators, in Proc. OTEC Biofouling, Corrosion and Materials Workshop, Rosslyn, Va., 1979, U.S. Department of Energy, Washington, D.C., 101.

30. **Dexter, S. C.,** Influence of substrate wettability on the formation of bacterial slime films on solid surfaces immersed in natural seawater, in Proc., 4th Intl. Cong. Marine Corrosion and Biofouling, Antibes, France, 137, 1976.

31. **Janda, J. and Work, E.,** A colorimetric estimation of lipopolysaccharides, *FEBS Lett.,* 16(4), 343, 1971.

32. **Mason, J. R., Somerville, H. J., and Pirt, S. J.,** Estimation of biomass with an immobilized cell biocatalyst by means of a luminol assay for haem, *J. Appl. Chem. Biotech.,* 28, 770, 1978.

33. **Atkinson, B. and Daoud, I. S.,** Diffusion effects within microbial films, *Trans. Inst. Chem. Eng.,* 48, 245, 1970.

34. **Harremoes, P.,** Half order reactions in biofilm and filter kinetics, *Vatten,* 2, 123, 1977.

35. **Kornegay, B. H. and Andrews, J. F.,** Characteristics and kinetics of biological film reactors, in FWPCA Final Report, Research Grant WP-01181, Department of Environmental Systems Engineering, Clemson University, Clemson, S.C., 1967.

36. **Zelver, N.,** Biofilm development and associated energy losses in water conduits, Master's thesis, Rice University, Houston, Tex., 1979.

37. **Characklis, W. G., Nimmons, M. J., and Picologlou, B. F.,** Influence of fouling biofilms on heat transfer, *Heat Transfer Eng.,* 3(1), 23, 1981.

38. **Bryers, J. D. and Characklis, W. G.,** Early fouling biofilm formation in a turbulent flow system: overall kinetics, *Water Res.,* 15, 483, 1980.

39. **Costerton, J. W., Geesey, G. G., and Cheng, K. J.,** How bacteria stick, *Sci. Am.,* 238, 86, 1978.

40. **Dexter, S. C., Sullivan, J. D., Williams, J., and Watson, S.,** Influence of substrate wettability on the attachment of marine bacteria to wetted surfaces, *Appl. Microb.,* 30(2), 298, 1975.

41. **Fetkovich, J. G., Granneman, G. N., Mahalingam, L. M., and Meier, D. L.,** Studies of biofouling in OTEC plants, in Proc. 4th Conf. OTEC, New Orleans, La., VII15, 1977.

42. **Knudsen, J. G.,** Apparatus and techniques for measurement of fouling of heat transfer surfaces, *Condenser Biofouling Control,* Garey, J. F., Jorden, R. M., Aitken, A. H., Burton, D. T., and Gray, R. H., Eds., Ann Arbor Science, Ann Arbor, Mich. 143, 1980.
43. **LaMotta, E. J.,** Internal diffusion and reaction in biological films, *Environ. Sci. Technol.,* 10(8), 765, 1976.
44. **McCoy, W.,** Immunofluorescence as a technique to study marine biofouling bacteria, Directed Research Project-M699, University of Hawaii, 1979.
45. **Trulear, M. G.,** Dynamics of biofilm processes in an annular reactor, Master's thesis, Rice University, Houston, Tex., 1980.
46. **Trulear, M. G.,** Cellular reproduction and extracellular polymer formation in the development of biofilms, Ph.D. thesis, Montana State University, Boseman, Mont., 1983.
47. **Watson, S. W., Novitsky, T. J., Quinby, H. G., and Valois, F. W.,** Determination of bacterial number and biomass in the marine environment, *Appl. Environ. Microbiol.,* 33, 940, 1977.
48. **Characklis, W. G., Trulear, M. G., Stathopoulos, N. A., and Chang, L. C.,** Oxidation and destruction of biofilms, in *Water Chlorination,* Jolley, R. L., Ed., Ann Arbor Science, Ann Arbor, Mich., 349, 1980.
49. **Bakke, R.,** Unpublished results.

Chapter 9

# A CONSTANT-DEPTH LABORATORY MODEL FILM FERMENTER

**Adrian Peters and Julian W. T. Wimpenny**

## INTRODUCTION

The ubiquity and economic importance of microbial film has been discussed both by Characklis and Wimpenny elsewhere in this volume. Basically, almost any surface in an aqueous environment can become colonized first by bacteria. The resulting microbial film is a complex heterogeneous ecosystem with bacteria held in a matrix of polymeric substances produced by one or more of the species involved.

It is generally assumed that the film becomes spatially differentiated soon after attachment and growth of the pioneer species. For instance, the latter, responsible for forming the film, may pave the way for a second species which, while not able to colonize the surface alone, finds a particular niche within the film environment. This spatial order will be dominated by solute gradients throughout the film. There is a viscous sublayer of water associated with the film surface, and while this may help protect the film from shear effects, it probably also slows diffusion of solutes into it. Oxygen access in particular is likely to be an important limiting factor in developing film. Organisms in the upper layers of the film can theoretically use all the available oxygen, producing steep oxygen gradients and an anoxic region in the deeper zones. Facultative anaerobes or even anaerobes in this oxygen-depleted area will switch to anaerobic metabolism, lowering the pH of the film and possibly producing gas, which may destabilize the entire structure causing it to be sloughed away, and thus allowing the whole process of colonization to begin again. This brief scenario for film development serves to illustrate that both temporal and spatial organization occurs in natural microbial film.

In natural ecosystems, the bacterial film can become much more complex, especially where high eukaryotic organisms colonize it. Algae and cyanobacteria form tough mats which produce oxygen when illuminated, and protozoa can graze on the bacteria. The film will also collect large amounts of inorganic matter: silts, clays, etc., and animal and vegetable detritus.

The importance of microbial film and its complexity as an ecosystem make it a most interesting area for ecological study. However, this very complexity makes *in situ* investigation extremely difficult, suggesting that laboratory model systems are needed if the fundamental processes of microbial film are to be understood.

Table 1, revised from Wimpenny, Lovitt and Coombs,[44] lists the different types of model-film fermenter which have been developed. Most of these try to reproduce a specific type of film or are designed to investigate specific film processes. Models include trickling filters and effluent treatment films, marine fouling, dental plaque, microbial attachment to fermenter vessel walls, adhesion to various surfaces and systems to investigate film reaction kinetics. Others bring part of the natural environment into the laboratory, creating a microcosm. These have been used to investigate, for instance, oxygen profiles in algal mats.

The majority of these models do not attempt to reduce or restrict the complexity of the natural system to a level where unequivocal experiments on film structure and function can be performed. In his chapter, Characklis describes model reactors which allow control of all the relevant variables. These reactors were designed to mimic a specific film type: in general, the growth of film in tubular reactors such as pipelines or water-cooling towers. In this chapter, we advocate the use of a generalized microbial film fermenter capable of operating under quasi-steady-state conditions.

## Table 1
## MODEL FILM FERMENTERS

| Fermenter type | Organism | Nutrient source | Nutrient conc (mg/ℓ) | Surface | Film thickness (mm) | Comment | Ref. |
|---|---|---|---|---|---|---|---|
| Flat plate | Mixed | Glucose, mineral-salts | 25—1000 | Plastic | 0.48—1.4 | Simulation of trickling filter | 1 |
| Flat plate | Mixed | Glucose, mineral-salts | Up to 10,000 | Glass or aluminun | 0.073 or 2.0 | Kinetics of fixed film reactors; constant film thickness | 2 3 |
| Rotating disc | Mixed | Sewage | N.D. | N.D. | 0.194 | 4-ft-diameter disc for effluent treatment | 4 |
| Rotating disc | Mixed | Various effluents | 5.9—365 (BOD) | N.D. | 0.210 | Sectored rotating discs; describes model for substrate uptake kinetics | 5 |
| Rotating disc | Mixed | Sewage | 3.2 total organic carbon | Fiberboard sealed with polyurethane and polyethylene | Up to 1 | Two rotating disc reactors; film studied using light and electron microscopy | 6 |
| Rotating disc | Mixed | Sucrose, mineral-salts, artificial sea water | 100 | Plexiglass® | N.D. | Treatment of saline wastewater | 7 |
| Annular reactor | Mixed | Glucose, mineral-salts | 27—440 | Plastic | Up to 0.25 | Kinetics of fixed film reactors | 8 |
| Annular reactor | Mixed | Glucose, mineral-salts | 2—200 | Plastic | Up to 0.5 | Growth and uptake kinetics in fixell film reactor | 9 |
| Annular reactor | Mixed | Glucose, mineral-salts | N.D. | Plastic | 0.15 | Processes of biofilm development | 10 |

| Annular reactor | *Pseudomonas aeruginosa* | Glucose, mineral-salts | 1.9—16.8 | Plastic | N.D. | Activity of *P. aeruginosa* in biofilms | 11 |
|---|---|---|---|---|---|---|---|
| Tubular reactor | Mixed | Sucrose, yeast extract | 6.0 Sucrose 6.0 yeast extract | Plastic | N.D. | Scanning electron microscopy of film development | 12 |
| Tubular reactor | Mixed | Glucose, tryptic soy broth | 10 Glucose, 10 TSB | Glass | 0.08 | Rates of biofilm formation | 13 |
| Tubular reactor | Mixed | Glucose, mineral salts | 44.26 | Alloy | N.D. | Detachment of biofilm by a chelating agent | 14 |
| Tubular reactor | Mixed | Casitone, glycerol yeast autolysate, mineral-salts | 13—262 | PTFE | 0.0153—0.913 | Denitrification and oxygen uptake in films | 15 |
| Rotating tube | Mixed | Sewage | 200 (BOD) | Perspex coated with Kieselguhr | 1.15—1.5 | Oxygen penetration measured at 0.2 $\mu$m | 16 |
| Tubular sampling device | Mixed | Wastewater | N.D. | Polyvinyl chloride | N.D. | Electron microscopy of wastewater biofilm | 17 |
| Packed bed | Mixed | Glucose, mineral-salts | 180—330 | Plastic | 0.705 | 4-inch tube plus table tennis balls: model trickling filter | 18 |
| Packed column | Mixed | Sewage | 250—300 | Plastic | Up to 2.0 | Trickling filter model | 19 |
| Packed column | Mixed | Molasses | Up to 3000 | Plastic tube or sheet | 0.118—2.55 | Trickling filter model | 20 |
| Packed column | Mixed | Acetate | 2.2—7.2 | Glass | Up to 0.25 | Solute gradient and film thickness | 21 |
| Packed column | Mixed | D-galactose, mineral-salts | 0.03—3 | Glass beads | N.D. | Trickling filter model | 22 |
| Inclined plane | 4 Different mixed cultures | Glucose, mineral-salts | 30—130 | — | 0.5 | Trickling filter model | 23 |

## Table 1 (continued)
## MODEL FILM FERMENTERS

| Fermenter type | Organism | Nutrient source | Nutrient conc (mg/ℓ) | Surface | Film thickness (mm) | Comment | Ref. |
|---|---|---|---|---|---|---|---|
| Glass slide or cover slip | Mixed | Sewage | 50 | Glass | 0.076—0.096 | Film reaction kinetics | 24 |
| Glass slide or cover slip | Mixed | Nutrient broth | — | Glass | 0.021 | Limiting film thickness | 25 |
| Glass slide or cover slip | Mixed | Nutrient broth | N.D. | Glass | 0.2 | Oxygen profiles in film using microelectrodes | 26 27 |
| Glass slide or cover slip | 30 Pure species | Various | N.D. | Various glasses | 0.001 | Survey of microbial adhesion to surfaces | 28 |
| Glass slide or cover slip | Mixed | Seawater | 4—5 (total organics) | Glass | 0.0012 | Microbial adhesion to surfaces | 29 |
| Glass slide or cover slip | Mixed | Seawater | N.D. | Glass, ground glass | 0.00123—0.0056 | Role of bacteria in marine fouling | 30 31 |
| Glass or cover slips | Mixed | Seawater plus or minus 10 mg glucose or nutrient broth | | Glass | 0.02—0.06 | Stack of cover slips in laminar flow of nutrient | 31 |
| Glass slide | *Pseudomonas* sp. | Glucose | N.D. | Glass | N.D. | Rate and extent of surface colonization over 6 hr | 32 |
| Glass fermenter | Mixed or *S. natans* | Various | 5—100 | Glass | 0.004—0.0122 | Film attachment and wall growth | 33 |
| Glass fermenter | 8 species | Nutrient broth | N.D. | Glass | 0.001 | Comparison of attachment to fermenter walls | 34 |

| Continuously stirred or fluidized bed | Mixed | Methanol | 16—57 | Glass | 0.04—1.207 | Fluidized bed reactor | 35 |
|---|---|---|---|---|---|---|---|
| Continuously stirred or fluidized bed | Mixed or a pure yeast culture | Various | N.D. | Glass or steel filament balls or plastic | 0.23—0.69 | Fluidized bed reactor for effluent treatment or for mass culture | 36 37 2 |
| Artificial mouth | Cariogenic streptococci | Complex including synthetic saliva | N.D. | Enamel (extracted teeth) | N.D. | Changes in pH and Eh in in vitro plaques | 38 |
| Artificial mouth and test tubes | Pure cariogenic streptococci | Cycle of complex media | N.D. | Enamel acrylic porcelain glass nichrome stainless steel | N.D. | Cariogenicity and plaque formation in in vitro model systems | 39 |
| Misc. | Mixed | Nutrient broth | 20 | Epoxyresin | N.D. | Electron microscopy of a slime layer | 40 |
| Misc. | Mixed with *Legionella pneumphila* | Domestic hot water supply | N.D. | Copper, stainless steel, aluminum, glass beads, silicone rubber, natural rubber | N. D. | Growth of *L. pneumophila* films in model hot water system | 41 |
| Misc. | Mixed or pure *Nitrobacter* | Tapwater plus 100 mg each of $NH_4^+$ $NO_2^-$ or $NO_3^-$ | Membrane | 0.2—0.5 | 0.2—0.5 | Film of filtered packed cells | 42 |
| Misc. | Pure cultures | Complex | N.D. | PTFE | 0.3 | Constant-depth PTFE pan fermenter | 43 |

*Note:* N.D., not determined; BOD, biological oxygen demand; PTFE, (poly) tetra-fluoroethylene.

The model described by Coombe et al.[43] goes some way toward providing the facilities needed if the development and maintenance of any film is to be investigated. It consists of six (poly)tetra-fluoroethylene (PTFE) film pans, each machined to a depth of 300 μm. These rotate under a stationary wiper blade to provide a constant-depth film. The system is sterilized and gas and media regimes selected. The model was first used to simulate dental plaque formation (see Chapter 10) and although the principles behind the design were sound, some aspects of the design were operationally unsatisfactory.

A major problem was that only six samples could be taken during the course of a single experiment. This was not felt to be enough to give statistically significant results in some types of experiment, in particular for determining growth kinetics of microbial film. Variance between samples could not be attributed to a specific process, for example increased biomass, since the experimental error is unquantified. Secondly, the areas allowing film growth, which were grooves of 1, 2, or 3 mm in each film pan, were too small, providing insufficient biomass for further analysis. Thirdly, the scraper bar was not a knife section but had a leading edge, which allowed a significant amount of film to grow under the scraper bar. The medium inlet was placed too close to the film pans and could thus become blocked with film. The final problem was due to the alloy material used to construct the body of the fermenter. This corroded easily, in severe cases blocking the medium outlet port.

This model has now undergone further development in an attempt to alleviate these problems, giving a model film fermenter which enables all experimental variables to be finely controlled. This design allows an investigation of the spatial and temporal organization of any film to be undertaken. As understanding improves, other factors can be introduced, so that the model itself moves closer to reality.

In this chapter, the model will be described fully and preliminary results presented.

## DESIGN AND CONSTRUCTION OF THE CONSTANT-DEPTH FILM FERMENTER

### General Design Features

If a laboratory-model film fermenter is to be useful, it must be capable of giving answers not obtainable easily from natural film. This requirement determines the minimum design features of any model fermenter. Before describing in full the thin-film fermenter, it is worth setting out these features.

A variety of surfaces may become colonized by microbial film, and characteristic properties of a given film may be determined by the nature of the substratum material. Thus, it is important that any surface can be selected for use in the model and that changes can readily be made.

One of the biggest problems facing the microbial ecologist is sampling techniques. All too often film sampling consists of scraping the colonized surface, completely destroying the spatial order within the film. One consequence of such a practice might be that any anaerobes present which are suddenly exposed to air in this way may lose viability and escape detection. Any model film fermenter must have facilities to enable discrete, representative, and reproducible samples to be taken. Such samples must also include the substratum so there is as little disturbance as possible to the delicate structure of the film.

In natural habitats, film generally proliferates until it becomes destabilized, when the whole structure may become detached. Although steady-state films may never arise under normal circumstances, it is advantageous to be able to maintain such a film in a model fermenter. This makes investigation of spatial order in the film easier. The constant-depth film fermenter maintains a steady-state by physically removing growth once a predetermined depth of film has been generated. This depth should be reproducible and easy to set accurately in the first place.

For a model to be useful, all parameters must be controllable. Thus, any chosen regime of nutrient flow, atmosphere, shear rate, and temperature must be easy to set and maintain. It must also be easy to achieve and maintain sterility so that a pure culture or preselected film community may be investigated without unwelcome intruders!

Key aspects of film structure are the solute gradients which will always be present. Recent advances in microelectrode techniques make it possible to measure $pO_2$, pH, Eh, and some other solutes accurately and with a high degree of spatial resolution. The latter may range from less than 1 $\mu$m, in the case of polarographic oxygen electrodes, increasing to about 30 $\mu$m for pH electrodes. Any film fermenter design should allow the deployment of these electrodes, either while the film is still in the fermenter or by taking suitable samples without disturbing film structure.

Other important techniques for detailed study of film structure and physiology are transmission and scanning electron microscopy, X-ray microanalysis, and freeze sectioning. These permit the distribution of elements, species, cell viability, enzymes, substrates, and metabolic products to be determined throughout the film. Facilities must be provided to enable samples to be taken in a form which is appropriate to these techniques. Once agian, it is most important that the spatial order of the film is not disturbed during sampling, fixation, or freeze sectioning.

## Constructional Details
### Evolution of a Concept

The constant-depth film fermenter developed in Cardiff, Wales attempted to fulfill all these requirements. The first fermenter consisted of a rotating (poly) tetra-fluoroethylene (PTFE) disk approximately 40 mm in diameter. This was in contact with a stationary PTFE scraper blade which was spring-loaded to maintain contact with the disk. Set flush within this disk are six PTFE sample pans each 10 mm in diameter. These are fitted with short stainless steel pins which project from the side of the main disk. The pin allows easy removal of the sample pans through a port in the polycarbonate lid. New pans can also be placed in the disk once a sample has been removed. On early models of the fermenter, these pans had a single machined recess in which film could grow; this meant that only six samples could be taken during any one experiment, allowing no scope for assessing reproducibility of results. Problems also arose because accurate machining of PTFE is impossible and depths of individual pans varied by as much as 100 $\mu$m. The surface of these pans was very rough, since it proved impossible to remove marks left by the cutting tool.

In later models, some of the problems mentioned above were overcome in the following way. The main PTFE disk is 80 mm in diameter. Each 20 mm diameter sample pan is fitted with five PTFE plugs, each 5 mm in diameter, recessed to the required depth with a special tool (Figure 1). These plugs are easy to produce from any material and can be accurately ground down to give a flat surface. Any degree of roughness can then be given to the finished surface. With six film pans, it is only possible to take six samples during any one experiment; however, as can be seen, these are divided into five subsamples of known area and film depth. Thus the reproducibility of results obtained from any single sample can be assessed. Since all the plugs are recessed in the same way, the accuracy of the depth does not depend on machining of individual plugs but on the accuracy of the tool used to recess them. As this is turned from stainless steel, it can be made with great precision. Figure 2 is a top view of the PTFE disc showing film pan configuration and the position of the scraper blade and medium inlet. Nutrient flow is directed at the center of the disk, where it is distributed over the surface by the scraper blade. This was found to be the best way of achieving nutrient flow; earlier models were prone to blockages in the feed pipe since it was positioned very close to the film pans. Film growth occurs only on the surface of the plugs, since excess growth is wiped away by the scraper blade. Film pans and plugs are

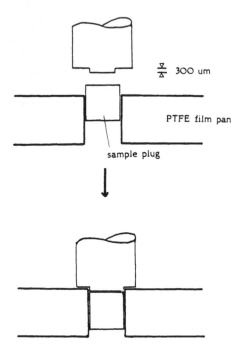

FIGURE 1. Technique used to recess sample plugs.

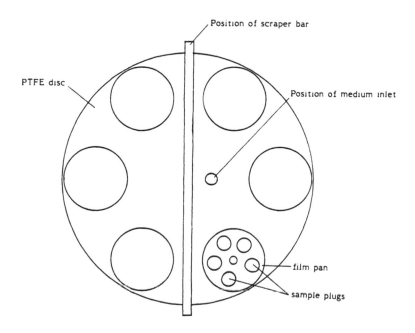

FIGURE 2. Top view of main PTFE disc showing film pan configuration.

held together by the "natural spring" in the PTFE. Sample plugs made of materials other than PTFE should also be held securely in the sample pans for the same reason. Plugs could be made by machining or they could be cast using the holes in the sample pans as molds. Cast plugs would then have to be ground down to give a flat surface.

*The Final Design*

The body of the fermenter was initially machined from a piece of aluminum alloy which

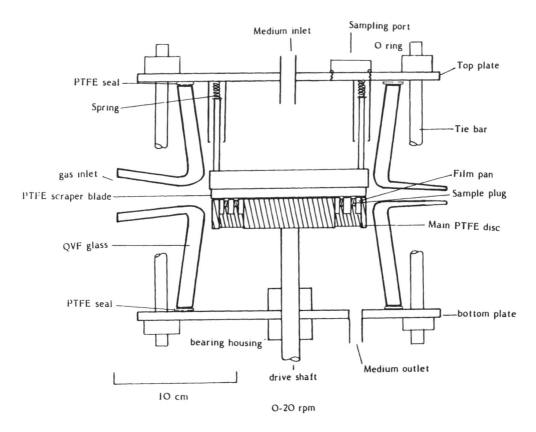

FIGURE 3.    The glass-bodied film fermenter.

proved susceptible to corrosion. A glass-bodied fermenter with stainless steel top and bottom plates was developed (Figure 3). This overcame the corrosion problem as well as being less expensive to produce. The QVF glass section (Dow Corning) has an internal diameter of 150 mm and two side arms for gas inlet and a thermistor probe. The stainless-steel base plate holds a bearing housing with a double-shaft seal. This takes the stainless-steel drive shaft to which the PTFE disk is attached. Drive is direct, powered by a small electric motor and gearbox. A speed control unit allows rotational speeds from 1 to 10 rpm to be selected. A port for waste medium is also provided in the base plate. The top plate houses the medium inlet and sample ports, and the stainless-steel surface can be flamed, allowing better aseptic sampling. The sprung scraper bar assembly is suspended from the top plate, all parts being constructed of stainless steel except for the thin PTFE blade. The main PTFE disc has a diameter of 140 mm and holds 15 sample pans, each with 5 plugs, giving 75 samples in all. The stainless-steel pins described in earlier models have been superseded by a small threaded hole in the center of each pan. A threaded stainless-steel tool is used to remove and replace pans. The fermenter is maintained at the right temperature with a heating belt wrapped around the body, although it is easy to envisage a double-walled vessel allowing water jacketing. Gas inlet and thermistor probe ports would then be repositioned in the top plate.

## EXPERIMENTS WITH THE FILM FERMENTER

### Isolation of Film Organisms
*Initial Selection*

A glass slide fermenter was used to isolate organisms from river water. This fermenter consisted of a 1 $\ell$ vessel with a holding volume of 500 m$\ell$. Agitation was achieved using a magnetic stirrer bar. Six microscope slides were held vertically around the edge of the vessel, suspended in the liquid flow. Water (20 $\ell$) collected from a small, fast-flowing, freshwater stream, was recirculated through the fermenter pumped at 10 m$\ell$/min with a Watson and Marlowe HR peristaltic pump. No nutrients were added to the river water.

After 2 weeks the glass slides were supporting thick film growth; microscopic examination showed that this consisted of a mixture of filamentous algae and protozoa together with many bacteria. The film was removed by scraping, and then homogenized in isotonic saline. The homogenate was diluted by serial dilution and plated onto tryptic soy agar. A total of 19 distinct bacterial types were isolated. No attempt was made to identify these at this stage. Eukaryotes were not selected or further examined.

### Isolation of Six Film-Forming Bacteria

The constant-depth film fermenter was used to select 6 bacterial types from the 19 already isolated. The fermenter was inoculated with 0.2 m$\ell$ of each of the 19 isolates in 24 hr broth culture.

### Media Composition

Naturally occurring film exists under conditions of extremely low nutrient concentrations. For this reason a very dilute medium was used containing (per liter distilled water):

| | |
|---|---|
| $KH_2PO_4$ | 100 mg |
| $MgSO_4$ | 20 mg |
| $CaCl_2$ | 1 mg |
| $NH_4Cl_2$ | 10 mg |
| Glucose | 5 mg (50 mg/$\ell$ for the initial recirculation) |

After inoculation, 500 m$\ell$ of medium was recirculated through the fermenter. This medium contained a tenfold increase in glucose concentration to maintain growth during the recirculation period, which continued for 24 hr before the main medium was connected.

### Experimental Conditions

Operational parameters were set as follows:

| | |
|---|---|
| Rotational velocity | 5 rpm |
| Nutrient flow rate | 2 m$\ell$/min |
| Temperature | Ambient |
| Gas flow | Constant flow of air (unquantified) |

The fermenter was run for 3 weeks during which time a thick film developed. On inspection this proved to be purely bacterial. Samples were taken and homogenized before being plated out as before. The six most numerous bacterial types were selected for use in all further experiments.

### Identification of Organisms

An attempt was made to classify the six bacteria selected using the API 20B kit for the identification of aerobic heterotrophs. It proved impossible to identify the isolates to species level, but genus classification for four of the six organisms was obtained. These were as follows:

> 2 *Bacillus* spp.
> *Aeromonas* sp.
> *Azotobacter* sp.

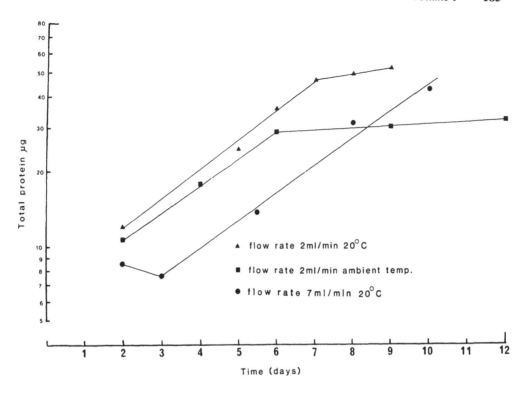

FIGURE 4. Film growth, measured as total protein, under three different operating conditions.

The other two organisms, one a Gram-negative coccus which does not utilize any carbohydrate and the other a Gram-positive nonspore-forming rod, gave variable results to some of the tests and have yet to be identified.

The organisms identified are commonly isolated from freshwater, and interestingly, the *Azotobacter* produces an extracellular polyuronide as a capsular material around the organism. This gives rise to colonies with a slimy appearance and may be important in film structure where the polymer could enmesh other organisms. Carbohydrate utilization by the azotobacter indicate that it may be *Azotobacter vinelandii* although this is not certain.

## Growth of Film Consortium
### Preliminary Experiments
These were designed to investigate the operating characteristics of the thin-film fermenter. Film growth rates were measured at different temperatures and nutrient flow rates. Steady-state operation of the fermenter was also assessed. Medium composition and operational details were as already described and 24 hr broth cultures of the 6 organisms were used as inoculum.

Growth was measured by analyzing samples for total protein content. Plugs were carefully removed from the sample pan with the film still intact on the surface. Each plug was boiled in 0.5 mℓ 1M NaOH for 5 min. Protein analysis was then carried out using the Lowry method. All five subsamples were treated in the same way so that reproducibility of results could be checked.

The dry weight of the film was also used as a measure of growth. Plugs were carefully removed from the sample pan. The plugs were then dried at 105°C for 24 hr and then until constant weight was observed. After cleaning in an alkaline detergent and drying for 1 hr, the plugs were reweighed. Substraction of these weights gave the dry weight of the film.

Growth curves for three experimental runs under different operating conditions are shown in Figure 4. The log of total protein is plotted against time, each point being the average

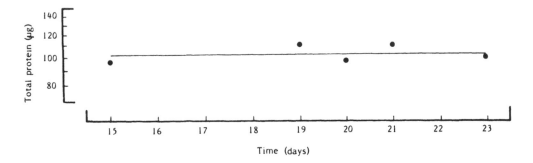

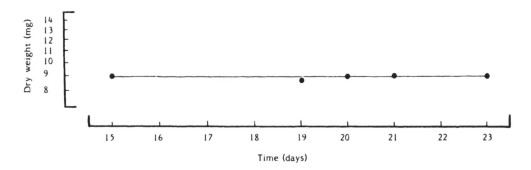

FIGURE 5. Film at steady state measured in terms of biomass. The top graph shows total protein, the lower, dry weight.

result from five subsamples. With a nutrient flow rate of 2 mℓ/min at ambient temperature, an apparent steady state was reached after 6 days. This film, while filling the pans, was transparent. Maintaining the temperature at a constant 20°C gave a curve with the same growth rate but no steady state was reached. Increasing the flow rate to 7 mℓ/min again had no effect on the growth rate but produced a visably denser film. A steady state was not reached during the experiment.

### Towards a Steady State

Because only six samples were available during any one experiment, it was impossible to cover film growth from the time of inoculation until a steady state was obtained. A further experiment allowed film to grow for 2 weeks before any samples were taken. Because previous results had shown that the Lowry method of protein measurement was reproducible, only two of the five subsamples were used for total protein measurement. Growth on two other plugs was assessed using dry weight measurement, while the fifth plug was used to measure the partial pressure of oxygen within the film. Experimental conditions were identical to those described above with a nutrient flow rate of 7 mℓ/min at 20°C.

Figure 5 shoes the resultant growth curves measured by total protein and dry weight. Both methods confirm that it was possible to maintain a steady state over the period of the experiment. The protein curve can be plotted on the same axes as the curve obtained for early film development under identical operating conditions. The resulting plot (Figure 6) indicates that the two sets of results agree.

### Oxygen Penetration in Microbial Film

#### The Oxygen Microelectrode

Polarographic oxygen microelectrodes have a typical tip diameter of less than 10 μm.

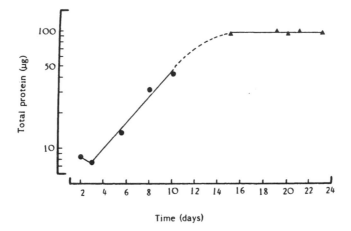

FIGURE 6.   Composite growth curve.

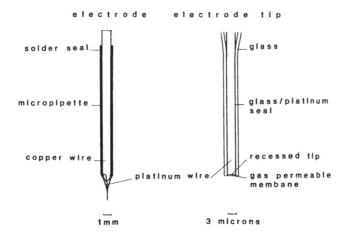

FIGURE 7.   Oxygen microelectrode construction showing the entire electrode and details of the electrode tip. (From Coombs, J. P. and Peters, A. C., *J. Microbiol. Methods*, 3, 201, 1985. With permission.)

They are available commercially but are very expensive, and several methods for producing them have been described in the literature. The electrodes used in this study were made by the relatively simple method described by Coombs and Peters.[45] These electrodes are reliable, inexpensive, and have a tip diameter of less than 5 μm. They consist (Figure 7) of a length of platinum wire soldered to a piece of copper wire. The platinum is electrolytically etched until it has a diameter of about 3 μm with a long, gently tapering tip. This is heat sealed inside a glass micropipette. The protruding tip of platinum wire is again etched until the electrode has a slightly recessed tip. A gas-permeable membrane of cellulose acetate is applied to the tip; this makes the electrode response more stable. Because of the recessed tip, the electrode has a spatial resolution of less than 1 μm with a 90% response time of 2 to 3 sec.

*Positional Control of Electrodes Using the BBC Microcomputer*

The microbial film under investigation was only 300 μm thick, supported on a solid substratum. When probing this film with electrodes, it was vital that the precise position if

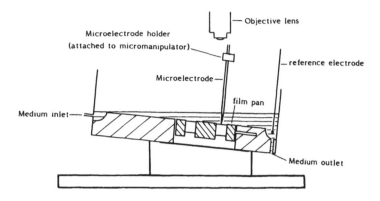

FIGURE 8.   Inclined sample holder for oxygen electrode studies in film.

the electrode tip was known and controlled carefully, so that the electrode could not be driven into the hard surface below the film. A BBC Model B microcomputer was interfaced with hardware which controlled a stepper motor. This motor was attached to the fine control of a Prior micromanipulator in which the electrode was mounted. This arrangement allowed positional control of the electrode tip to 0.6 μm. An advantage of using complex hardware to control the motor when the BBC could have been programmed to do the same was that only a small amount of simple code was needed in a BBC program to send data to the motor hardware telling it what to do. The system allows the starting position of the electrode to be set, and the electrode can then be moved in any number of steps until a maximum, preset depth is reached. The electrode will then return to its starting point.

*Data Logging Using the BBC Microcomputer*

The oxygen microelectrode was connected to a Transidyne chemical microsensor (Model 1201, Transidyne General Corporation) with a silver/silver chloride electrode as reference. The microsensor output was interfaced with the BBC microcomputer via the analogue digital converter which gave the computer digital data it could handle. The A/D converter output on the BBC was found to vary by as much as 15% as the temperature of the machine fluctuated. This was due to poor quality components used to regulate the reference voltage. The section of circuit responsible was cut out and replaced by a new circuit which cured the problem. Software on the BBC collected data every 10 msec when the electrode was stationary in a position where the $pO_2$ was to be measured. This data was paired with the electrode position which was also known. All data was stored on floppy disc and could be output either numerically or graphically on the VDU or printed out on a dot-matrix printer.

*Experimental Procedure*

Because of the need to maintain sterility and because the sample pans were rotating, it was impossible to measure $pO_2$ in the film while the sample was still in place in the fermenter. A special chamber was constructed to hold the sample when $pO_2$ determinations were being made.

***The Sample Holder***

Figure 8 shows details of the chamber used to hold the film sample. The sample pan, with plugs and film still in place, was fitted in the base so that the pan was flush with the surface. Nutrients were pumped through the chamber so the film was immersed. The whole chamber was held in the field of view of a binocular microscope with 40× magnification. The film pan was tilted at a slight angle so that the tip of the electrode, which was held

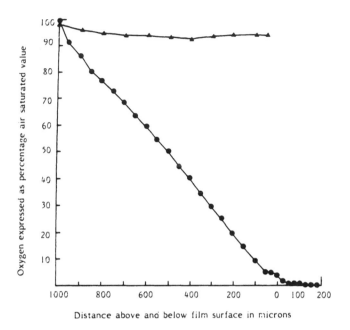

FIGURE 9.    Oxygen profiles in film 300 μm thick (●), and in air saturated medium above a clean film pan (▲).

normal to the film surface, could be observed through the microscope. Nutrients were recirculated using a peristaltic pump while temperature was held at 25°C.

### Calibration of the Electrode

The electrode was calibrated immediately before a series of measurements were made. First, the electrode was placed in a sample of medium which had been sparged with compressed air and the output adjusted to read 100% $pO_2$. Next, the electrode was placed in nitrogen-sparged medium, the zero being adjusted to read 0% $pO_2$. It is essential to make these adjustments when the electrode is in the same media as that used to irrigate the film since measured $pO_2$ varies with chloride ion concentration. All temperatures must be maintained with $+/-$ 2.5°C as electrode output varies with temperature.

### Taking Measurements

After the electrode was calibrated, the film was drained and a small piece of tissue was used to draw as much liquid from the film surface as possible. The electrode was then lowered until it just made contact with the film. This could be observed through the microscope. The electrode was raised until it was 1000 μm above the film surface. Then nutrient flow was reconnected and the film irrigated with air saturated medium to a depth of 1000 μm. The whole procedure took only a few seconds so film shrinkage was kept to a minimum. The program which lowered the electrode and measured $pO_2$ was then activated and the oxygen profile measured. One profile measuring oxygen at 25 μm intervals until the electrode tip was 50 μm above the film surface, and then at 5 μm intervals for a further 300 μm, took about 3 min. All results were automatically stored on floppy disc.

### Results

Oxygen was measured in the medium above the film surface and to a depth of 250 μm in the 300 μm film using the technique described. The total depth of the profile was kept to 250 μm to eliminate the chance of electrode damage. Figure 9 shows the average of five

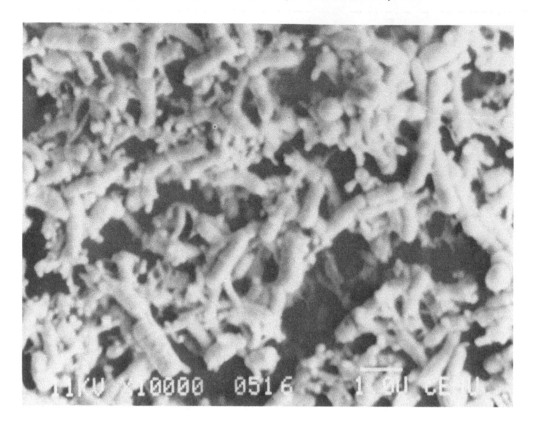

FIGURE 10.    Scanning Electron Micrograph of the surface of 300 μm film. Bar = 1 μm.

such profiles. The control was produced using the above procedure but with an empty film pan in the sample holder. The results show that the film is respiring rapidly and becomes anaerobic some 30 to 50 μm below the film surface. Although the film was irrigated with air-saturated medium, it is not an open system, so the oxygen concentration in the medium above the film is changing with time as well as with depth. Further experimental work is needed to investigate oxygen penetration in a steady-state situation.

### The Electron Microscopy of Microbial Film
*Film Fixation and Observation*

Samples were fixed by carefully immersing a sample pan in 2% glutaraldehyde in sodium cacodylate buffer for 30 min. The sample was then washed in fresh buffer for 30 min before being dehydrated by placing sucessively in 10%, 50%, 80%, 90%, and 100% ethanol for 20 min. each. The film did not attach firmly to the PTFE plugs, so extreme care had to be taken to ensure that none of the steps resulted in detachment of the film. Further investigations indicate that vapor fixing in formaldehyde and acrolein in a water-saturated atmosphere prior to fixing in gluteraldehyde improved adherence. Samples were then dried in a critical point dryer and sputter coated with gold before examination using a Joel scanning electron microscope.

*Results*

Figure 10 is a scanning electron micrograph of the surface of a steady-state film. Chains of rod-shaped organisms can be seen, and there is some evidence of the polymeric compounds which bind the film which appear as thin dehydrated sheets when fixed.

# DISCUSSION

The film fermenter described in this chapter owes its origin to the view that a model system rather than a microcosm should be chosen where the basic aim is to investigate fundamental and ubiquitous properties of film rather than to concentrate on the unique properties of a single film. Such an approach is bound to have advantages and disadvantages.

A distinctive feature of this film fermenter is that film is maintained at a constant depth. Although this moves away from the cyclic pattern of attachment growth and sloughing found in many natural films, there are advantages in constant-depth film. It has been shown that it is possible to enter a long-term quasi-steady-state where some properties of the film do not change significantly with time. Once a steady state has been achieved, perturbing the system becomes easy and unequivocal results can be obtained. By maintaining a constant volume of film in each film pan, reproducibility can be assessed and the deployment of fragile microelectrodes is made easier. The fermenter also allows aseptic operation and control of all environmental conditions.

## Applications of Constant-Depth Film Fermenters in Fundamental Studies of Film Biology

The following headings indicate the sorts of experiments that could be done.

### The Attachment of Microbes to Surfaces

A major area of interest is the process by which microbes attach to insoluble surfaces.[46] The film fermenter described here provides numerous separate film surfaces, thus allowing the comparison of many different attachment substrates simultaneously. The effects of solutes and other environmental factors on attachment can also be examined under reproducible conditions.

### Growth Rate Determinations

The number of film surfaces available in the fermenter allow measurements of growth rate to be determined accurately. Each of the five film plugs may be assayed separately, giving a statistically respectable value for growth after a particular time.

### Rates of Formation of Cell Products

Cell products like matrix polysaccharides, etc., can be monitored in each film pan using appropriate chemical methods. Again the effect of environment on productivity can be easily assessed in this film fermenter because of its innate reproducibility.

### Film Diffusivity

The use of radioactive tracers should allow diffusion rates to be measured in film formed in each of the pans. Films exposed to the tracer can be removed and freeze-sectioned in the horizontal dimension.[47] Sections can then be counted in the usual way. It ought also to be possible to section the film vertically and prepare autoradiograms of the isotope distribution as was done for bacterial colonies by Reyrolle and Letellier.[48]

### Rates of Solute Uptake

It should be easy to determine rates of solute uptake in films produced in the film fermenter. *In situ* determinations are possible in the whole fermenter by measuring concentrations in the influent medium and comparing this with concentrations in the effluent. Individual film pans may be removed and incubated in the presence of specific substrates. Film could be placed in respirometer flasks to measure oxygen removal rates directly in the presence or absence of added substrates.

*Successional Changes in Film Biology*

It is important to recognize the differences between the pioneer strains that initially colonize the substratum and the organisms that appear later to make a climax or steady-state population. This is simply done at the most basic level by removing film and homogenizing it, as described above, followed by counting each of the species present.

*Spatial Order Within the Film*

One of the most interesting problems in film biology is the way in which the film differentiates according to environmental circumstances at each point within the structure. This information is quite hard to obtain. The following approaches may be worth pursuing. Horizontal freeze-sectioning is a possibility. Here the film is frozen on to a cryotome stub and sectioned serially. Each section is gently homogenized and then plated out to measure viable counts of each species.

Vertical freeze-sectioning, coupled with immunofluorescent staining, could be used to assess species heterogeneity. We believe that this could constitute a most powerful tool once the initial problems of preparing fluorescent labels for each member of the film community have been overcome.

Spatial differentiation can be investigated using transmission electron microscopy provided that the individual species are recognizably different.

*Chemical Differentiation*

The distribution of chemical species across film is another important area for investigation. As this chapter has shown, it is comparatively easy to deploy oxygen microelectrodes to determine the distribution of this solute. Other microelectrodes are available. These can include pH and redox potential electrodes, sulfide electrodes, as well as a growing family of other specific-ion electrodes.

The distribution of radioactive tracers has already been mentioned. A combination of high activity tracers plus freeze-sectioning is worth considering as long as sensitive detection methods are available.

Element distribution is another important area for investigation. This is comparatively easy to do for some elements using electron microprobe analysis.

These represent just a few of the ways in which the biology of microbial film can or could be investigated. Using all these techniques should throw considerable light on the physiology of microbial film.

Not all applications of the film fermenter are as academic as these. Microbial films are responsible for so much economic damage that a brief discussion of economic applications of film fermenters seems important.

## Specific Applications of Constant-Depth Film Fermenters in Economic Microbiology
*Research into the Structure and Physiology of Anaerobic Corrosion Film*

Steel structures used in industrial environments are prone to damaging attack by sulfate-reducing anaerobic corrosion films. The relevant microbes attach to the metal surface under appropriate environmental conditions. The growth of pioneer species leads to oxygen depletion in the film adjacent to the metal followed by the growth of sulfate-reducing bacteria. Anaerobic activities, due in part to sulfide accumulation, then causes pitting corrosion in the structure. The economic losses due to sulfide corrosion, sustained by the oil industry in the U.K. alone, run to hundreds of millions of pounds per annum.[49] If losses incurred by the chemical and manufacturing industry, the gas supply industry, and marine interests are also considered, the figures become so large that research on the structure and activity of corrosion films is seen as an absolute necessity to the relevant *industry*.

*Films Harboring Legionella*

There is at present a major interest in controlling the build up of microbial films in water piping in large buildings, in particular hospitals and hotels. Outbreaks of legionnaires' disease have been shown to be due to contamination of air-conditioning systems and internal water pipe work. It is established that the pathogen exists in association with other film-forming species on the walls of pipes and other regions in contact with circulating fluids.[41] The availability of secure film fermenters capable of operating under totally aseptic conditions will allow fundamental investigations into the film structure and behavior, and possibly more important, will allow testing of treatments designed to kill pathogenic cells and eliminate biofilms.

*Dental Plaque*

Dental caries normally forms beneath an accumulation of food debris and microbial biomass called dental plaque. There are a number of dental plaque models available as discussed by Tatevossian (Chapter 10) in this volume. Although many of these reflect specific oral situations quite well, their design does not lead to the same reproducibility as is possible with the constant-depth film fermenter.

*Films in the Effluent Treatment Industry*

Films form on insoluble substrata used as trickling filters in the treatment of polluting effluents. The films are composed of microbes in a matrix consisting mainly of polysaccharide polymers. Metabolic activities within these films oxidize organic effluents to harmless products, including carbon dioxide and water under aerobic conditions, or to other simple fermentation products where the filters are operated anaerobically. The films are regarded by effluent engineers as simple catalysts and little is known about the physiology of these structures.

*Marine Fouling*

This is initiated by a microbial film which attaches to exposed surfaces. The film fermenter provides a good way of following, once more under carefully controlled conditions, the initiation and early development of these films. Again, specific antifouling treatments can be monitored using the film fermenter.

*The Electricity Supply Industry*

Film growth significantly reduces the transfer of heat across heat exchangers. Knowledge of the attachment and development of the relevant organisms is a worthwhile goal, as is the ability to develop treatment processes under well-defined conditions.

## ACKNOWLEDGMENTS

Adrian Peters was funded by the Natural Environment Research Council, which we gratefully acknowledge.

## REFERENCES

1. **Maier, W. J., Behn, V. C., and Gates, C. D.,** Simulations of the trickling filter process, *J. Sanit. Eng. Div. Am. Soc. Civ. Eng.,* 93, 91, 1967.
2. **Atkinson, B., Daoud, I. S., and Williams, D. A.,** A theory for the biological film reactor, *Trans. Inst. Chem. Eng. (London),* 6, T245, 1968.
3. **Atkinson, B. and Daoud, I. S.,** The analogy between microbial "reactions" and heterogeneous catalysis, *Trans. Inst. Chem. Eng. (London),* 48, T245, 1970.
4. **Borchardt, J. A.,** Biological waste treatment using rotating discs, *Biotechnol. Bioeng. Symp.,* 2, 131, 1971.
5. **Famularo, J., Mueller, J. A., and Mulligan, T.,** Application of mass transfer to rotating biological contactors, *J. Water Pollut. Control. Fed.,* 50, 653, 1978.
6. **Kinner, N. E., Balkwill, D. L., and Bishop, P. L.,** Light and electron microscopic studies of microorganisms growing in rotating biological contactor films, *Appl. Environ. Microbiol.,* 45, 1659, 1983.
7. **Klemetson, S. L. and Lang, M. E.,** Treatment of saline wastewaters using a rotating biological contactor, *J. Water Pollut. Control Fed.,* 56, 1254, 1984.
8. **Kornegay, B. H. and Andrews, J. F.,** Kinetics of fixed film biological reactors, *J. Water Pollut. Control Fed.,* 40, 460, 1968.
9. **La Motta, E. J.,** Kinetics of growth and substrate uptake in a biological film system, *Appl. Environ. Microbiol.,* 31, 286, 1976.
10. **Trulear, M. G. and Characklis, W. G.,** Dynamics of biofilm processes, *J. Water Pollut. Control Fed.,* 54, 1288, 1982.
11. **Bakke, R., Trulear, M. G., and Characklis, W. G.,** Activity of *Pseudomonas aeruginosa* in biofilms: steady state, *Biotechnol. Bioeng.,* 26, 1418, 1984.
12. **McCoy, W. F., Bryers, J. D., Robbins, J., and Costerton, J. W.,** Observations of fouling biofilm formation, *Can. J. Microbiol.,* 27, 910, 1981.
13. **Bryers, J. D. and Characklis, W. G.,** Processes governing primary biofilm formation, *Biotechnol. Bioeng.,* 24, 2451, 1982.
14. **Turakhia, M. H., Cooksey, K. E., and Characklis, W. G.,** Influence of a calicum specific chelant on biofilm removal, *Appl. Environ. Microbiol.,* 46, 1236, 1983.
15. **Strand, S. E., McDonnell, A. J., and Unz, R. F.,** Concurrent denitrification and oxygen uptake in microbial films, *Water Res.,* 19, 335, 1985.
16. **Tomlinson, T. G. and Snaddon, D. H. M.,** Biological oxidation of sewage by films of microorganisms, *Int. J. Air Water Pollut.,* 10, 865, 1966.
17. **Eighmy, T. T., Maratea, D., and Bishop, P. L.,** Electron microscopic examination of wastewater biofilm formation and structural components, *Appl. Environ. Microbiol.,* 45, 1921, 1983.
18. **Monadjemi, P. and Behn, V. C.,** Oxygen uptake and mechanism of substrate purification in a model trickling filter, *Proc. 5th Int. Water Pollut. Res. Conf., Paper No. II-12,* Pergamon, Oxford, England, 1971.
19. **Bruce, A. M. and Merkens, J. C.,** Recent studies of high rate biological filtration, *Water Pollut. Control,* 69, 113, 1970.
20. **Rincke, G. and Wolters, N.,** Technology of plastic medium trickling filters, *Proc. 5th Int. Water Pollut. Res. Conf.,* Paper number II-15, Pergamon, Oxford, England, 1971.
21. **Rittman, B. E. and McCarty, P. L.,** Model of steady-state biofilm kinetics, *Biotechnol. Bioeng.,* 22, 2343, 1980.
22. **Rittman, B. E. and Brunner, C. W.,** The nonsteady-state biofilm process for advanced organics removal, *J. Water Pollut. Control Fed.,* 56, 874, 1984.
23. **Vaughan, G. M. and Holder, G. A.,** Substrate removal in the trickling filter process, *J. Water Pollut. Control Fed.,* 56, 417, 1984.
24. **Lee, E. J., De Witt, K. J., Bennett, G. F., and Brockwell, J. L.,** Investigation of oxygen transfer to slime as a surface reaction, *Water Res.,* 10, 1011, 1976.
25. **Sanders, W. M.,** Oxygen utilization by slime organisms in continuous culture, *Air Water Pollut. Int. J.,* 10, 253, 1966.
26. **Bungay, H. R., Whalen, W. J., and Sanders, W. M.,** Microprobe technique for determining diffusivities and respiration rates in microbial slime systems, *Biotechnol. Bioeng.,* 11, 765, 1969.
27. **Whalen, W. J., Bungay, H. R., and Sanders, W. M.,** Microelectrode determinations of oxygen profiles in microbial slime systems, *Environ. Sci. Technol.,* 3, 1297, 1969.
28. **Zvyagintsev, D. G.,** Adsorption of microorganisms by glass surfaces, *Microbiology,* 28, 104, 1959.
29. **Zobell, C. E.,** Marine microbiology: a monograph of hydrobacteriology, in *Marine Microbiology,* Chronicia Botanica Company, Waltham, Mass., 1946.
30. **Wood, E. J. F.,** Investigation on underwater fouling. II. The biology of fouling in Australia, *Australian J. Mar. Freshwater Res.,* 1, 85, 1950.

31. **Pedersen, K.,** Method for studying microbial biofilms in flowing water systems, *Appl. Environ. Microbiol.,* 43, 6, 1982.
32. **Nelson, C. H., Robinson, J. A., and Characklis, W. G.,** Bacterial adsorbtion to smooth surfaces: rate extent and spatial pattern, *Biotechnol. Bioeng.,* 27, 1662, 1985.
33. **Dias, F. F., Dondero, N. C., and Finstein, M. S.,** Attached growth of *Sphaerotilus* and mixed populations in a continuous flow apparatus, *Appl. Microbiol.,* 16, 1191, 1968.
34. **Larsen, D. H. and Dimmick, R. L.,** Attachment and growth of bacteria on surfaces of continuous culture vessels, *J. Bacteriol.,* 88, 1380, 1964.
35. **Mulcahy, L. T. and LaMotta, E. J.,** Mathematical model of the fluidised bed biofilm reactor, Report No. Env. E. 59-78-2, Environmental Engineering Program, Department of Civil Engineering, University of Massachussetts, Amherst, 1978.
36. **Atkinson, B. and Davies, I. J.,** Completely mixed microbial film fermenter. Method for overcoming washout in continuous fermentation, *Trans. Inst. Chem. Eng. (London),* 50, 208, 1972.
37. **Atkinson, B. and Knights, A. J.,** Microbial film fermentors their present and future applications, *Biotechnol. and Bioeng.,* 17, 1245, 1975.
38. **Russell, C. and Coulter, W. A.,** Continuous monitoring of pH and Eh in bacterial plaque growth on a tooth in an artificial mouth, *Appl. Microbiol.,* 29, 141, 1974.
39. **Jordan, H. V. and Keyes, P. H.,** *In vitro* methods for the study of plaque formation and carious lesions, *Arch. Oral Biol.,* 11, 793, 1966.
40. **Jones, H. C., Roth, I. L., and Sanders, W. M., III.,** Electron microscopic study of a slime layer, *J. Bacteriol.,* 99, 316, 1969.
41. **Schofield, G. M. and Wright, A. E.,** Survival of *Legionella pneumophila* in a model hot water distribution system, *J. Gen. Microbiol.,* 130, 1751, 1984.
42. **Williamson, K. and McCarty, P. L.,** Verification studies of the biofilm model for bacterial substrate utilization, *J. Water Pollut. Res. Fed.,* 48, 281, 1976.
43. **Coombe, R. A., Tatevossian, A., and Wimpenny, J. W. T.,** Bacterial thin films as *in vitro* models for dental plaque, in *Surface and Colloidal Phenomena in the Oral Cavity: Methodological Aspects,* IRL Press Ltd., London, 1982, 239.
44. **Wimpenny, J. W. T., Lovitt, R. W., and Coombs, J. P.,** Laboratory model systems for the investigation of spatially and temporally organised microbial ecosystems, *Symp. Soc. Gen. Microbiol.,* 34, 67, 1983.
45. **Coombs, J. P. and Peters, A. C.,** A simple method for making microelectrodes to measure oxygen, *J. Microbiol. Methods,* 3, 199, 1985.
46. **Fletcher, M. and Marshall, K. C.,** Are solid surfaces of ecological significance to aquatic bacteria?, *Adv. Microbial Ecol.,* 6, 199, 1982.
47. **Wimpenny, J. W. T. and Parr, J. A.,** Biochemical differentiation in large colonies of *Enterobacter cloacea, J. Gen. Microbiol.,* 114, 487, 1979.
48. **Reyrolle, J. and Letellier, F.,** Autoradiographic study of the location and evolution of growth zones in bacterial colonies, *J. Gen. Microbiol.,* 114, 483, 1979.
49. **Hamilton, W. A.,** Sulphate-reducing bacteria and anaerobic corrosion, *Annu. Rev. Microbiol.,* 39, 195, 1985.

Chapter 10

# FILM FERMENTERS IN DENTAL RESEARCH

**A. Tatevossian**

## INTRODUCTION

Films that form on dental tissues have been described since early in the history of science. In 1683 Antonie van Leeuwenhoek[1] described small animalcules which he had scraped off his own teeth. He can thus be credited with discovering that human teeth are coated with a film of bacteria, now called dental plaque. However, Leon Williams[2] had considerable difficulty in obtaining recognition of the same fact by his colleagues in dental research. As recently as the early 1960s, a common view among dental researchers was that the teeth and oral soft tissues were covered with a film of saliva and food debris, together with the epithelial cells shed from the cheek, tongue, and gum margins. A symposium[3] in 1969 made a significant contribution to our understanding of the nature of dental plaque.

When a tooth surface is cleaned, it instantly acquires a surface coating of irreversibly adsorbed salivary proteins, termed the acquired pellicle (often shortend to ''pellicle''). This conditioning film is always the substrate to which bacteria, suspended in saliva, become adsorbed (Figure 1). The composition of this conditioning film has been investigated with some difficulty; scraping the tooth surface or demineralizing its outer enamel has yielded enough material to analyze the aminoacid composition[4-9] of the acquired pellicle. This conditioning film contains salivary phosphoproteins with high affinity for dental enamel, indicating a degree of selective adsorption in its formation.[10,11] However, a lower affinity for adsorption does not preclude the presence of several other salivary proteins, enzymic, and antibody activity in the film.[12,13] Salivary amylase, firmly adsorbed lysozyme activity, and IgA was consistently found;[14] albumin and IgG were frequently found. The carbohydrate composition of experimental salivary pellicles was different from that of the saliva from which the film derived.[15,16]

The adsorption of oral bacteria to the pellicle is a slow process, taking 3 to 4 hr before bacteria are adsorbed in significant numbers. These are essentially selectively adsorbed streptococci, particularly *Streptococcus sanguis*,[17,18] and coaggregated actinomycetes. The sparsity of these species in saliva, where *S. salivarius* predominates, is a striking confirmation of this selectivity, but in vitro models have failed to demonstrate selectivity or its basis. Although salivary bacterial counts can be relatively high, ($10^8$/m$\ell$ on average), indicating a rich source of inoculum for the teeth, the sparsity of the early bacterial coating remains unexplained in the context of the rapid kinetics of the process in vitro, where surface saturation is typically found within the first hour. In vitro a small proportion of the bacteria in solution can be instantaneously and irreversibly adsorbed in sufficient numbers to result in the subsequent growth of these bacteria into a confluent film.

The influence of the conditioning layer of pellicle on the initial adsorption of bacteria to the tooth is not well defined. Several studies with *S. sanguis*, *S. mitior*, *S. mutans*, and *Bacillus* sp. indicate increased adsorption compared with uncoated substrate.[19-22] However, it has also been reported that *S. mutans* adsorption is decreased by pellicle.[23-26] The specificity of the initial adsorption has been ascribed to lectin-type interactions,[27] although a purely physico-chemical mechanism based on charge interactions between pellicle and the coated surface of approaching bacteria has also been proposed.[28] Adsorption of *S. mutans* to hydroxyapatite may depend on the zeta potential of the hydroxyapatite surface and chemical treatment to increase the zeta potential of the surface reduced adsorption.[29] The surface

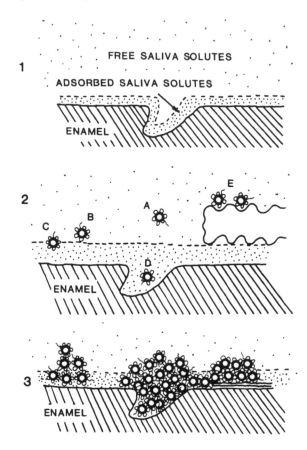

FIGURE 1.    The formation of a pellicle and dental plaque on a cleaned tooth surface. Bacteria are shown as open circles and their adsorbed or elaborated coatings are shown as fine loops. Areas of surface roughness may not have been denuded of bacteria by cleaning and are also more likely to carry adsorbed bacteria. (1) An adsorbed layer, the acquired pellicle, forms instantaneously when a cleaned enamel surface is exposed to saliva. (2) The initial reversible phase of adsorption: A, salivary bacteria (both free and desorbed from the surface); B, reversibly adsorbed bacteria; C, D, gradual incorporation of bacteria and surface growth in discrete colonies; E, bacteria associated with epithelial cells shed from the gum or cheek. (3) Growth of colonies and elaboration or adsorption of extracellular polymers aid further surface accretion of reversibly adsorbed cells and their retention, signaling a transition to a permanent attachment to the surface. Coalescence of isolated attached colonies leads to confluence and subsequent growth as a film.

potential of the bacteria also affects their adsorption, but the zeta-potential effect seems to depend on the species of streptococcus examined, or even differs between strains of the same species and can be altered by growth conditions.[30] Surface area and roughness are two factors which have been ignored in most studies of bacterial adsorption to dental enamel or hydroxyapatite, but there is some evidence that these factors may be important.[31] Surface roughness is known to be important for adsorption in other film-forming systems, particularly where hydrodynamic shear forces are present.[32] Extrapolation from other systems to the mouth may be dangerous. The rate of bacterial adsorption to a surface is directly proportional to their concentration in suspension for systems described by Bryers and Characklis;[33] however, the selectivity of tooth adsorption processes described above does not reflect concentrations of the different species in saliva. The medium used for growing cells may have an effect on their adsorption properties which may vary depending on the strain of streptococcus or actinomycetes used.[34,35] Adsorption of oral streptococci may be increased by glycerol teichoic acids[22] and facilitated by specific interbacterial aggregations which include lectin-type interactions between surface components of coaggregating species.[36] The formation of insoluble extracellular polyglucan may enhance the initial adsorption of *S. mutans*,[37] but there is no reason to believe that this is the primary or essential determinant of adsorption in this or other species.[38,39] The role of bacterial surface appendages such as fibrils and fimbriae has recently been discussed.[40,41] The presence of liquid/solid/air interfaces causes an accumulation of larger numbers of bacteria than present at a liquid/solid interface or in the bulk liquid[42] and, it has been suggested that this may account for the particular distribution of early plaque accumulation adjacent to gum margins.[39] Hydrophobic interaction has also been proposed as a mechanism for bacterial adsorption to pellicle[43] since a significant density of hydrophobic groupings can be predicted from the amino-acid composition of pellicle; on the other hand, the wettability of the surface, as measured by the contact angle or the critical surface tension has been proposed as a guide to its adsorptive potential for bacteria.[44] Fresh isolates of oral streptococci were all found to be highly hydrophobic.[45] Both their hydrophobicity and their adsorption to hydroxyapatite (the bulk of tooth mineral) were reduced by repeated subculture of *S. mutans*, serotype c strains, or *S. sanguis*. The hydrophobicity of several oral species was examined by Gibbons and Etherden,[46] who concluded that although there was a correlation with the extent of their adsorption to saliva-coated hydroxyapatite, such a mechanism lacked specificity. In contrast, no correlation between hydrophobicity and adsorption to acrylic or hydroxyapatite was found for six strains of *S. sanguis*, and, in any event, the apparent hydrophobicity of bacteria can be changed by altering the suspending medium without necessarily altering their adsorption characteristics.[141]

The process of bacterial adsorption to pellicle remains to be explained. Although both the proposed specific and nonspecific mechanisms have their adherents, it is probable that both mechanisms are involved to some extent and that others remain to be discovered. In all probability, several mechanisms are needed for bacterial substantivity in the hostile conditions found on the teeth, where washout in saliva is a constant threat.

The retention and proliferation of pioneer organisms on the teeth, and the subsequent development of a climax population and the ecological successions which occur in the formation of "mature" plaque, have been very extensively reviewed[3,17,47] and are beyond the scope of this chapter. It is well established that the early colonizers are mainly streptococci and actinomycetes, and that once an average film thickness of about 100 μm is attained, there is a succession to Gram-negative and filamentous species. Interrelationships between species based on nutritional commensalism (e.g., utilization of streptococcal lactate by *Veillonella* sp., which in turn may produce hydrogen as an energy source for *Vibrio sputorum;* *p*-aminobenzoate produced by *S. sanguis* can support growth of *S. mutans;* oral diphtheroids may facilitate the establishment of *B. melaninogenicus* by producing vitamin K) and inhibition (e.g., hydrogen peroxide or bacteriocins produced by several oral species are inhibitory to sensitive organisms) resulting in palisades of single or mixed species are known to occur, and interactions between species during growth affect the final composition and properties of the plaque. The chemical composition of the free aqueous phase in the film is quite distinct and easily distinguishable from that of saliva even though the film, often less than 100 to 300 μm thick, is constantly bathed in saliva.[48] This suggests that gradients of nutrient and metabolic products may occur in such films and contribute to their pattern of ecological succession, but virtually nothing is known about this aspect of dental plaque ecology.

## OBJECTIVES FOR USING FILM FERMENTERS IN DENTAL RESEARCH

### Previous Studies

The complexity of the oral environment and ethical problems associated with the study of disease processes in human subjects inevitably directed attention to the development of

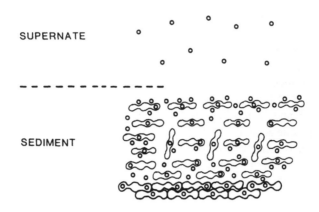

FIGURE 2.   Distribution of bacteria in saliva supernatant and sediment. The shed epithelial cells harbor dense clusters of bacteria on and in them, while the supernatant is free of eukaryote cells and harbors a lower bacterial population density.

laboratory models. The objectives of many of these studies were to isolate factors which influence the initiation and progress of dental caries.

### Models to Study the Processes Involved in Dental Caries

Several in vitro models contained bacterial films either by design or as a consequence of including bacteria in the incubation vessels. Dietz used a reaction vessel containing a tooth[48] which was suspended in saliva and a mixed salivary bacterial flora. Interest was focused on demineralization of the teeth and on the chemical composition of the solution in which dissolution of the tooth occurred. The nature or properties of the bacterial films which formed on teeth in such a system were not examined. Several such models which include a bacterial film and are potentially useful for study have been described.[49-52] Their contribution to an understanding of the caries process was reviewed by Slade.[53] In these and other recent variants, an emphasis on understanding caries has mitigated against deriving useful data on the bacterial films they contain. However, these studies often give an insight into the properties of bacterial films, often in mixed culture, and deserve mention.

### Models to Study Fermentation in Dental Plaque

Sugar metabolism in mixed salivary bacterial populations has been investigated by several workers in order to identify fermentation products made under different conditions and their relevance to the mechanism of tooth tissue demineralization.[54-58] Kleinberg[59] has reviewed many of these studies. Although these were aqueous systems, the suspended bacteria were distributed in a nonhomogeneous way, often associated with clumps of epithelial cells shed into the saliva (Figure 2). Each cell in this system represents a densely coated "film fermenter", and the metabolic activity in these systems is mainly associated with these clumps, since the activity associated with the dispersed bacteria in the fluid phase is negligible.[60] The salivary systems share some of the attributes of dental plaque: a broadly similar distribution of bacterial species, the major fermentation product (lactate), a fall and then a rise in pH following exposure to a limited source of carbohydrate. These systems also differ significantly from dental plaque in the spatial distributon of the bacteria, in inoculum density, and in the chemical composition of the aqueous environment of the bacteria. These differences may not be large enough to preclude using the salivary systems to predict the general metabolic pathways and processes in films of dental plaque. However, they do not allow conclusions to be drawn about the magnitude and direction of the exchange processes taking place at the interfaces between saliva and dental plaque or between dental plaque and the

1.

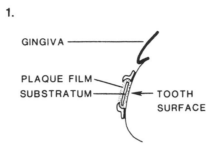

2.

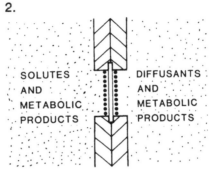

FIGURE 3.   Growth of a film of dental plaque in the mouth and its use for the study of fermentation in vitro, summarized from several studies.

hard or soft dental tissues. The spatial organization of dental plaque slows down the rates of diffusion of solutes across these boundaries, and gradients of solutes are likely to develop in such unstirred environments. The film of dental plaque also imposes a directionality on the exchange processes at these interfaces, which is lacking in the salivary systems.

A useful alternative approach to this problem is to grow a film of dental plaque in the mouth on a membrane and to remove the intact composite film with as little disturbance as possible to allow investigations of the properties of the film in vitro (Figure 3). This system was used[62] to differentiate between the rates of acid production at the saliva and enamel interfaces of intact plaque films and to monitor the kinetics of the production of volatile and nonvolatile organic acids as a function of carbon source concentration.[63] There was found to be a great deal of variability in such films, even when collected from the same site in the same subject; the film cohesion was such that the loss of parts of the film during the in vitro work compromised a quantitative study of the rates of diffusion of solutes or the kinetics of the process involved.

*Models for the Formation of Films of Oral Bacteria on the Soft and Hard Tissues in the Mouth*

Pioneering studies by Gibbons and coworkers[64,65] established that the soft and hard tissues in the mouth were coated with bacterial films whose species composition depended on the nature of the surfaces.[17] Films forming on hard dental tissues (dental plaque) were shown to be dense layers of bacteria whose species content differed considerably from that found in saliva (but see Denepitiya and Kleinberg[61]) and which showed a relatively reproducible ecological succession.[17,66] The importance of a selective nonreversible adsorption process was established as the first essential requirement for film formation and a profusion of simple in vitro systems and more complex models, sometimes requiring prior exposure to intraoral conditions, was developed for the study of factors which may affect the initial adsorption process.[67-76] These studies have contributed significantly to an understanding of the initial colonization of solid surfaces in a wide variety of ecological niches (reviewed by Wimpenny, Lovitt, and Coombs).[77] A number of recent scientific meetings have considered this topic in detail but have shown that there is a gap between physico-chemical treatments of the problem and experimental models. Physico-chemical treatments address themselves to adsorption processes according to the colloidal theory of particle interactions, propounded by Deryagin, Landau, Verwey, and Overbeek (DLVO theory) or its variants. The theory considers the interactions when particles with the configuration of idealized smooth spheres approach a flat substrate of homogeneous composition.[78,79] In real ecological problems, however, the surface limits of the bacteria are often blurred and their symmetry distorted by extracellular matrix components, by appendages, and by coatings of material adsorbed to their surfaces from the solvent phase. These complications make it hard to apply DLVO theory to such problems. The surfaces to which the bacteria adsorb are also hard to define physicochemically (e.g., their surface energy cannot be measured directly) and cannot easily be assessed with respect to physical or chemical homogeneity, roughness, and other factors which are relevant theoretically. The constants used in physico-chemical calculations are also somewhat arbitrary and of doubtful validity when used in mixed, chemically nonhomogeneous media containing ions and macromolecules which may compete with, or augment, the bacterial adsorption process.

*Models for Studying Microbial Interaction in Dental Plaque*

The bacteria initially colonizing the teeth establish the conditions which allow the growth and substantivity of species which in some cases could not otherwise establish in this habitat. This interaction may be chemical (e.g., coaggregation between *A. viscosus* and *S. sanguis* is blocked by D-galactose and lactose; hydrogen peroxide produced by *S. sanguis* under aerobic conditions inhibits a wide range of oral bacteria), metabolic (e.g., *Veillonella* sp., which depend on lactate producers), or physical (e.g., competition for available sites; threshold concentration of species in saliva affect their chances of colonizing tooth surfaces). The gnotobiotic rat has been a useful experimental model for investigating ecological factors which influence the formation of films of dental plaque.[80] Film fermenters have been used to examine the effects of bacterial antagonism, or of commensal dependence, on the success of mixed species in establishing growing films on surfaces of teeth.[81,82] These studies illustrate the complexity of factors influencing the growth of bacterial communities on tooth, tongue, or cheek surfaces. In reducing the variables which exist in real ecosystems like the mouth, such studies isolate parameters to investigate whose significance in the end may be hard to assess, especially if complex factors excluded from these models could have had an interactive effect on the system. This, of course, is one of the key objections to modeling.

*Models for Examining the Physical, Chemical, and Biological Factors which Affect the Diffusional Properties of Thin Bacterial Films*

Few dental studies have considered the bacterial thin film from the point of view of its diffusional properties or of the factors which affect the growth, substantivity, and chemical profile of the film as a whole ecosystem. The factors affecting growth, substantivity, and the effect of diffusion on the chemical profile of the film have been discussed for microbial fermenters in stirred conditions.[83,84] However, the current work using oral isolates grown in a thin-film fermenter is at a developmental stage with little more than preliminary data to indicate its relevance or application to analytical studies or to the modeling of dental plaque.[85] There is recent interest in films forming on the walls of vessels used for growing bacterial cultures and on solid surfaces inserted into continuous cultures of oral bacteria, which can be grown as stable mixed bacterial populations.[86] The films formed may demonstrate an ecological succession somewhat analogous to that described for dental plaque in vivo.

### Current Objectives for the Use of Dental Film Fermenters

In the light of the information above, a partial list of problems amenable to investigation using laboratory models is given.

1. The importance of substratum surface on initial pellicle formation and on bacterial adsorption
2. The attachment to and early growth of pioneer species on the pellicle
3. The temporal succession of species from earliest colonizers to the climax population
4. The spatial order of species across the dental plaque layer
5. The aerobic and anaerobic metabolism of organisms in plaque
6. The instantaneous concentrations and fluxes of solutes at different points in the plaque under steady-state or transitional conditions
7. Interactions between species in plaque
8. Kinetic parameters of cells growing in plaque
9. The effects of nutrient composition on plaque film growth
10. The effects of gas composition on plaque formation
11. Changes in substratum composition which may affect the caries process
12. Causes and effects of the transient changes on film behavior
13. The importance of eukaryotic cells in plaque formation and function

## WHAT FEATURES ARE ESSENTIAL OR DESIRABLE IN FILM FERMENTERS THAT MODEL THE MOUTH?

It is not immediately obvious which features need to be included in a film fermenter in order to model aspects of the mouth environment. Most workers have evolved the simplest and most practicable experimental models, without discussing the pros and cons of the conditions they have chosen. The following discussion outlines the problems to which suitable answers are needed when considering the design of particular models.

### Nutrient Medium

There is little experimental evidence concerning the influence of substances other than sugars or mucin on the formation and maintenance of dental bacterial films. An important question is whether the medium should give maximal growth rates for the microorganisms under investigation, or whether it should simulate one or other of the three fluid phases in the mouth. These phases are:

1. Saliva, which is itself a mixture, in different and variable proportions, of fluid from the parotid, submandibular, sublingual, and the minor salivary glands, each of which differs significantly in composition from the others. It is probable that unstimulated saliva represents the best starting point in designing a synthetic saliva and such a formulation was described by Shellis.[87] Proteins, immunoglobulins, antibodies, and other salivary constituents are not easily included in artificial salivary substitutes. Among these constituents are some poorly characterized components which aggregate specific bacterial species and which may be expected to affect their filming properties.[88,89] Saliva is a poor carbon source and only sporadically contains enough carbohydrate to allow substantial bacterial glycolysis. These episodes of carbohydrate excess, caused by eating or drinking, can be mimicked in model systems.[90] Cycling between carbon excess and carbon limitation needs further research to show whether the inclusion of such cycling processes is essential or affects the results obtained from models. Evidence discussed in Chapter 2 (this volume) suggests that cycling can play an important part. The composition of bulk saliva may be different from that present

on the cheeks where the mucous secretions of the minor salivary glands exert a local effect; the tongue reservoir of bacteria is also probably influenced by the secretions of the minor glands on the hard and soft palates against which the tongue normally presses in the "physiological rest position".

2.  Plaque fluid, which differs in composition from saliva, is the millieu of the bacteria in dental plaque. It contains higher than salivary concentrations of potassium, sodium, calcium, magnesium, inorganic and organic phosphates, aminoacids, albumin, lysozyme, and immunoglobulin G, but comparable concentrations of free fluoride, hydrogen ions, and of the secretory immunoglobulin A.[91] There are several uncharacterized components in plaque fluid including proteins coaggregated with IgA, lactoferrin, or lysozyme. The biological effects of these on bacterial aggregation and adsorption to surfaces have been characterized[92,93] as modifying the filming properties of oral bacteria in model fermenters. Although the inclusion of these components in media for models can be difficult, not least in obtaining purified preparations, their effects can be significant. Thus, the finding that unusually high potassium concentrations boosts glycolytic activity has been one indicator that small differences in composition of the medium may generate different results in some model systems.[94] In addition, effects on extracellular glucan and surface polymer formation,[95,96] on interbacterial antagonism in mixed cultures,[97] and on physical and adsorptive properties of bacteria have all been documented.[45,46,98,99]

3.  Crevicular fluid, which is a transudate from extracellular tissue fluid and varies considerably in composition, depending on the health of the gingival (gum) tissues. This fluid, present between the tooth and gingiva, may contain blood plasma constituents at higher concentrations than in the plasma or tissue fluid from which it seems to originate.[100] The protein composition of crevicular fluid appears to make a major contribution to that of plaque fluid since the major protein in both fluids is serum albumin, with a greater concentration in crevicular than in plaque fluid.[91]

The complexity of these possible reference solutions for basing decisions on the choice of nutrient medium for dental models, and the problems they pose, has led to the use of commercially available bacterial culture media. Although complex and variable in detailed composition, such media, or the well-characterized synthetic media, offer the best prospects for reproducibility in replicate experiments (Table 1).

## Composition of the Gas Phase

The gas phase in the mouth has not been well characterized. Some measurements of oxygen tension in intraoral air shows it to vary between 3 to 120 mmHg (0.4 to 16 kPa) depending on location and, most likely, also on the method used.[101,102] The salivary oxygen tension as it enters the mouth is 20 to 140 mmHg (2.6 to 18.4 kPa),[103] the latter value being above that of the gland tissues in which saliva was formed, suggesting that air contamination may be a serious problem in such studies. Apparently, a significant oxygen uptake from saliva by salivary bacteria occurs in the short time interval between the secretion of saliva into the mouth and its removal by swallowing. The oxygen tension in dental plaque, measured with a monopolar probe-type oxygen electrode of 1 mm tip diameter, was 23 to 61 mmHg (about 3 to 8 kPa). The oxygen tension depended on the intraoral location of the plaque, and fell to 0 in some samples after 5 days of abstention from oral hygiene. During this time plaque thickness should increase and the ecological succession leading to anaerobic regions being formed should be in progress, since the highest measured values for the oxygen tension in dental plaque samples were considerably lower than in air.[103] Similar ideas were proposed by Stralfors 20 years earlier, basing his conclusions on *a study of columns of gel-entrapped oral bacteria in vitro.*[104] He calculated that at about 300 µm below the gel surface, anaerobic conditions could prevail if the surface of the film was in contact with air (160 mmHg or 21 kPa).

There appear to be no reliable estimates of the carbon dioxide tension in the mouth, and, as with oxygen, gas composition is likely to vary, depending on the duration and nature of mouth movements, which regulate exchange with ambient air. There is therefore no consensus as to atmospheric conditions that best represent the mouth environment; air, 95% air with 5% carbon dioxide, 95% nitrogen with 5% carbon dioxide, have all been used, with air as the most popular choice (Table 2).

## Choice of Substratum

A wide range of substrata have been used for growing dental films in vitro (Table 3). Studies using whole teeth may be thought to be the most appropriate models. However, aspects of the adsorption mechanism and of the surface chemistry relevant to the process of film formation may be more conveniently studied by using more homogeneous and better-characterized materials. To a large extent, the question of which material to use depends on whether it influences the adsorption of macromolecules and bacteria in a crucial way. There appears to be no consensus about this. Although there is evidence that the surface chemistry affects the types of macromolecules which may initially adsorb to it,[105,106] there is little difficulty in demonstrating the adsorption of oral bacterial isolates to glass,[67] mylar strips,[68] epoxy resin,[69] hydroxyapatite beads[70] and even (poly)tetra-fluoroethylene (PTFE).[107] It has been stated that the adsorbed macromolecular layers on a surface exert an effect over more than 10 nm, which is approximately the distance associated with the secondary minimum necessary for most attractive interactions between a surface and a particle in its proximity.[108] Surface roughness, however may affect this value. Rothen[109] has concluded from an extensive series of studies that the successive deposition of about 50 molecular layers of adsorbed macromolecules no longer allows the characteristics of the original substratum to affect surface-particle interactions. Surface energy may be an important determinant of filming characteristics,[110] and critical surface tension, obtained by the extrapolation of plots relating the surface tension of a series of liquids and their contact angles, may dictate whether a surface is likely to be bioadhesive or not.[106,111] At best, these considerations are qualitative and represent only a part of the relevant information. For example, not only the teeth, but also adjacent soft tissues and the oral bacteria are covered by a salivary pellicle. A method has recently been described for measuring the contact angles of liquids on a bed of bacteria.[112] If one knows the contact angle, the apparent surface free energy of the bacterial cell surfaces can be derived, but it is unclear how this is affected by a coating of adsorbed salivary components. If such coatings impart a surface energy to these surfaces similar to that on enamel pellicles, it would be hard to explain the very different bacterial species found colonizing enamel and the soft tissues on the basis of this value.

## Type of Inoculum

Both pure and mixed bacterial cultures have been used in model film systems. It is clear, however, that the origin, age, and preparation of the inoculum are variables whose effects on the formation of films in vitro need further research. Repeated subculture, growth conditions, and media composition can all influence the filming properties of oral bacteria.[45,46,86,98] While almost all the bacteria examined in dental model systems originated from the mouth (Table 4), their "freshness" varied from a sample of dental plaque removed immediately before use to bacteria obtained from type culture collections and transferred through numerous subcultures before use. Inoculum density may affect the initial adsorption process and several reports indicate that the process is concentration dependent.[29,33] Duration of exposure of the substratum to the inoculum has been arbitrary, varying from a single-batch exposure, to continuous flow of a pure culture of bacteria, or to saliva containing an indigenous flora (Table 4).

## Table 1
## NUTRIENT MEDIA USED IN DENTAL MODELS

| Type | Carbon Addition | Control of Parameters | | Medium flow rate | pH | Ref. |
|---|---|---|---|---|---|---|
| | | Cycling of media | | | | |
| Human saliva | Bread or other | — | | + | 6 | 48 |
| Bacto-tryptone | 0.5% Glucose | — | | + <br> 5—7 drops/min | 6.4 — 6.8 | 49 |
| Tryptone, yeast extract (thioglycollate) | 0.5% Glucose | — | | + | 6.4 | 118 |
| Tryptone, yeast extract calcium phosphate | 0.25% Glucose | — | | + <br> 0.1—0.3 ml/min | 6.2 | 119 |
| Tryptone, yeast extract | 0.25% Glucose | — | | + | 6.3—6.5 | 120 |
| Krebs-Henseleit | Glucose | — | | 3—6 ml/day <br> + | 7.4 | 121 |
| Nutrient broth, yeast extract | 0.5% sucrose <br> or 0.5% glucose <br> or 0.5% fructose | — | | + | 7.0 | 50, 122 |
| Tryptone soy, 0.1 M phosphate model saliva | 0.1% Sucrose | + <br> + <br> + | | + <br> + <br> + | * <br> 7.0 <br> * | 123, 134 |
| Brain-heart, thioglycollate, model saliva | 2 % Sucrose <br> or 2 % glucose <br> or 2 % fructose <br> or 2 % glucose syrup <br> or 2 % fructose/glucose | + | | + <br> 1 ml/hr | * | 124 |
| Human saliva (4 × concentrated) | 1% Sucrose | — | | — | 7.0 | 125 |
| Beef and yeast extracts, peptone, model saliva | | — | | + <br> 0.1 ml/min (medium) <br> 0.05 ml/min (model saliva) | 8 | 126—129 |
| Brain-heart, model saliva | 2% Sucrose | + | | + <br> 1 ml/hr (medium) <br> 0.5 ml/hr (model saliva) | * | 90 |

|  |  |  |  |  |  |
|---|---|---|---|---|---|
| Model saliva | 0.5% Glucose or 0.5% sucrose | + | + | 6.8 | 81 |
| Model saliva, bactocasitone and yeast extract | 5% Sucrose or 0.05% glucose | + | + 0.5 mℓ/hr | 6.8 7.0 | 130 |
| Human saliva | 1.5—5% Glucose or equivalent as solid foods | − | + 0.5 mℓ/hr | * | 52, 131 |
| Trypticase, soy broth | 0.2% Glucose or 0.2% sucrose | − | + 7.5—11 mℓ/hr | * | 132 |
| Casamino, yeast extract | 0.1% Sucrose | − | + 0.1 mℓ/min | 7.0 | 85, 135 |
| Model saliva | 0.1% Sucrose | + | + 5 mℓ/hr | 6.4 | 136—138 |
| Todd-Hewitt, phosphate, model saliva | − | + | + | 7.0 | 139 |
|  | 0.1 *M* Sucrose | + | + | 7.0 |  |
|  | − | + |  | 7.0 |  |

*Note:*  +, yes;  −, no;  *, not stated.

## Table 2
## GAS PHASES USED IN DENTAL MODELS

| Gas composition | Flow (F) or diffusion (D) | Ref. |
|---|---|---|
| Air | D | 48—50, 52, 54, 118—123, 125, 131, 132, 136—138 |
| 95% air/5% $CO_2$ | F | 81, 90, 124, 130 |
| 85% $N_2$/10% $O_2$/5%$CO_2$ | F | 85 |
| 95% $N_2$/5% $CO_2$ | F | 135, 139 |
| Air (port available for gas input) | D | 126—129 |

## Table 3
## SUBSTRATUM AND FILM PARAMETERS USED IN DENTAL MODELS

| Surface used | Film growth parameters | | | Ref. |
|---|---|---|---|---|
| | Duration | °C | Sampling | |
| Tooth sections | Not given | 37 | — | 48 |
| 2 Teeth in contact | 8 weeks | Room | — | 49, 120 |
| 1 or 2 teeth | <8 weeks | 35 | Plate counts | 118, 119 |
| 1 Tooth | 6—8 weeks | 37 | Smears and plate counts | 121 |
| 2 Teeth in contact | 7 days | 37 | — | 50, 122 |
| 2 Teeth in contact | "several months" | 37 | — | 54 |
| Several teeth | 20 days | 37 | — | 123 |
| 1 Tooth | 10 days | 37 | — | 124 |
| Glass slide | 1 hr | Room | Gram stain, O.D., glycolysis | 125 |
| Acid-etched tooth | 48 hr | 35 | pH and Eh, plate counts on effluent and film | 126—129 |
| Mylar film, enamel slices or teeth, filter paper | 2—5 weeks | 37 | — | 90 |
| 2 teeth in contact | 5 days | 37 | Total CFU | 81, 130 |
| 5 mm enamel squares, glass, "plastic" | 4 days | 37 | Visual score, smears, plate counts, pH response to 20% sucrose | 52, 131 |
| Glass-coated hydroxyapatite (HAP) | 51 hr | 37 | HAP dissolved for count of adsorbed cells | 132 |
| (Poly)tetrafluoroethylene, enamel | 4—10 days | 37 | Viable, total counts, protein, film thickness, effluent pH, and CFU | 85, 135 |
| Bovine teeth, platinum foil | 5 days | 37 | pH and F in film | 136—138 |
| Enamel slices | 28 days | 37 | — | 139 |

*Note:* —, no.

There remain two other factors which need further investigation. First, it is generally agreed that film formation depends on a transition from a reversible adsorption to a more permanent residence as a result of extracellular formation of matrix material such as polysaccharides. However, the time-scale for this transition in model systems is not known. Second, most models avoid a disturbance of the surface or generation of any detaching forces during the inoculation period, although it may be that this could distinguish between surface-related film growth and that arising from a loose entrapment of cells close to the surface.

## Table 4
## INOCULUM SOURCE AND PROCEDURES USED IN DENTAL MODELS

| Source of bacteria | Procedure | Duration of initial contact | Re-infection | Ref. |
|---|---|---|---|---|
| Mixed saliva | Gauze-filtered saliva, replaced every 24 hr | 24 hr | + | 48 |
| Dental plaque + mixed saliva | Extracted tooth washed then exposed to saliva | 30 min | + | 49, 118 |
| ATCC and salivary isolates | Tooth exposed to broth culture | 24 hr | − | 119 |
| Mixed saliva | Cloth impregnated by inclusion in broth inoculated with saliva | * | − | 120 |
| Plaque left on extracted tooth | Tooth washed in sterile saline before insertion into model | * | − | 121 |
| Cultures of human oral isolates | Two successive batch-cultures before using medium for inoculum | * | − | 50, 54, 122 |
| Cultures of rat/hamster isolates | Teeth into 18 hr batch-culture | * | + Daily till visible film | 123 |
| *Streptococcus mutans* (Ingbritt) | Batch- or chemostat-grown, flowing over sample | * | + | 124 |
| Mixed saliva | Glass rotated into air and saliva (2 min cycle) | 1 hr | + | 125 |
| ATCC, NCTC, Manchester University Collection | Pure cultures or a mixture of pure cultures fed onto teeth in model | 4 hr | − | 126—129 |
| *S. sanguis* mixed saliva | Culture medium fed onto specimens | * | − | 90 |
| *S. sanguis* and *S. mitior* LPA-1 | Batch-grown, add 1.5 m$\ell$ ($10^8$ cells/m$\ell$) to teeth | * | − | 81, 130 |
| Dental plaque and mixed saliva | Plaque dispersed into homologous saliva and added per 5 mm enamel block | 1 hr | + Repeated 3 times | 52, 131 |
| Mixed saliva | 1 g HAP beads/125 m$\ell$ saliva in 90% $N_2$/10% $CO_2$ | Overnight | − | 132 |
| Dental plaque | Plaque homogenized in Tween® 80, added to teeth in tryptone soy broth | 8 hr | + Repeated after 24 hr | 134 |
| NCTC dental plaque | 40 hr culture in tryptone soy broth or solid biomass | 6 hr | + Dead volume over film | 85, 135 |
| *S. mutans* FA 1 | "Pellicle" formed on enamel, then teeth in broth culture adjusted to 5 × $10^7$ CFU/m$\ell$ | 3 days | + Dead volume over film | 136—138 |
| *S. mutans* C67-1 | Todd-Hewitt broth culture flowing over specimen in model | 24 hr | − | 139 |

*Note:* +, yes; −, no; *, not stated; ATCC, American Type Culture Collection; NCTC, National Collection of Type Cultures.

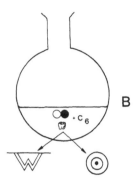

FIGURE 4.   Schematic view of the model of Dietz. See Figures 19 and 20 for keys to symbols.

### Cycling or Continuous Culture Conditions

The intraoral environment is characterized by fluctuations in the composition of the bulk phase saliva and in the availability of carbon, nitrogen, mineral, and other nutrients. There are corresponding fluctuations in pH, associated with an acidification during periods when sufficient metabolizable carbohydrates are present. It is not known how these fluctuations affect either the development or physiological growth or function of dental films. Frequent ingestion of sugary foods accelerates the accumulation of dental plaque[113] and alters the proportion of *S. mutans* in it.[114] The aciduric characteristics of *S. mutans* and the adsorptive advantage due to its ability to produce sticky extracellular polysaccharide have been proposed as explanations for these findings. These effects have been seen in systems using pure or limited mixed cultures, where ecological competition is not so important. Work is needed to compare effects of cycling vs. continuous culture on film formation or properties. It has been suggested that the mouth is essentially a continuous-culture vessel and that use of continuous culture techniques improves the homogeneity of the system because the mean residence time of the bacteria which are present in the culture remains constant.[115,116] In contrast, the generation of fluctuations in nutrient supply was incorporated into the design of the model used by Dibdin et al.[90] However, most models have used the technically simpler and cheaper expedient of a continuous nutrient flow (see also earlier section about the study of dental caries).

### Eukaryote Cells

A cursory examination of saliva is enough to indicate the importance of shed epithelial cells both for their contribution to the bulk of the solid matter in saliva (i.e., saliva sediment, separated by centrifugation) and also their contribution to the bacterial population. Most of the bacterial metabolic activity is associated with shed epithelial cells and these cells can be seen under a microscope to harbor dense clumps of bacteria. Few if any epithelial cells can be seen within the dental plaque film. Shed epithelial cells and migrating blood polymorphonuclear white cells from the crevice between the gum margin and the tooth structure, which rapidly disrupt on entry into the mouth or in saliva,[117] could both contribute to plaque film formation. What effects these host cells have on the processes of film formation is purely conjectural since the problem has not as yet been investigated in models.

## AVAILABLE MODEL SYSTEMS

Dietz[48] devised a model (Figure 4) according to the following criteria: (1) the tooth in the model should be immersed in saliva, (2) the saliva should be as fresh as possible and should be periodically agitated, (3) acidity arising from bacterial metabolism should not be

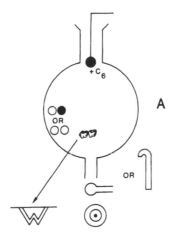

FIGURE 5. Schematic view of model of Pigman and co-workers. See Figures 19 and 20 for keys to symbols.

so high that it readily decalcifies the tooth. The model contained a tooth section, mounted in a glass chamber which could be viewed under a microscope. The tooth section was immersed in 50 mℓ gauze-filtered saliva donated by two caries-prone subjects. Nutrient was supplied by placing white bread or various carbohydrates in pinhead sized samples onto the tooth and replacing the nutrient with a fresh sample every 8 hr. The saliva was agitated for 15 to 20 sec every hour. Every 24 hr, the entire apparatus was washed with soap and water, rinsed in distilled water, and the whole procedure repeated. Smears taken from the tooth surfaces and plated out invariably showed streptococci and lactobacilli growing as "plaque". Yeasts, staphylococci, micrococci, Gram-positive bacilli, and filamentous microorganisms were also found. Unfortunately, interest focused on examining the tooth sections for caries and thus no useful information was available on the film itself, other than it was present.

Pigman and coworkers[49] described a model for producing caries in teeth and monitoring its progress in situ by using X-rays. Decalcification in their system was generally found to be over the entire surface rather than at discrete sites, as seen naturally, so that X-rays were not useful in this context. The design of the fermenter (Figure 5) allowed either a continuous flow over the teeth under investigation, or else their phased immersion in the nutrient medium using a syphon to drain the chamber periodically. Plaque orginally present on the teeth was disturbed as little as possible during extraction and the subsequent removal of any attached gum tissue. This was carried out in isotonic saline and the teeth were then mounted singly or in pairs in acrylic boxes using dental modeling compound. The medium used was 0.5% bactotryptone, 0.5% glucose, and 0.5% sodium chloride, pH 6.4 to 6.8. The medium was allowed to drop at a rate of "several drops per minute" onto the biting (occlusal) surfaces of the teeth. The inoculum was a mixture of the resident flora on the tooth at the time of extraction, and of salivary bacteria which were supplied in 0.5 mℓ of pooled saliva freshly obtained from 3 to 5 subjects. Every 3 to 4 days, the teeth were removed from the apparatus which was cleaned and reassembled. The teeth were also washed but not mechanically cleansed in a flow of water and returned to the chamber. When the pH in the effluent was compared with that on the tooth surfaces, there was a difference in their range: pH 4.8 to 5.4 in the effluent vs. 4.8 to 6.2 on the teeth. Within 24 hr of starting each experiment, a white, somewhat sticky material deposited on the tooth surfaces and on the surfaces of the container and outlet tube. The plaque-like deposits on the teeth could not be washed away in a stream of water. In a later report,[118] the protocol included a daily brushing of the teeth and reinoculation with saliva. The effect of different glucose concentrations in the medium

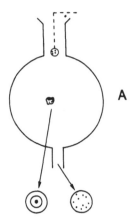

FIGURE 6.     Schematic view of model of Wagg and co-workers. See
Figures 19 and 20 for keys to symbols.

on decalcification of dentine or enamel and on the bacterial population on the teeth was
investigated.[118] The latter involved prior brushing of the teeth and using the cleaned tooth
crown, dipped into nutrient medium, as a source of inoculum for enumerating the relative
numbers of streptococci and lactobacilli remaining on the tooth surface.

Details of specimen handling and sampling in these experiments cast doubt on the origin
of the microorganisms that were examined. Superinfection of tooth sites already occupied
by salivary bacteria with air-borne, water-borne, or even toothbrush-borne bacteria could
have occurred during dismantling and toothbrushing procedures. However, the apparatus
could be used without such manipulations and was capable of maintaining pure cultures of
lactobacilli, streptococci, clostridia, and neisseria for up to 25 weeks with only an occasional
contaminant reported.[119] A modified version of the apparatus was used to test the effects of
fluorides, sarcosinates, and toothpastes on tooth decalcification.[120]

Wagg and coworkers[121] considered that medium flow rates in previous models were too
rapid compared with in vivo rates and designed a wicking system to reduce the rate of
nutrient flow to 3 to 6 m$\ell$ per day (Figure 6). A freshly extracted tooth was used in the
model, the inoculum being the natural flora adhering to the extracted tooth after washing it
in sterile saline. The tooth was irrigated with sterile Krebs-Henseleit solution, pH 7.4.
Although a syphon arrangement was included in the effluent line, the length of time that
the tooth was covered by the medium was short compared with the period needed for enough
fluid to collect in the syphon to cover the tooth. The tooth surface was monitored by taking
loopfuls of deposits and plating out on various media. The unsupplemented medium caused
most bacterial species to disappear from the tooth surface, but lactobacilli of the casei-
plantarum group survived the 6 to 8 week experimental period.

Rowles and coworkers[54] described an apparatus designed to produce dental caries in vitro
(Figure 7). Sterilized teeth were mounted on polyethylene tubing with epoxy resin and held
in the appartus in a configuration which formed points of contact between the teeth, as found
naturally. They were inoculated with mixed saliva organisms and incubated for "several
months". Films growing on the teeth caused demineralization, as judged by light microscopy,
7 weeks after inoculum with infected carious dentine. No details of media, method of
inoculation, and so on were given and it seems that there was no interest in the bacterial
films formed, except for the comment that both aerobic and anaerobic bacteria survived in
the system. Sidaway[50,122] reported that the medium used by Pigman did not support the
growth of more fastidious organisms and therefore used an infusion broth or Lab-Lemco to
supplement a medium containing peptone, yeast extract, and glucose. Using the model

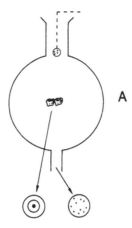

FIGURE 7.   Schematic view of model of Rowles and co-workers. See Figures 19 and 20 for keys to symbols.

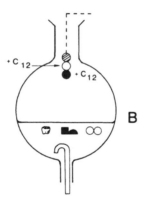

FIGURE 8.   Schematic view of model of Jordan and Keyes. See Figures 19 and 20 for keys to symbols.

described by Rowles, et al.,[54] she inoculated eight teeth with one or a group of organisms previously grown in broth culture. Medium flow was maintained for 7 days. At 3 days, half the samples were removed and, after noting the amount of visible deposit, cultures were taken to check for the presence of inoculated species and for contamination. The films were more easily removed from the enamel surfaces than would be the case with a natural plaque. The filming ability of a range of oral isolates in the presence of sucrose, fructose, or glucose supplement was reported, and demineralization of the tooth surfaces could be demonstrated in 7 days with some species.

Jordan and Keyes[123] defined their requirements for a model system as follows: (1) exposure to sufficient essential nutrients to allow limited growth; (2) intermittent short-term exposure to high levels of metabolizable carbohydrate; (3) relatively long rest periods in a saliva-like medium; (4) frequent removal of bacteria not adsorbed to or growing on the surface as well as the removal of products of fermentation, simulating clearance in saliva by swallowing. Several models were tested, including batch cultures in test tubes with a wire or glass rod immersed into the medium, a system used by several other investigators.[71-74] They also described a closed chamber housing teeth onto which media were metered in a repeating cycle controlled by solenoid valves in both entry and drainage lines (Figure 8). Up to six solutions could be controlled and the chamber could be filled to a preset level, controlled

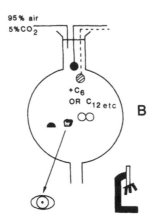

FIGURE 9.   Schematic view of model of Wilson. See Figures 19 and 20
for keys to symbols.

by an internal sensor. The teeth could be bathed for periods from 6 min to 6 hr before the
system was drained. Such a model represents a phasic batch culture whose constituents from
one solution would be washed out by the solution cycled next. The apparatus was kept at
37°C using a warming mantle and the teeth, mounted on stainless-steel blocks, were inoculated with 10 mℓ of a batch culture of a streptococcus on successive days until visible
plaque had formed on them. The earliest plaque could be seen after 3 days of repeated
inoculation. Sucrose was essential for plaque formation and other sugars which were tested
could not substitute for it. After 20 days, the teeth were completely covered with a heavy
deposit of plaque which, in contrast to the natural film, was mostly removed by a stream
of water, leaving some firmly attached bacteria close to the surface of the teeth.

Wilson[124] (Figure 9) described an experimental plaque and caries model which used a
mixture of nutrient medium and an artificial saliva whose make-up was not described in
detail but which contained hog gastric mucin. A constant inoculum of batch- or chemostat-
grown *S. mutans* was fed to the substratum housed in a chamber through which 95% air/
5% carbon dioxide was passed. The apparatus produced a "thick" plaque of *S. mutans* in
10 days at 37°C. Electron microscopy and chemical analyses for carbohydrate polymers
were reported.

Dolan and coworkers[125] grew films on agar-coated glass slides which were rotated regularly
between saliva and air (Figure 10). The deposit was examined by measuring its optical
density and also by measuring the change in pH following immersion of the slide into glucose
or sucrose solution at 37°C under aerobic or anaerobic conditions. Optimum conditions for
their system necessitated coating the slides with brain-heart infusion agar and rotating them
into fourfold concentrated salivary sediment. A known inhibitor of plaque formation, chlorhexidine, was shown to inhibit the formation of deposits in this system. Low and high power
microscopy of the slides after Gram staining showed a mixed population which reflected
that found in the salivary sediment from which it was formed. There seemed therefore to
be no selective process apparent in the adsorption step. The glycolytic activity was considerably lower in these films than in dental plaque. The authors attributed this to the difference
in the bacterial population density in the two films.

Russell and Coulter[126] pointed out that little kinetic information had emerged about bacterial
growth and activity from the previous attempts at modeling the mouth. They devised a
system based on that of Naylor et al.[124] but with an additional port to allow deployment of
an electrode assembly opposite the tooth (Figure 11). They removed preexisting pellicle
from the teeth by immersing the teeth in 1 *M* HCl for 5 min and had therefore pre-etched

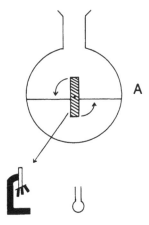

FIGURE 10. Schematic view of model of Dolan and co-workers. See Figures 19 and 20 for keys to symbols.

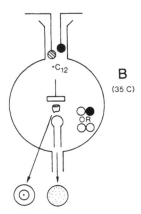

FIGURE 11. Schematic view of model of Russell and co-workers. See Figures 19 and 20 for keys to symbols.

the teeth to produce a roughened surface before their insertion into the fermenter. The teeth were inoculated with mixed human saliva, which was allowed to fall onto the teeth for 4 hr at a rate of 0.05 m$\ell$/min. A nutrient broth based on Lab Lemco beef extract, yeast, peptone, and the artificial saliva described by Muhler (which utilized hog gastric mucin), was then delivered at twice the first rate. The apparatus was housed in a warm room at 35°C. Continuous monitoring of pH was via a glass electrode inserted through a drilled hole along the long axis of the tooth, so that the electrode glass bulb substituted for the enamel along the surface of the tooth. Remote junction reference and platinum Eh electrodes were inserted from the port opposite to the tooth. The medium was collected at the bottom of the fermenter housing and went to waste via a covered outlet which avoided back-contamination. Smears of the effluent from plaque grown in medium without added carbohydrate showed masses of neisseria and streptococci as well as veillonellae, diphtheroids, and fusobacteria after 24 hr incubation. When 1% w/v sucrose was added to the medium, the effluent became predominantly streptococcal after 24 hr. Cycling between sucrose-supplemented and unsupplemented media led to pH perturbations which were associated inversely with changes in Eh. The deposit on the teeth lowered the redox potential progressively as it thickened, but episodes of low pH and correspondingly high Eh were believed to explain both the loss of

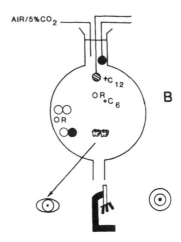

FIGURE 12.    Schematic view of model of Dibdin and co-workers. See
Figures 19 and 20 for keys to symbols.

anaerobes and the transition to a mainly streptococcal film. Later studies examined the
development of pure and mixed bacterial cultures and some factors affecting their formation.[127,128] The relative distribution of different species in the effluent after inoculation with
saliva was broadly comparable to that on the tooth surfaces.[127] The tenacity of the films
formed in the unsupplemented medium was poor compared with that when sucrose was used
as a supplement, the latter films being predominantly streptococcal. When pure cultures
were used[82] as inocula, some species produced little or no visible surface growth or pH and
Eh changes. Using a mixed inoculum containing up to four bacterial species, these authors
examined the nature of the association between *S. sanguis* (a lactate producer) and *Veillonella
parvula,* a commensal dependent on a source of lactate. Conditions which would allow
coexistence of these two in the film were not obtained. In pure culture, veillonella were
unable to produce a film on the teeth unless reducing agent and lactate were included in the
medium. Inclusion of *Neisseria pharyngis* and an oral diphtheroid provided such an environment and promoted the growth of the veillonella in the surface-associated film. However,
the addition of *S. sanguis* to the system, even after establishment of such a community,
resulted in washout and the elimination of the veillonella from the film unless diphtheroid
species showing antagonism to *S. sanguis* were included. Sucrose was considered essential
for the formation of tenacious films on the teeth in this model[128] and saliva was neither a
promoter nor inhibitor of surface adsorption and growth, in contrast to other reports cited
above. A later report explored the use of chlorhexidine as an inhibitor of plaque formation.[129]

The apparatus described by Dibdin and coworkers[90] was designed to grow artificial plaques,
allowing several replicates to be run simultaneously (Figure 12). A programmable sequence
of nutrient supply was possible with the aid of a multichannel peristaltic pump. The apparatus,
based on "Quickfit"® glassware, housed two teeth making a contact point or else other
surfaces as disks. Nutrient broth was supplied at 1 m$\ell$/hr and an artificial saliva[87] at 0.5
m$\ell$/hr. The medium collected at the bottom of each flask and was removed with a peristaltic
pump. Pure cultures of oral streptococci or mixed salivary bacteria were inoculated by
injection through a sample port and incubations were carried out at 37°C in a gas phase of
95% air/5% carbon dioxide, chosen to approximate the composition of expired air. Plaques
were formed within a few days on all the substrates tried, and incubation was continued for
2 to 5 weeks. The films appeared to be morphologically similar to natural dental plaque on
the basis of stained sections. Effects of nutrition on interactions between *S. mitior* and *S.
mutans* were investigated.[81] The behavior of these two organisms was quite different in agar

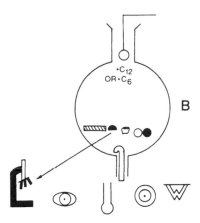

FIGURE 13.   Schematic view of model of Yaari and Bibby. See Figures
19 and 20 for keys to symbols.

from that in the model system. In agar, *S. mitior* inhibits *S. mutans* by producing hydrogen
peroxide unless glucose or sucrose is present when *S. mutans* inhibits the growth of *S. mitior*
by producing acid. No bacteriocins or other factors appear to be involved. In the model,
however, *S. mitior* was shown to inhibit *S. mutans* in the absence of glucose or sucrose.
Sucrose addition was accompanied by a further increase in the proportion of the *S. mitior*
population. Addition of glucose, however, led to mutual antagonism and fewer total numbers
were recovered than in unsupplemetned synthetic saliva. The effect of the sequence of
inoculation of these two species was examined when a continuous supply of artificial saliva
was supplemented 1 hr in every six with sucrose or glucose. The possibility that the species
which are already present on a surface may inhibit the attachment of others was put forward
to explain the lower numbers of *S. mutans* which were found in di-association with preexisting
*S. mitior* on tooth surfaces.[130] The apparatus has been developed further by Donoghue and
her coworkers to include continuous monitoring of the pH of the film on the teeth in the
fermenter and to allow samples to be taken for analysis of the fluid adhering to the tooth
assembly. Several replicate units were used simultaneously to examine the interaction between *S. rattus* (mutans) BHT and *S. mitior* LPA-1, which produces hydrogen peroxide.[142]

Yaari and Bibby[131] used a carousel slide projector to house several replicates of enamel
slices sandwiched between glass slides and exposed to the medium in the space between
the two slides. Hanging above the carousel was a reservoir of mixed human saliva which
had been pooled from two donors and frozen in dry ice at −78°C. The saliva was slowly
thawed using a small heating element inserted into it and the thawed saliva dropped onto
the rotating carousel bathing each sample with fresh drops of saliva at approximately body
temperature. The constant flow of inoculum caused a film of plaque to develop not only on
the tooth edges but on the entire apparatus. Smears stained with Gram stain were used to
prepare differential counts of the bacteria present in saliva and in the films. The latter could
be sampled repeatedly during the run. Different foods and sugars were stirred into the carousel
tray and their effects on the tooth sections were determined microscopically. In principle,
this model appears to have many advantages over some of those already mentioned, particularly in regard to its relative simplicity (Figure 13). In practice, however, film formation
was rather inconsistent, and films were difficult to collect for examination. A later report[52]
described attempts to improve the reproducibility and the amount of film developed in the
model by using an initial inoculum of dental plaque which was allowed contact with the
specimens for 1 hr before saliva flow started. Plaque was visible after 1 day, increasing in
thickness to a maximum by the fourth day. The film acidity was thought to be the reason

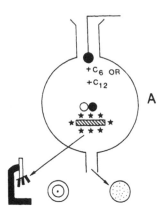

FIGURE 14.   Schematic view of model reactor of Sudo. See Figures 19 and 20 for keys to symbols.

for obtaining a mainly streptococcal biomass by the end of the experiment. However, the presence of some Gram-negative rods and filamentous organisms encouraged the authors to speculate that their films more closely resembled natural in vivo plaque than those produced in other laboratory models. A comparison of wax, glass, and enamel as substrata for film formation revealed that enamel harbored the most florid growth.

A column reactor consisting of hydroxyapatite-coated glass beads was described by Sudo.[132] The initial adsorption of mixed salivary bacteria was carried out in batch culture for 2 hr at 37°C; prereduced medium was then added to the saliva/beads mixture and incubated overnight. The beads were then washed free of nonadsorbed cells and transferred to a glass column through which trypticase soy broth was pumped with a peristaltic pump (Figure 14). Nutrient medium was used with or without supplements of 0.2% glucose or sucrose, or with hemin and menadione. Bacteria on the beads could be recovered by removing the apatite layer with lactic acid or by sonication. The effluent was also monitored and a comparison was made between the organisms in the effluent and those which persisted in the reactor. This allowed some estimate of both the proportion of cells which may have been shed from the reactor and those growing in the effluent in transit to the collecting point. Streptococci were the dominant species in the reactor but lactobacilli were also consistently found, though in smaller numbers. Unexpectedly, *S. salivarius* dominated the streptococcal count of initial flora after adsorption from batch cultures. The addition of supplements to the culture medium, as well as the method of cultivation and harvesting of the cells used for the initial adsorption, affected the result. In contrast to the findings of Russell and Coulter, the addition of sucrose to the medium increased both streptococci and veillonella adsorbed from the batch cultures. Inclusion of hemin and menadione increased the numbers of nearly all organisms. Scanning electron microscopy of beads after adsorption of bacteria from the batch cultures showed surface growth as closely packed microcolonies. The addition of lysine phosphate to the column reactor converted it from one where the surface community of *S. salivarius* was minimal to one in which this species dominated. The density of growth in the reactor was thought to contribute to the apparent difference from the batch mode in terms of the surface colonial species. A microbial succession may have occurred in the reactor system, judging by the changes in the effluent bacterial population, though this was not investigated further.

Leech and coworkers[133] examined deposition onto glass slides of five oral bacteria in pure and mixed bacterial suspensions (Figure 15). Although all species tested were deposited preferentially at the solid/liquid/air interface, their ranking in terms of the density of deposit could be altered by changing the liquid composition. Under *all experimental conditions*, however, *S. sanguis* and *Actinomyces viscosus* had a greater tendency to attach than did *S. mutans*, *S. salivarius* and *Lactobacillus casei*.

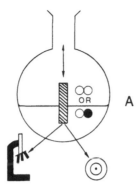

FIGURE 15. Schematic view of model of Leech and co-workers. See Figures 19 and 20 for keys to symbols.

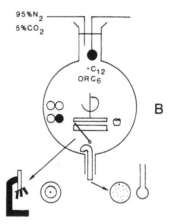

FIGURE 16. Schematic view of model of Coombe and co-workers. See Figures 19 and 20 for keys to symbols.

Singer and Buckner[134] used a modified version of the apparatus and methodology described by Jordan and Keyes[123] to grow plaque cultures in a cycle of nutrient broth, buffered sucrose solution, and artificial saliva. Interest was in the toxicity of products from such cultures towards mouse fibroblast cells and no useful microbiological information was collected.

Coombe and coworkers[85,135] reported on the use of a thickness-limited fermenter for growing pure and mixed culture films on six replicate PTFE pans 300 μm deep (Figure 16). The pans were swept by a PTFE paddle housed in a chamber maintained at 37°C. The gas phase was 85% nitrogen, 10% oxygen, 5% carbon dioxide, or 95% nitrogen with 5% carbon dioxide. Nutrient medium based on casamino acids, yeast extract, and 0.1% sucrose was spread over the pans by the rotating paddle and removed by peristaltic pumps. Four representative oral streptococci were grown as films in pure culture but the limiting depth of 300 μm was not achieved in any of the film pans, despite florid growth in the effluent from the fermenter. Transmission and scanning electron microscopy of the films, viable and total counts of films and effluent, film thickness measurements by light microscopy and pan protein content were used to monitor film formation and some of the factors which affected the process. The density of the initial coverage of the pans by adsorbed bacteria was not a critical factor for growth of thin films and the shear or cohesive forces in films of pure or mixed plaque inocula may have affected film dimensions, which were usually <100 μm

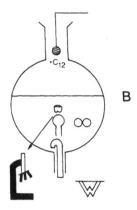

FIGURE 17.    Schematic view of model of Klimek and co-workers. See
Figures 19 and 20 for keys to symbols.

thick, except near the walls of pan recesses. Cohesive forces between elements in the film
were considered important for the formation and retention of the bacterial films.

Klimek and coworkers[136,137] developed a model containing up to 40 replicate bovine tooth
enamel blocks on which bacterial attachment and film growth was followed after precoating
the enamel for 3 days with a pellicley conditioning surface derived from an artificial saliva-
containing bacterial mucin. The enamel blocks were then transferred to a thioglycollate broth
culture of *S. mutans* FA1 which was neutralized every 24 hr. After 3 days exposure to this
culture, a thin bacterial film had formed and the blocks were transferred to a chamber into
which test solutions or artificial saliva could be pumped, at a flow rate of 5 m$\ell$/hr (Figure
17). During the 5 days following cyclical exposure to these solutions, the pH of the inner
layer of the film was continuously monitored in some samples using an inbuilt micro-pH
glass electrode. After 5 days, a "plaque" about 6 mm thick was present above white chalky
areas of tooth enamel decalcification. The fluoride and phosphate concentrations in the film
were measured, and an attempt was made to decide whether enamel fluoride treatment and
its distribution in the enamel layer was affected by the presence of the film. The thickness
of the film was greater than that expected in natural plaque.

Noorda and coworkers[139] have used a modern version of Pigman's apparatus to follow
plaque-induced caries. The pH changes in the film following exposure to sucrose were
monitored and enamel decalcification beneath the film was evaluated by quantitative mi-
croradiography (Figure 18). An interesting model, designed to test predictions from a mathematical model for salivary clearance of solutes, was reported by Lagerlof and coworkers.[140]
Unfortunately, in its present form it cannot be sterilized, and therefore is of limited value;
however, it incorporates a cyclical emptying and filling of the chamber with a saliva substitute
and has other features which can generate more quantitative information about ongoing
events than many other models and warrants further adaptations to allow its use for study
of microbial films.

## AN EVALUATION OF THE USE OF FILM FERMENTERS IN DENTAL
## RESEARCH

To date, microbial model systems for developing caries in vitro have been of little or no
analytical value. The use of models for studies of dental bacterial films has been limited
mainly to preliminary evaluations of the possibilities without a great deal of systematic
experimentation to examine the variables which can and should be predetermined for studies
of such complex processes. Models which rely on the diffusion of air for a gas supply are

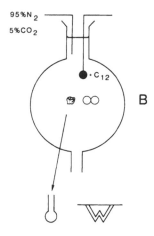

FIGURE 18. Schematic view of model of Noorda and co-workers. See Figures 19 and 20 for keys to symbols.

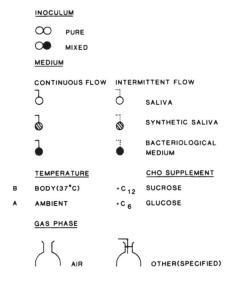

FIGURE 19. Symbols for figures.

too limited for use and, because of the absence of positive pressure within the chamber, are too easily contaminated during sampling (Table 2). Thus, only recent models or modifications of those lacking this facility are considered adequate tools. The potential to modify the medium is common to all models but those which allow a complex cycling between several media are more flexible (Table 1); the ability to vary the flow rate of the medium is an advantage and systems with flow rates of $<1$ m$\ell$/min are probably more realistic. The minimum flow rate to obtain films resembling dental plaque cannot be deduced from the resting salivary flow rates because the distribution of the saliva secreted from the major glands is uneven, collecting mainly under the tongue and in the cheek pouches under resting conditions. Similarly, it is not possible to assess the drainage or clearance rate of solutes at discrete sites in the mouth, and therefore to suggest whether models should incorporate dead space volumes (e.g., model of Coombe et al.) or allow more complete renewal of the biofilm aqueous phase composition by inflowing medium (e.g., models of Russell or Dibdin and coworkers).

FIGURE 20.    Symbols for figures.

Models which allow sampling for kinetic measurements are essential for monitoring the formation of the films (Table 3). However, the technology for such kinetic studies is at present very limited, so that apart from sampling for microbial enumeration or measurements with pH or Eh electrodes located in the model, there seems to be no satisfactory way to measure the rate of growth or distribution of surface-associated bacterial films. Any sampling which removes a part of the biofilm disturbs the film structure, and therefore unacceptably perturbs the system under study. The continual presence of an inoculum either as an inflow or in the dead-space of the model is a desirable feature. The former is potentially available in all models but the latter facility is not available in most models (see Table 4).

## SUMMARY AND CONCLUSIONS

The past 20 years have been a most productive period in furthering our understanding of the wide range of microbial surface attachment mechanisms found in nature. Many aspects of attachment and of the properties of oral bacterial films can be pursued with simple homogeneous in vitro systems, and this approach has revealed strategies open to microorganisms both for attachment to host oral tissue and for the evasion of host defense mechanisms. The behavior of intact communities of pure or mixed bacteria in a film, however, reveals other properties of the species which cannot be predicted from simpler studies. For example, ecological successions in dental plaque and the gradual transition to a calcified mass of dead bacteria known as calculus or tartar are both hard to predict from known or inferred data and difficult to study without more complex film-forming models. The evolution of such models has been limited by the questions to which the investigators sought answers. Recent interest in the attachment to and growth on solid surfaces should be a stimulus to the use of such models to ask increasingly complex ecological questions. Whether the models will be representative of the situations they attempt to reproduce is not clear at this stage, but the long list of factors which await study indicates this to be a fruitful area for more pioneering research.

# REFERENCES

1. **Mikx, F. H. M.**, Het tandplaque-onderzoek van Antoni van Leewenhoek in 1683, *Ned. Tijdschr. Tandheelk.*, 90, 421, 1983.
2. **Williams, J. L.**, A contribution to the study of pathology of enamel, *Dent. Cosmos.*, 39, 169, 1897.
3. **McHugh, W. D.**, *Dental Plaque*, Livingstone, Edinburgh, 1970.
4. **Armstrong, W. G.**, Origin and nature of the acquired pellicle, *Proc. R. Soc. Med.*, 61, 923, 1968.
5. **Mayhall, C. W.**, Concerning the composition and source of the acquired pellicle of human teeth, *Arch. Oral Biol.*, 15, 1327, 1970.
6. **Sönju, T. and Rölla, G.**, Chemical analysis of the acquired pellicle formed in two hours on cleaned human teeth in vivo. Rate of formation and amino-acid analysis, *Caries Res.*, 7, 30, 1973.
7. **Mayhall, C. W.**, Amino acid composition of experimental salivary pellicles, *J. Periodont.*, 48, 78, 1977.
8. **Juriaanse, A.**, The adsorption of peptides and purified salivary proteins onto tooth enamel. A study on pellicle formation, Ph.D. thesis, State University at Groningen, the Netherlands, 1980.
9. **Öste, R., Rönstrom, A., Birkhed, D., Edwardsson, S., and Stenberg, M.**, Gas-liquid chromatographic analysis of amino-acids in pellicle formed on tooth surface and plastic film in vivo, *Arch. Oral Biol.*, 26, 685, 1981.
10. **Hay, D. I.**, The interaction of human parotid salivary proteins with hydroxyapatite, *Arch. Oral Biol.*, 18, 1517, 1973.
11. **Bennick, A., Wong, P., and Cannon, M.**, Structure and biological activities of salivary acidic proline-rich phosphoproteins, in *Calcium Binding Proteins and Calcium Function*, Wasserman, R. H., Corradino, R. A., Carafoli, E., Kretsinger, R. H., MacLennan, D. H., and Siegel, F. L., Eds., North-Holland, New York, 1977, 391.
12. **Örstavik, D. and Kraus, F. W.**, The acquired pellicle: immunofluorescent demonstration of specific proteins, *J. Oral Pathol.*, 2, 68, 1973.
13. **Örstavik, D. and Kraus, F. W.**, The acquired pellicle: enzyme and antibody activities, *Scand. J. Dent. Res.*, 82, 202, 1974.
14. **Kraus, F. W. and Mestecky, J.**, Salivary proteins and the development of dental plaque, *J. Dent. Res.*, 55, C149, 1976.
15. **Sönju, T., Christensen, T. B., Kornstad, L., and Rölla, G.**, Electron microscopy, carbohydrate analyses and biological activities of the proteins adsorbed in two hours to tooth surfaces in vivo, *Caries Res.*, 8, 113, 1974.
16. **Mayhall, C. W. and Butler, W. T.**, The carbohydrate composition of experimental salivary pellicles, *J. Oral Pathol.*, 5, 358, 1976.
17. **Gibbons, R. J. and van Houte, J.**, Bacterial adherence in oral microbial ecology, *Annu. Rev. Microbiol.*, 29, 19, 1975.
18. **van Houte, J.**, Bacterial adherence in the mouth, *Rev. Infect. Dis.*, 5, Suppl. 4, S659, 1983.
19. **Zahradnik, R. T., Propas, D., and Moreno, E. C.**, Effect of pellicle formation time on in vitro attachment and demineralization by *Streptococcus mutans*, *J. Dent. Res.*, 57, 1036, 1978.
20. **Hillman, J. D., van Houte, J., and Gibbons, R. J.**, Sorption of bacteria to human enamel powder, *Arch. Oral Biol.*, 15, 899, 1970.
21. **Gibbons, R. J., Moreno, E. C., and Spinell, D. M.**, Model delineating the effects of a salivary pellicle on the adsorption of *Streptococcus miteor* onto hydroxyapatite, *Infect. Immun.*, 14, 1109, 1976.
22. **Bolton, R. W.**, Adherence of oral streptococci to hydroxyapatite in vitro via glycerol teichoic acid, *Arch. Oral Biol.*, 25, 111, 1980.
23. **Mcgaughey, C., Field, D. B., and Stowell, E. C.**, Effect of salivary proteins on the adsorption of cariogenic streptococci by hydroxyapatite, *J. Dent. Res.*, 50, 917, 1971.
24. **Ruangsri, P. and Örstavik, D.**, Effect of the acquired pellicle and of dental plaque on the implantation of *Streptococcus mutans* on tooth surfaces in man, *Caries Res.*, 11, 204, 1977.
25. **van Houte, J.**, Bacterial adhesion in the mouth, in *Dental Plaque and Surface Interactions in the Oral Cavity*, Leach, S. A., Ed., Information Retrieval, London, 1980, 69.
26. **Liljemark, W. F. and Schauer, S. V.**, Competitive binding among oral streptococci to hydroxyapatite, *J. Dent. Res.*, 56, 157, 1977.
27. **Gibbons, R. J.**, On the mechanism of bacterial attachment to teeth, in *Saliva and Dental Caries*, Kleinberg, I., Ellison, S. A., and Mandel, I. D., Eds., Information Retrieval, New York, 1979, 267.
28. **Rölla, G.**, Formation of dental integuments — some basic chemical considerations, *Swed. Dent. J.*, 1, 241, 1977.
29. **O'Brien, W. J., Fan, P. L., Loesche, W. J., Walker, M. C., VI, and Apostolids, A.**, Adsorption of *Streptococcus mutans* on chemically treated hydroxyapatite, *J. Dent. Res.*, 57, 910, 1978.
30. **Olsson, J., Glantz, P.-O., and Krasse, B.**, Surface potential and adherence of oral streptococci to solid surfaces, *Scand. J. Dent. Res.*, 84, 240, 1976.

31. **Fan, P. L., O'Brien, W. J., and Walker, M. C., VI,** Surface morphologies of chemically treated hydroxyapatite, *J. Dent. Res.,* 58, 1566, 1979.
32. **Characklis, W. G.,** Fouling biofilm development: a process analysis, *Biotech. Bioeng.,* 23, 1923, 1981.
33. **Bryers, J. D. and Characklis, W. G.,** Processes governing primary film formation, *Biotech. Bioeng.,* 24, 2451, 1982.
34. **Wu-Yan, C. D., Tai, S., and Slade, H. D.,** Properties of *Streptococcus mutans* grown on synthetic medium:binding of glucosyltransferase and in vitro adherence, and binding of dextran/glucan, and glycoprotein and agglutination, *Infect. Immun.,* 23, 600, 1979.
35. **Peros, W. J. and Gibbons, R. J.,** Influence of growth medium on adsorption of *Streptococcus mutans, Actinomyces viscosus and Actinomyces naeslundii* to saliva-treated hydroxyapatite surfaces, *Infect. Immun.,* 32, 111, 1981.
36. **Cisar, J. O., Kolenbrander, P. E., and McIntyre, F. C.,** Specificity of coaggregation reactions between human oral streptococci and strains of *Actinomyces viscosus or Actinomyces naeslundii, Infect. Immun.,* 24, 742, 1979.
37. **Gibbons, R. J. and Nygaard, M.,** Synthesis of insoluble dextran and its significance in the formation of gelatinous deposits by plaque-forming streptococci, *Arch. Oral Biol.,* 13, 1249, 1968.
38. **Stäat, R. H., Langley, S. D., and Doyle, R. J.,** *Streptococcus mutans* adherence: presumptive evidence for protein-mediated attachment followed by glucan-dependent cellular accumulation, *Infect Immun.,* 27, 675, 1980.
39. **Abbott, A., Berkeley, R. C. W., and Rutter, P. R.,** Sucrose and the deposition of *Streptococcus mutans* at solid/liquid interfaces, in *Microbial Adhesion to Surfaces,* Berkeley, R. C. W., Lynch, J. M., Melling, J., Rutter, P. R., and Vincent, B., Eds., Ellis Horwood, Chichester, U.K., 1980, Chap. 6.
40. **Cisar, J. O., Vatter, A. E., and McIntire, P. C.,** Identification of the virulence-associated antigen on the surface of fibrils of *Actinomyces viscosus* T14, *Infect. Immun.,* 19, 312, 1978.
41. **Hogg, S. D., Handley, P. S., and Embery, G.,** Surface fibrils may be responsible for the salivary glycoprotein-mediated aggregation of the oral bacterium *Streptococcus sanguis, Arch. Oral Biol.,* 26, 945, 1981.
42. **Rutter, P. R. and Abbott, A.,** Study of the interaction between oral streptococci and hard surfaces, *J. Gen. Microbiol.,* 105, 219, 1978.
43. **Leach, S. A. and Agalamanyi, E. A.,** Hydrophobic interactions that may be involved in the formation of dental plaque, in *Bacterial Adhesion and Prevention Dentistry,* ten Cate, J. M., Leach, S. A., and Arends, J., Eds., IRL Press, London, 1984, 43.
44. **Baier, R. E.,** Conditioning surfaces to suit the biomedical environment: recent progress, *J. Biomech. Eng.,* 104, 257, 1982.
45. **Westergren, G. and Olsson, J.,** Hydrophobicity and adherence of oral streptococci after repeated subculture in vitro, *Infect. Immun.,* 40, 432, 1983.
46. **Gibbons, R. J. and Etherden, I.,** Comparative hydrophobicities of oral bacteria and their adherence to salivary pellicles, *Infect. Immun.,* 41, 1190, 1983.
47. **Hardie, J. M.,** The microbiology of dental caries, *Dent. Update,* 9, 199, 1982.
48. **Dietz, V. H.,** In vitro production of plaques and caries, *J. Dent. Res.,* 22, 423, 1943.
49. **Pigman, W., Elliott, H. C., Jr., and Laffre, R. O.,** An artificial mouth for caries research, *J. Dent. Res.,* 31, 627, 1952.
50. **Sidaway, A., Marsland, E. A., Rowles, S. L., and MacGregor, A.,** The artificial mouth in caries research, *Proc. R. Soc. Med.,* 57, 1065, 1964.
51. **Koulourides, T., Bodden, R., Keller, S., Manson-Hing, L., Lastra, J., and Housch, T.,** Cariogenicity of nine sugars tested with an intraoral device in man, *Caries Res.,* 10, 427, 1976.
52. **Bibby, B. G. and Huang, C. T.,** Some observations on in vitro dental plaques, *J. Dent. Res.,* 59, 1946, 1980.
53. **Slade, H. D.,** In vitro models for the study of adherence of oral streptococci, in *Immunological Aspects of Dental Caries,* Bowen, W. H., Genco, R. J., and O'Brien, T. C., Eds., Information Retrieval, Washington, D. C., 1976, 21.
54. **Rowles, S. L., Sidaway, D. A., MacGregor, A. B., and Marsland, E. A.,** An apparatus for the production of dental caries in vitro, *Arch, Oral Biol.,* 8, 311, 1963.
55. **Molan, P. C. and Hartles, R. L.,** The source of glycolytic activity in human saliva, *Arch. Oral Biol.,* 12, 1593, 1967.
56. **Tonzetich, J. and Friedman, S. D.,** The regulation of metabolism by the cellular elements in saliva, *Ann. N.Y. Acad. Sci.,* 131, 815, 1965.
57. **Kleinberg, I.,** Effect of varying sediment and glucose concentrations on the pH and acid production in human salivary sediment mixtures, *Arch. Oral Biol.,* 12, 1457, 1967.
58. **Newbrun, E.,** Sugar and dental caries: a review, *Science,* 217, 418, 1982.
59. **Kleinberg, I., Jenkins, G. N., Denepitiya, L., and Chatterjee, R.,** Diet and dental plaque, in *The Environment of the Teeth,* Ferguson, D. B., Ed., S. Karger, Basel, 1981, 88.

60. **Tatevossian, A. and Jenkins, G. N.**, The source of metabolic activity in human saliva, *Arch. Oral Biol.*, 14, 1121, 1969.
61. **Denepitiya, L. and Kleinberg, I.**, A comparision of the microbial compositions of pooled human dental plaque and salivary sediment, *Arch. Oral Biol.*, 27, 739, 1982.
62. **Gilmour, M. N. and Poole, A. E.**, The fermentative capabilities of dental plaque, *Caries Res.*, 1, 247, 1967.
63. **Geddes, D. A. M.**, Acids produced by human dental plaque metabolism in situ, *Caries Res.*, 9, 98, 1975.
64. **Gibbons, R. J. and van Houte, J.**, Selective bacterial adherence to oral epithelial surfaces and its role as an ecological determinant, *Infect. Immun.*, 3, 567, 1971.
65. **Gibbons, R. J. and van Houte, J.**, On the formation of dental plaques, *J. Periodont.*, 44, 347, 1973.
66. **Hardie, J. M. and Bowden, G. H.**, The normal microbial flora of the mouth, in *The Normal Microbial Flora of Man*, Skinner, F. A. and Carr, J. G., Eds., Academic Press, London, 1974, 47.
67. **Mukasa, H. and Slade, H. D.**, Mechanism of adherence of *Streptococcus mutans* to smooth surfaces, *Infect. Immun.*, 8, 555, 1973.
68. **Svanberg, M. and Loesche, W. J.**, The salivary concentration of *Streptococcus mutans* and *Streptococcus sanguis* and their colonization of artificial tooth fissures in man, *Arch. Oral Biol.*, 22, 441, 1977.
69. **Listgarten, M. A., Mayo, H. F., and Tremblay, R.**, Development of dental plaque on epoxy resin crowns in man, *J. Periodont.*, 46, 10, 1975.
70. **Lie, T.**, Growth of dental plaque on hydroxyapatite splints. A method of studying early plaque morphology, *J. Periodont. Res.*, 9, 135, 1974.
71. **McCabe, R. M., Keyes, P. H., and Howell, A., Jr.**, An in vitro method for assessing the plaque-forming ability of oral bacteria, *Arch. Oral Biol.*, 12, 1653, 1967.
72. **Bladen, H., Hageage, G., Pollock, F., and Harr, R.**, Plaque formation in vitro on wires by Gram-negative oral microorganisms *(Veillonella)*, *Arch. Oral Biol.*, 15, 127, 1970.
73. **Bass, G. E., Dillingham, E. O., and Powers, L. J.**, Quantitative studies of in vitro inhibition of *Streptococcus mutans* plaque formation by organic amines, *J. Dent. Res.*, 54, 968, 1975.
74. **Dummer, P. M. H. and Green, R. M.**, A comparison of the ability of strains of streptococci to form dental plaque-like deposits in vitro with their cariogenicity in gnotobiotic rats, *Arch. Oral Biol.*, 25, 245, 1980.
75. **Clayton, J. A. and Green, E.**, Roughness of pontic materials and dental plaque, *J. Prosthet. Dent.*, 23, 407, 1970.
76. **Norman, R. D., Mehra, R. V., Swartz, L., and Phillips, R. W.**, Effects of restorative materials on plaque composition, *J. Dent. Res.*, 51, 1596, 1972.
77. **Wimpenny, J. W. T., Lovitt, R. W., and Coombs, J. P.**, Laboratory model systems for the investigation of spatially and temporally organised microbial ecosystems, in *Microbes in Their Natural Environments*, Slater, J. H., Whittenbury, R., and Wimpenny, J. W. T., Eds., Cambridge University Press, 1983, 67.
78. **Friberg, S.**, Colloidal phenomena encountered in the bacterial adhesion to the tooth surface, *Swed. Dent. J.*, 1, 207, 1977.
79. **Gregory, J.**, Physical aspects of particle aggregation and adhesion, in *Dental Plaque and Surface Interactions in Oral Cavity*, Leach, S. A., Ed., IRL Press, London, 1980, 7.
80. **Mikx, F. H. M., van der Hoeven, J. S., König, K. G., Plasschaert, A. J. M., and Guggenheim, B.**, Establishment of defined microbial ecosystems in germ-free rats. 1. The effect of the interaction of *Streptococcus mutans* or *Streptococcus sanguis* with *Veillonella alcalescens* on plaque formation and caries activity, *Caries Res.*, 6, 211, 1972.
81. **Donoghue, H. D., Dibdin, G. H., Shellis, R. P., Rapon, G., and Wilson, C. M.**, Effect of nutrients upon *Streptococcus mutans* BHT and *Streptococcus mitior* LPA-1 growing in pure culture on human teeth in an artificial mouth, *J. Appl. Bacteriol.*, 49, 295, 1980.
82. **Coulter, W. A. and Russell, C.**, pH and Eh in single and mixed culture bacterial plaque in an artificial mouth, *J. Appl. Bacteriol.*, 40, 73, 1976.
83. **Atkinson, B. and Fowler, H. W.**, The significance of microbial film in fermenters, *Adv. Biochem. Eng.*, 3, 223, 1974.
84. **Matson, J. V. and Characklis, W. G.**, Diffusion into microbial aggregates, *Water Res.*, 10, 877, 1976.
85. **Coombe, R. A., Tatevossian, A., and Wimpenny, J. W. T.**, Bacterial thin films as in vitro models for dental plaque, in *Surface and Colloid Phenomena in the Oral Cavity: Methodological Aspects*, Frank, R. M. and Leach, S. A., Eds., Information Retrieval, London, 1981, 239.
86. **Marsh, P. D., Hunter, J. R., Bowden, G. H., Hamilton, I. R., McKee, A. S., Hardie, J. M., and Ellwood, D. C.**, The influence of growth rate and nutrient limitation on the microbial composition and biochemical properties of a mixed culture of oral bacteria grown in a chemostat, *J. Gen. Microbiol.*, 129, 755, 1983.
87. **Shellis, R. P.**, A synthetic saliva for cultural studies of dental plaque, *Arch. Oral Biol.*, 23, 485, 1978.
88. **Kashket, S. and Guilmette, K. M.**, Further evidence for the non-immunoglobulin nature of the bacterial aggregating factor in saliva, *Caries Res.*, 12, 170, 1978.

89. **Rundegren, J. and Ericson, T.,** An evaluation of the specificity of salivary agglutinins, *J. Oral Pathol.,* 10, 261, 1981.

90. **Dibdin, G. H., Shellis, R. P., and Wilson, C. M.,** An apparatus for the continuous culture of micro-organisms on solid surfaces with special reference to dental plaque, *J. Appl. Bacteriol.,* 40, 261, 1976.

91. **Tatevossian, A.,** The chemistry of plaque fluid, in *The Environment of the Teeth,* Ferguson, D. B., Ed., S. Karger, Basel, 1981, 66.

92. **Eggert, F. M.,** The nature of secretory agglutinins and aggregating factors. IV. Complexes between non-mucin glycoproteins, immunoglobulins and mucins in human saliva and amniotic fluid, *Int. Arch. Allergy Appl. Imunol.,* 62, 46, 1980.

93. **Pollock, J. J., Iacono, V. J., Bicker, H. G., Mackay, B. J., Katona, L. I., Taichman, L. B., and Thomas, E.,** The binding, aggregation and lytic properties of lysozyme, in *Microbial Aspects of Dental Caries,* Stiles, H. M., Loesche, W. J., and O'Brien, T. C., Eds., Information Retrieval, Washington, D.C., 1976, 325.

94. **Marsh, P. D., Williamson, M. I., Keevil, C. W., McDermid, A. S., and Ellwood, D. C.,** The influence of sodium and potassium ions on acid production by washed cells of *Streptococcus mutans* Ingbritt and *Streptococcus sanguis* NCTC 7865 growing in a chemostat, *Infect. Immun.* 36, 476, 1982.

95. **Spinell, D. M. and Gibbons, R. J.,** Influence of culture medium on the glucosyltransferase and dextran binding capacity of *Streptococcus mutans* 6715 cells, *Infect. Immun.,* 10, 1448, 1974.

96. **Ellwood, D. C., Baird, J. K., Hunter, J. R., and Longyear, V. C.,** Variations in surface polymers of *Streptococcus mutans, J. Dent. Res.,* 55, C42, 1976.

97. **Russell, C. and Tagg, J. R.,** Role of bacteriocin during plaque formation by *Streptococcus salivarius* and *Streptococcus sanguis* on a tooth in an artificial mouth, *J. Appl. Bacteriol.,* 50, 305, 1981.

98. **Vadeboncoeur, C. and Trahan, L.,** Comparative study of *Streptococcus mutans* laboratory strains and fresh isolates from carious and caries-free tooth surfaces and from subjects with hereditary fructose intolerance, *Infect. Immun.,* 40, 81, 1983.

99. **Örstavik, J.,** Influence of in vitro propagation on the adhesive qualities of *Strep. mutans* isolated from saliva, *Acta Odontol. Scand.,* 40, 57, 1982.

100. **Cimasoni, G.,** *The Crevicular Fluid,* Monographs in oral science, Vol. 3, S. Karger, Basel, 1974.

101. **Eskow, R. N. and Loesche, W. J.,** Oxygen tension in the human oral cavity, *Arch. Oral Biol.,* 16, 1127, 1971.

102. **Charlton, G.,** The oxygen tension of saliva within the parotid duct and on the floor of the mouth of humans, *J. Dent. Res.,* 41, 512, 1962.

103. **Globerman, D. Y. and Kleinberg, I.,** Intraoral $P_{O_2}$ and its relation to bacterial accumulation on the oral tissues, in *Saliva and Dental Caries,* Kleinberg, I., Ellison, S. A., and Mandel, I. D., Eds., Information Retrieval, New York, 1979, 275.

104. **Stralfors, A.,** An investigation of the respiratory activities of oral bacteria, *Acta Odontol. Scand.,* 14, 153, 1956.

105. **Sonju, T. and Skjorland, K.,** Pellicle composition and initial bacterial colonization on composite and amalgam in vivo, in *Microbial Aspects of Dental Caries,* Stiles, H. M., Loesche, W. J., and O'Brien, T. C., Eds., Information Retrieval, Washington, 1976, 133.

106. **Baier, R. E. and Glantz, P.-O.,** Characterization of oral in vivo films formed on different types of solid surfaces, *Acta Odontol. Scand.,* 36, 289, 1978.

107. **Maetani, T., Miyoshi, R., Nahara, Y., Kawazoe, Y., and Hamada, T.,** Plaque accumulation on teflon-coated metal, *J. Prosthet. Dent.,* 51, 353, 1984.

108. **Rutter, P. R. and Vincent, B.,** The adhesion of microorganisms to surfaces: physicochemical aspects, in *Microbial Adhesion to Surfaces,* Berkeley, R. C. W., Lynch, J. M., Melling, J., Rutter, P. R., and Vincent, B., Eds., Ellis Horwood, Chichester, U.K., 1980, chap. 4.

109. **Rothen, A.,** Surface film techniques, in *International Revues of Cytology,* Vol. 80, Bourne, G. H., Danielli, J. F., and Jeon, K. W., Eds., Academic Press, London, 1983, 218.

110. **Baier, R. E.,** Conditioning surfaces to suit the biomedical environment: recent progress, *J. Biomech. Eng.,* 104, 257, 1982.

111. **Zisman, W. A.,** Reduction of the equilibrium contact angle to liquid and solid constitution, *Adv. Chem.,* 43, 1, 1964.

112. **Busscher, H. J., Weerkamp, A. H., van der Mei, H. C., van Pelt, A. W. J., de Jong, H. P., and Arends, J.,** Measurement of the surface free energy of bacterial cell surfaces and its relevance for adhesion, *Appl. Environ. Microbiol.,* 48, 980, 1984.

113. **Carlsson, J. and Egelberg, J.,** Effect of diet on early plaque formation in man, *Odontol. Rev.,* 16, 112, 1965.

114. **Morhart, R. E. and Fitzgerald, R. J.,** Nutritional determinants of the ecology of the oral flora, *Dent. Clin. North Am.,* 20, 473, 1976.

115. **Griffith, C. J. and Melville, T. H.,** Growth of oral streptococci in a chemostat, *Arch. Oral Biol.,* 19, 87, 1974.

116. **Mikx, F. M. H. and van der Hoeven, J. S.**, Symbiosis of *Streptococcus mutans* and *Veillonella alcalescens* in mixed continuous cultures, *Arch. Oral Biol.*, 20, 407, 1975.
117. **Wright, D. E.**, The source and disintegration rate of leukocytes in saliva from caries-free and caries-active subjects, *Br. Dent. J.*, 106, 278, 1959.
118. **Pigman, W., Hawkins, W. L., Watson, J., Powell, R., and Gaston, C.**, The effect of the concentration of D-glucose on the attack of the tooth substances in the artificial mouth, *J. Dent. Res.*, 34, 537, 1955.
119. **Pigman, W., Gilman, E., Powell, R., and Muntz, L.**, The action of individual bacterial strains on human teeth under in vitro conditions, *J. Dent. Res.*, 36, 314, 1957.
120. **Pigman, W. and Newbrun, E.**, Evaluation of anticaries agents by the use of the artificial mouth, *J. Dent. Res.*, 41, 1304, 1962.
121. **Wagg, B. J., Melville, T. H., and Hartles, R. L.**, A technique for studying the microorganisms associated with extracted teeth when under continuous irrigation, *Br. Dent. J.*, 103, 121, 1957.
122. **Sidaway, D. A.**, The bacterial composition of natural plaque and the in vitro production of artificial plaque, in *Dental Plaque*, McHugh, W. D., Ed., Livingstone, Edinburgh, 1970, 225.
123. **Jordan, H. V. and Keyes, P. H.**, In vitro methods for the study of plaque formation and carious lesions, *Arch. Oral Biol.*, 11, 793, 1966.
124. **Naylor, M. N., Wilson, R. F., and Melville, M. R. B.**, Mono- and di-saccharide solutions and the formation of plaque in vitro, in *Dental Plaque*, McHugh, W. D., Ed., Livingstone, Edinburgh, 1970, 41.
125. **Dolan, M. M., Murphy, C. V., Kavanagh, B. J., and Yankell, S. L.**, Development of an in-vitro plaque model from human salivary sediment suspensions, *Arch. Oral Biol.*, 17, 147, 1972.
126. **Russell, C. and Coulter, W. A.**, Continuous monitoring of pH and Eh in bacterial plaque grown on a tooth in an artificial plaque, *Appl. Microbiol.*, 29, 141, 1975.
127. **Coulter, W. A. and Russell, C.**, A miniature continuous culture system for controlled production of simulated bacterial dental plaque, *Arch. Oral Biol.*, 21, 333, 1976.
128. **Russell, C. and Coulter, W. A.**, Plaque formation by streptococci, in an artificial mouth and factors influencing colonization, *J. Appl. Bacteriol.*, 42, 337, 1977.
129. **Coulter, W. A. and Russell, C.**, The effect of chlorhexidine on plaque development in an artificial mouth, *Microbios*, 16, 21, 1976.
130. **Donoghue, H. D., Hudson, D. E., Perrons, C. J., Dibdin, G. H., Rapson, G., Shellis, R. P., and Wilson, C. M.**, Effect of inoculation sequence and nutrients upon *Streptococcus mutans* BHT and *Streptococcus mitior* LPA-1 growing on human teeth in an aritificial mouth, *J. Appl. Bacteriol.*, 54, 23, 1983.
131. **Yaari, A. and Bibby, B. G.**, Production of plaques and initiation of caries in vitro, *J. Dent. Res.*, 55, 30, 1976.
132. **Sudo, S. Z.**, Continuous culture of mixed oral flora on hydroxyapatite-coated glass beads, *Appl. Environ. Microbiol.*, 33, 450, 1977.
133. **Leech, R., Marsh, P. D., and Rutter, P.**, The deposition of oral bacteria at the solid/liquid/air interfaces, *Arch. Oral Biol.*, 24, 379, 1979.
134. **Singer, R. E. and Buckner, B. A.**, Characterisation of toxic extracts of in vitro cultured human plaque, *J. Periodont. Res.*, 15, 603, 1980.
135. **Coombe, R. A., Tatevossian, A., and Wimpenny, J. W. T.**, Factors affecting the growth of thin bacterial films in vitro, in *Bacterial Adhesion and Preventive Dentistry*, ten Cate, J. M., Leach, S. A., and Arends, J., Eds., IRL Press, Oxford, 1984, 193.
136. **Klimek, V. J., Hellwig, E., and Ahrens, G.**, Der Einfluss von Plaque auf die Fluoridstabilität in Schmelz nach Application von Aminfluorid in künstlichen Mund, *Dtsch. Zahnaerztl. Z.*, 37, 836, 1982.
137. **Klimek, J., Hellwig, E., and Ahrens, G.**, Fluoride taken up by plaque, by the underlying enamel and by clean enamel from three fluoride compounds in vitro, *Caries Res.*, 16, 156, 1982.
138. **Klimek, J., Hellwig, E., and Ahrens, G.**, Movement of plaque fluoride under cariogenic conditions, *Caries Res.*, 17, 315, 1983.
139. **Noorda, W. D., De Koning, W., Purdell-Lewis, D. J., and van de Poel, A. C. M.**, De 'Kunstmond', een model voor plaque-en cariesonderzoek, *Ned. Tijdschr. Tandheelk.*, 91, 9, 1984.
140. **Lagerlof, F., Dawes, R., and Dawes, C.**, Salivary clearance of sugar and its effects on pH changes by *Streptococcus mitior* in an artificial mouth, *J. Dent. Res.*, 63, 1266, 1984.
141. **Russell, C.**, Personal communication, 1985.
142. **Donoghue, H. D.**, Personal communication, 1985.

Chapter 11

# GEL-PLATE METHODS IN MICROBIOLOGY

## Julian W. T. Wimpenny, Paul Waters, and Adrian Peters

## INTRODUCTION

Martinius Beijerinck, the Dutch microbiologist, was almost certainly the father of plate diffusion methods in analytical microbiology.[1] It was Beijerinck who gave the term "auxanography" to the microbiological literature. Auxanography, literally meaning "grow-drawing", described the use of gel-stabilized systems in which nutritional patterns were identified by placing small quantities of a nutrient on or in an agar layer and noting whether specific organisms grew better near these zones.

Gel-diffusion methods have been applied to two main areas of microbiology since that time. These include antibiotic sensitivity testing and immunodiffusion techniques. Testing the activity of inhibitory agents has proved a speedy and efficient tool especially to medical microbiologists who need rapid, reasonably precise information to help them choose appropriate antibiotics for particular microbial infections. The agent is placed in a well or a cup in the agar or more recently on filter paper "spiders" or discs on the agar surface. The latter is seeded uniformly with a lawn of the test organism. After incubation, the diameter of the inhibition zone is measured and decisions made about the spectrum of antibiotic inhibition of the isolate. Another version of the same test is to cross streak a number of bacterial strains against an unknown antibiotic-producing isolate. The width of each of the inhibition zones gives a lot of information with the minimum of effort. Immunodiffusion methodology allows an assessment of zones of precipitation when antigen and antibody diffuse towards one another in an agar substratum and is a powerful tool, not only in medical microbiology and immunology, but in many other biological disciplines.

The power of diffusion methods in microbiology has not really been properly explored outside these two areas. In particular, there is a clear case for applying diffusion methods to establishing the habitat range of bacterial species, but also in investigating possible interactions between more than one biologically active solute molecule at the same time. Such investigations can profit by the use of multidimensional techniques. A glimpse of the efficacy of multidimensionality is apparent in the pioneering work of Baas-Becking and his colleagues[2,3] who were investigating the biology of estuarine microbes. Through a large range of individual measurements, they were able to chart the responses of estuarine organisms to pH and Eh plotted one against the other. The two-dimensional map shows the limits of growth of each group of bacteria and indicates the great range of responses possible. This work was published in the fifties and has been cited fairly often; however, the multidimensional approach has not been pursued much since then. Perhaps the reason is that to perform such experiments in the laboratory can be extremely time-consuming if it relies on the preparation of numerous individual culture tubes, each having different solute concentrations. For example, a two-dimensional map for a range of ten concentrations of two solutes would need the preparation, inoculation, and subsequent monitoring of one-hundred culture vessels. As this chapter will show, two-dimensional gradient plates provide a simple and accurate answer to these problems. As we also make clear, the use of gradient plates allows multidimensionality to be further explored at least to the simultaneous variation of four factors, five if time is treated as a dimension as well.

## SINGLE-DIMENSION DIFFUSION PLATES

Agar diffusion methods were given precision by Szybalski and his colleagues[4-6] who developed the elegant but simple wedge plate technique used to determine the sensitivity of microbes to antibiotics and to select mutants resistant to these agents. Here, a circular plate was propped up a few millimeters at one side, and a measured amount of plain nutrient agar medium was added to the plate. Once this layer had set, the plate was restored to its normal flat position and a second layer containing the antibiotic was poured over it. After a period during which the antibiotic was allowed to equilibrate vertically, the plate was inoculated by streaking in the direction of the gradient with one or more strains of microorganisms.

After incubation, the region in which growth was observed gave a good indication of the sensitivity of the organism to the inhibitory agent. It is generally easy to determine the concentration profile since it tends to be linear. The error of the method may be determined in replicate experiments. The Szybalski technique does not rely on molecular diffusion to establish the horizontal gradient; however, it is assumed that the inhibitor can diffuse sufficiently far vertically to equilibrate in this plane during preincubation. Szybalski noted problems here with slowly diffusing antibiotics and performed his experiments with the layers reversed. Results were expressed as the mean of the length of the growth zones for both types of plate. Of course, if the streak lengths were the same, as was often the case, this confirmed the contention that vertical equilibration had taken place. For accurate quantitative work, the concentration of antibiotic in the upper layer was set at ten times the value in the lower. Under these conditions, the antibiotic gradient was linear from one to ten concentration units across the plate. The technique proved a useful way to detect the growth of resistant mutant organisms, which appeared as individual colonies in the regions of higher antibiotic concentration. Bryson and Szybalski[6] used gradient plates to investigate antibiotic resistance and cross-resistance in strains of *Escherichia coli*. Incidentally, Szybalski was aware that square petri plates would have been more suitable than the circular variety, however, he was not at that time able to obtain them commercially.

Four years after Szybalski published this method, it was used by Sacks[7] to prepare pH gradient plates. Sacks also used circular petri dishes. He added KH$_2$PO4 to a final concentration of 0.1 $M$ to one layer and an equal concentration of K$_2$HPO4 to the second layer after the first had set. The gradients were shown to be reasonably stable and to range from about pH 5.6 at one side of the plate to pH 7.8 at the other. Sacks demonstrated synergism between antibiotic activity and pH for streptomycin. Thus, the latter was more active at alkaline than at acid pH values for a number of species. In these experiments, antibiotic at a uniform concentration was incorporated into a thin surface agar layer.

The single-dimensional plate has been developed over a period by Davenport in his studies of yeast physiology.[8] Weinberg[9] used a similar gradient plate when investigating synergism between metal ions and antibiotics. He referred to his plates as "double-gradient" systems. These were actually one-dimensional plates in which the metal was incorporated into the first wedge while the second contained the antibiotic.

## SOME TWO-DIMENSIONAL GRADIENT SYSTEMS

There have only been a small number of systems developed for determining the responses of microbes to two factors at once. Such two dimensional maps seem potentially useful, yet these techniques have never been fully exploited in spite of the original work by Baas-Becking and his colleagues[2,3] on the sensitivity of estuarine bacteria to pH and redox potential discussed briefly in the introduction.

### Systems for the Investigation of Algal Growth

Experimental systems were devised to grow photosynthetic algae in the two dimensions

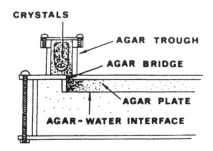

FIGURE 1.  The first Caldwell two-dimensional steady-state gradient plate. (Courtesy of D. E. Caldwell.)

of light and temperature.[10-12] In the model developed by Van Baalen and Edwards,[12] organisms were grown on a 25 × 25 cm square agar slabs or in separate containers on an aluminum plate in which a thermal gradient was established, using a cooling bath on one side and a heating bath at the other side of the plate. Hot or cold water was circulated through holes drilled along the edges of the plate. A light gradient was generated at right angles to the temperature gradient using a 20 watt fluorescent strip light placed about 4 in. above the front edge of the plate. This system was capable of generating a range of temperatures from 10.5 to 30.5 °C and of light from 35 to 500 fc. The utility of this device was illustrated using cultures of *Agmenellum quadruplicatum* and of *Centrocerus clavulatum* in separate experiments.

### The Caldwell Steady-State Diffusion Plates

The potential power of gradient plate techniques began to be discerned following the publication of papers describing a steady-state two-dimensional gradient plate by Caldwell and Hirsch and by Caldwell, Lai, and Tiedje in 1973.[13,14]

The Caldwell plate consists of a Perspex chamber in which two wells at right angles to one another feed the edges of a square block of agar with solutes dissolving from a crystalline supply. One surface of the agar block is washed with an aqueous solution which acts as a sink for the solutes diffusing into the block. The gradients should stabilize because the solute concentrations in the reservoirs are always saturated, at least while any crystals remain.

Because the method relies on molecular diffusion to generate the gradients, there is a practical upper limit to the size of the diffuson field to allow equilibration to take place in a reasonable time. Caldwell and his colleagues developed a number of variants of the steady-state diffusion plate. Each will be discussed separately as there are significant differences in design.

The first design (Figure 1) generated a diffusion field consisting of an 18 cm square agar block which was 1.3 cm thick. This model incorporated two troughs 19 × 7 × 3.3 cm, which fed two adjacent sides of the agar slab. These troughs contained the diffusible solute at saturating concentrations (usually as an excess of crystalline substrate). In order to operate as a steady state system, which was the essence of the design, the slab face opposite to the solute feeds was washed with an aqueous solution which acted as a sink for solute diffusion. In this design, cells were incorporated into the main agar slab after the troughs had been filled with solutes immobilized in agar. A third layer of sterile agar was poured over the main slab to provide a "seal" between the inoculum and the sink solution. In operation, solutes entered the diffusion field from above while the sink solution passed over the agar slab from below.

This model was tested first by constructing a pH gradient using sodium bicarbonate as source and acidified water as sink solution. It turned out that the solute front only extended some 3.1 cm into the agar slab in the first 34 hr and movement was extremely slow after this time. These results suggested that the slab field was really too large and this view was reflected in subsequent designs. Nevertheless, some interesting experiments were performed. *E. coli* was grown in glucose/bicarbonate gradients and duly grew best at a particular zone on the slab, signifying optimum concentrations for both substrates. Cultures of *Hyphomicrobium* sp. B-522, *Thiopedia* sp., and *Rhodomicrobium vannielii* could be separated in gradients of sodium sulfide and methylamine hydrochloride. *R. vannielii* cells showed characteristic morphological variations depending on whether they were growing actively or in a resting stage. Acetate vs. mud (a source of natural nutrients) were used to isolate characteristic species from a forest pond!

A second form of the plate used a much smaller diffusion field 2.5 cm square. In this case, the source diffusion chambers were built into the structure and fed the slab from adjacent edges below it. The slab itself was sterile acting simply to generate a diffusion field, while the inoculum was incorporated in a thin layer on its surface. Wash solution flowed over the top of the system. Diffusion patterns were predicted mathematically and tested using radioactive acetate as one source.

The final form of the steady-state two-dimensional diffusion plate is shown in Figure 2a and b. This is a more elegant version of the second model discussed above.

## THE TWO-DIMENSIONAL WEDGE-PLATE TECHNIQUE

### General Description

Two-dimensional diffusion systems could clearly be very powerful and a number of techniques, including the Caldwell plate, were investigated by my own group. The Caldwell plate was finally rejected in spite of its obvious elegance for three reasons: (1) a long time was needed to establish a gradient system by diffusion, (2) the equipment was relatively complex, and (3) the number of systems that would be needed to accomplish a large work program. We felt that it ought to be possible to use nonsteady-state systems and early experiments centered on agar plates with source troughs cut into the agar at right angles. These were not satisfactory because, once more, diffusion was so slow. We then developed two-dimensional gradient plates using the Szybalski wedge-plate technique,[15-17] believing that it ought to be possible by pouring four separate layers, to establish two sets of gradient wedges at right angles to one another. The methods used are described in detail[17] but are paraphrased here.

### Plate Production and Performance

*Preparation of the Plates*

In the U.K. we have used 10 cm square Sterilin wettable petri dishes. Wettable plates are slightly easier to pour than the normal plastic variety. We have also used 9 cm square plates in the U.S. These were nonwettable, and greater care must be taken when the first layer is poured to ensure that the surface is covered evenly. For 9 or 10 cm square dishes, we decided that the complete system should be about 6 mm thick. The edge of each dish is therefore propped up 3 mm with standard glass or metal rod (Figure 3). Layers that are 12 and 15 m$\ell$ of medium per layer are needed to achieve these dimensions in the 9 and 10 cm plates, respectively. For plate preparation, it is essential to use a perfectly flat surface, preferably a leveling board adjusted carefully with a spirit level. This should be set up in a sterile inoculating cabinet with easy access to a water bath which should be kept at 70°C.

Medium is prepared in 12 or 15 m$\ell$ aliquots in capped culture tubes. We commonly pour 24 plates at a time. Since 4 layers are needed per plate, 96 tubes must be prepared. For pH/NaCl gradient plates, the pH gradient is poured first. Tubes of medium containing 1.5 to 2.5% w/v agar are treated with $H_2SO_4$ just before the first layer is poured. Precise details of acid and base additions will depend on the pH profile required and the composition of

a

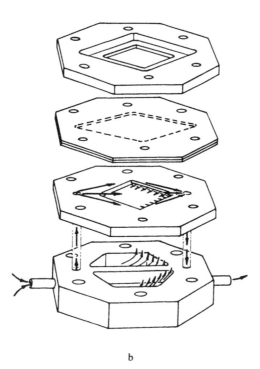

b

FIGURE 2.    The latest version of the Caldwell two-dimensional steady-state gradient
plate. (a) Assembled; (b) exploded view. (Courtesy of D. E. Caldwell.)

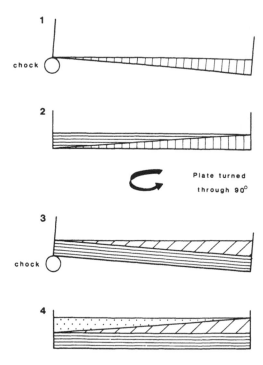

FIGURE 3.   The production of two-dimensional wedge plates. For pH/
NaCl gradients, an acid wedge is first made in the square plate, one of
whose sides is raised by 3 mm. After this layer has set, the plate is restored
to the horizontal position and the alkaline wedge pored. The plate is rotated
through 90° and the process repeated with the salt and then the salt-free
layers.

the medium itself. Using 15 mℓ per layer of a nutrient broth based medium, we added 0.22
mℓ 1.0 *N* sulfuric acid and 0.44 mℓ of 1.0 *N* NaOH to the first and second layers. For the
growth of phototrophic species on a malate/mineral-salts/vitamin mixture, 12 mℓ of medium
required 0.3 mℓ acid and 0.8 mℓ of alkali, respectively. On adding the acid or base, the
tubes should be well mixed and the layers quickly poured. A bunsen burner is used to flame
the agar surface to collapse any bubbles which may appear. Plates set quite quickly and
they can then be returned to the horizontal position before pouring the second layer.

If a salt gradient is chosen in the second dimension, this should be poured as a third layer
with the plate raised 3 mm in a direction at right angles to the first gradient. When using
high salt concentrations, the agar must be poured hot (at least 70 °C), since salt causes agar
to set quickly. The salt/agar medium tube should be quickly mixed, poured, and distributed
over the agar surface before it sets. After the layer has set, the final layer of medium should
be poured with the plate restored to its flat position. The plates may then be dried for an
hour and they should then be stored for 24 hr to allow the layers to equilibrate vertically.

*Plate Inoculation*
Three techniques have been tried and found suitable for particular groups of organism.

**A Semisolid Spread Inoculum**
An overnight culture (0.1 mℓ) is added to 9 mℓ of medium containing 0.2% w/v agar;
1.0 mℓ of this is added to the plate after mixing and spread with a flamed glass spreader.
This technique, though not perfect, proved suitable for a range of *aerobic heterotrophic*
species.[17]

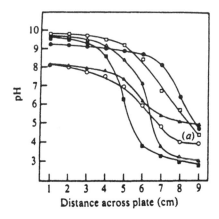

FIGURE 4. Selection of appropriate pH gradients. Acid and alkali wedges were prepared with the following additions: ○, 0.22 mℓ 1 *M* H₂SO₄ + 0.44 mℓ 1 *M* NaOH [standard system, curve (a)]; ●, 0.22 mℓ 1 *M* H₂SO₄ + 0.50 mℓ 2 *M* NaOH; □, 0.50 mℓ 0.5 *M* H₂SO₄ + 0.75 mℓ 1 *M* NaOH; ■, 1.00 mℓ 0.50 *M* H₂SO₄ + 0.75 mℓ 1 *M* NaOH; △, 0.75 mℓ 0.5 *M* H₂SO₄ + 0.75 mℓ 1 *M* NaOH; ▲, 1.00 mℓ 0.1 *M* acetic acid + 2.0 mℓ 0.1 *M* NH₄OH. (From Wimpenny, J. W. T. and Waters, P., *J. Gen. Microbiol.*, 130, 2921, 1984. With permission.)

## A Semisolid Layer Inoculum

When cultivating a range of phototrophic bacteria anaerobically in the light, the surface-spread inoculum gave very poor results with the organisms growing only slowly. After experimentation, it was found that cells grew very well if they were incorporated into a 5.0 mℓ aliquot of medium containing 0.8% w/v agar. The agar should be kept at 45°C until just before adding the inoculum, mixing carefully to avoid bubble formation, and pouring over the entire agar surface. There is little room to maneuver here as the layer sets quickly once it is in contact with the agar surface!

## The Simple Spread Plate

Experiments with luminous bacteria revealed that neither the semisolid spread nor the semisolid layer inocula were suitable. In this case 0.5 mℓ of an overnight broth culture were spread over the plate surface. This gave very satisfactory results.

## Choice of Gradients
### pH Gradient

For most applications, pH ranged from 3.5 to 4.0 at one edge of the plate to 8.0 to 9.0 at the opposite side. This is a large range of hydrogen ion concentrations and the gradients were never linear. Nonlinearity was due to the way in which gradients were established using strong acid or strong alkali to adjust the pH as well as to the buffer composition of medium. We have not explored the possibility of producing other, perhaps more linear, gradients, using, for example, a wide range of buffering agents with a spread of pK values. There are significant advantages to sigmoid relationships: most cells grow around the physiologically neutral region so no discrimination is exerted here, and the region need not be spatially extensive. Various combinations of acid and alkali were investigated in the nutrient-broth-based system described in the literature[17] and these are reproduced here (Figure 4).

### NaCl Gradients

NaCl was chosen for the second dimension in our own work because it is of fundamental importance to microbes since aquatic environments range from pure freshwater to the salt-saturated habitats adjacent to salt pans, etc. The concentration of NaCl in the salt layer varies, of course, according to the application. For a general response of different bacteria to salt and pH, we chose to use 25% w/v NaCl; however, when investigating a number of strains of the phototrophic *Rhodopseudomonas spheroides*, the concentration was reduced to 10% w/v since this separated related strains better.

### Other Gradients

Almost any other solute may be used to establish gradients in the wedge plate. Among the points to be borne in mind, perhaps diffusivity is the most critical. Although lateral diffusion is unimportant and indeed rather counter-productive, since it tends to alter the chosen gradient, vertical equilibration is necessary to equilibrate the gel in this dimension. If this is too slow, it becomes difficult to determine exactly what solute concentration is present at the surface which is seen by the cell inoculum. A particular problem is the binding of some charged molecules to the agar gel polymers.

### Measuring the Gradients

Specific analytical methods will need to be developed for particular gradient systems. What follows is some of our own experience with pH and NaCl gradients.

### pH Values

pH can be determined in two ways. In the first we have used commercially available microcombination pH electrodes manufactured by Microelectrodes Inc., New Hampshire. Results recorded in millivolts were read against a calibration curve. It is also possible to use flat-ended pH electrodes in much the same way. The second approach is to remove samples from the plate. We commonly remove 7 to 8 mm discs of agar using cork borers. The center of the hole remaining is marked on the back of the plate so that pH can be plotted against distance. The agar disc is melted in 2.0 mℓ distilled water in the steamer. The samples are stored at 45°C and the pH measured using a small bulb combination pH electrode. This method has been compared with pH measured in standard agar media of different pH values at 45°C and found to give results that were within 0.1 pH unit of one another.

### NaCl Concentrations

These have been estimated using conductivity measurements. Discs (7 to 8 mm) were removed from the agar, melted in 10.0 mℓ distilled water, and the conductivity of the solution determined. Calibration curves were prepared by making standard plates with known NaCl concentrations, removing a number of discs from each, and determining the average conductivity of these discs in 10.0 mℓ water. It is possible to construct simple platinum probes which can be placed in the agar to measure conductivity directly. The probe can be calibrated as before, using agar containing uniform concentrations of NaCl.

### Other Solutes

Ordinary analytical methods should be applied to other possible gradients systems. Sampling using the agar-disc method seems a good way to prepare samples for measuring solute concentration. Probe measurements *in situ* have the virtue of convenience. Electrodes are available to measure pH, $pO_2$, Eh, sulfide, and so on. Such techniques are really simpler and more accurate than sampling, and if these probes are available they ought to be used.

### Gradient Reproducibility

Variations in pH and in NaCl gradients were determined in replicate plates after storing them for 24 hr at room temperature and another 24 hr at 30°C. These periods correspond to the way actual experimental plates are treated. Results for pH gradients (Figure 5) indicated some variation near the center of the plate where change in pH value is steepest. At pH 6, for example, the SEM was ± 0.54, a value that fell to ± 0.10 near pH 4 or ± 0.17 near

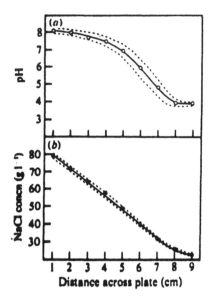

FIGURE 5. pH (a) and NaCl (b) gradients across each dimension of the two-dimensional diffusion gradient plate. Confidence limits are shown as broken lines. (From Wimpenny, J. W. T. and Waters, P., *J. Gen. Microbiol.*, 130, 2921, 1984. With permission.)

pH 8.5. Salt gradients were much more reproducible showing an SEM of ± 1.0 g/ℓ over the range 20 to 80 g/ℓ.

### Time-Dependent Gradient Changes

Growth experiments took a range of times up to 96 hr after pouring the plates. During this time, it is clear that lateral diffusion was taking place and experiments were undertaken to determine the extent of this factor. pH and NaCl concentrations were determined across inoculated and uninoculated plates as a function of time up to 96 hr after the plates were poured.

pH remained remarkably stable over this period in uninoculated plates (Figure 6); however, cell growth caused significant changes in pH particularly after prolonged incubation. *Bacillus cereus*, for example, made the acidic regions of the plate significantly more alkaline. We believe that these changes in pH in older cultures are unimportant in outlining the habitat domain of the organism. This is because the majority of the *increase* in cell numbers will have occurred early on, when numbers were low and the pH values relatively constant. This early growth will henceforth dominate the spatial pattern especially if it is recored early.

NaCl gradients were unaffected by microbial growth; however, they showed a time-dependent decay, as illustrated in Figure 6.

### Growth Pattern Reproducibility

The two-dimensional plate is not a completely precise system since there are several sources of error during plate preparation. It was necessary to see just how reliable a method it was. Four species of heterotrophic bacteria were examined, using the nutrient-broth-based system. Five plates were prepared for each. Maps of the growth zones and an indication of confidence limits as SEM values are shown in Figure 7.

### Mapping Growth Zones

There are several possible ways to map the plates after growth. Experience suggests that

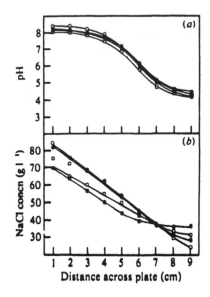

FIGURE 6.    Changes in pH gradient and in salt gradients as a function of time. ○, 24 hr; ●, 48 hr; □, 72 hr; ■, 96 hr. (From Wimpenny, J. W. T. and Waters, P., *J. Gen. Microbiol.*, 130, 2921, 1984. With permission.)

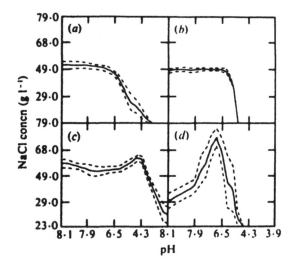

FIGURE 7.    Growth of four species of bacteria on pH/NaCl two-dimensional gradient plates. Broken line indicates the confidence limits for five samples. (a) *Bacillus subtilis*, (b) "*Flavobacterium arborescens*", (c) "*Serratia indica*", (d) "*Staphylococcus albus*". (From Wimpenny J. W. T. and Waters, P., *J. Gen. Microbiol.*, 130, 2921, 1984. With permission.)

all plates should be photographed under conditions that generate as much contrast as possible. This is particularly easy with pigmented strains such as *Serratia marcescens* or the phototrophic species, much less easy for unpigmented species producing scant or translucent growth. Experimenting with the intensity and angle of illumination and the type of background ought to suggest the optimum conditions for photography. A matte black background helps accentuate the more transparent growth. Color transparencies can be projected onto graph paper and the growth zone traced. It would be even better if it were traced directly onto a graphics tablet so that the data can be stored in machine-readable form on a computer.

A more laborious and prosaic approach, which we have used routinely, is to map the growth zone using an etched grid superimposed on the back of the plate.

Mapping, or even photography, unless it is very good, loses much of the information that can be seen on the plate. Thus, in some species, notably *B. cereus* and *Proteus vulgaris*, bands of growth seem to form in the acid direction on the plate. Periodic growth behavior is an important phenomenon and is discussed in more detail in another chapter.

One other answer to mapping problems is to use image-analysis techniques. Here it is possible to store all the information as a computer file, to superimpose images, to average them, and even to enhance images which may show very little contrast on the plate surface. We are investigating this approach at present.

### Some Results Using Two-Dimensional Gradient Plates

*Results with Some Aerobic Heterotrophs*

Once the method was developed, it was decided to test a range of aerobic heterotrophic bacteria on the plates. Some of these results have already been published.[17]

Two-dimensional gradient plates (pH/NaCl) were prepared as described above and incubated for 24 hr before inoculating. Organisms were maintained on tryptone soy agar (Difco) slopes and subcultured into nutrient broth (Difco) before use. The inoculum was diluted tenfold in semisolid nutrient broth containing 2% w/v Bacto-Agar, and 1.0 mℓ spread over the surface of the plate. Growth patterns are presented for 31 species and strains in Figure 8A. One example of a specific pattern is also shown (Figure 8B).

All the bacteria examined grew at some point in the gradients; however, in the case of *Chromobacterium violaceum*, growth was sporadic and occasionally completely absent, since the species was so sensitive to NaCl concentration. On the other hand, some bacteria, in particular Gram-positive cocci such as *Staphylococcus aureus* and *S. albus* could grow at the highest salt concentrations tested. It proved interesting to compare the responses within a single genus and the *Bacillaceae* were chosen for further study. Species like *B. mycoides* and *B. polymyxa* were sensitive to salt, *B. polymyxa* showing, in addition, a rather narrow range of pH tolerance. *B. megaterium*, in contrast, could grow at all the pH values tested and was reasonably insensitive to pH. *B. licheniformis* tolerated high salt concentrations but was sensitive to acidic pH values. Each bacillus had a characteristic profile on the plates, and this seemed to indicate that there is a place in systematics for multidimensional techniques. Profiles for coryneform organisms were much more uniform than those seen with the bacilli. This was especially clear on the pH axis. Among the Gram-negative rod shaped bacteria, *E. coli*, *Serratia marcescens*, and *Enteroacter cloacae* showed a broad tolerance to all pH values on the plate. *Proteus vulgaris*, a klebsiella, and *Pseudomonas aeruginosa* were all acid-sensitive, though the klebsiella could tolerate high salt concentrations. A miscellaneous group of Gram-negative bacteria were tested. All were quite salt sensitive. As has already been mentioned, *C. violaceum* is inhibited by low salt concentrations but so also is *Azotobacter vinelandii*. The relative pH sensitivity of these two species is quite different, however, and it is clear that their habitat domains overlap only slightly.

Two final points should be made about these results. First, it is clear in many cases that there is a "horn" of NaCl resistance which projects upwards on the plates at characteristic pH values, generally around neutrality. The connection between salt tolerance and pH is not immediately obvious, but seems nonetheless to be real. The second point to note is that the growth front on many of these two-dimensional plots show some curvature on the vertical axis. This indicates slight changes in the pH gradient as a function of salt concentration and is almost certainly a property of the plate and not a characteristic of the organism.

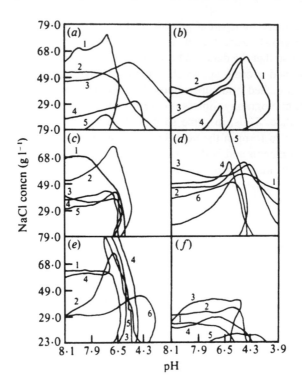

A

FIGURE 8.    A. Growth patterns observed with different organisms. (a) *Bacillus* species: 1, *B. licheniformis;* 2, *B. cereus;* 3, *B. megaterium;* 4, *B. macerans.* (b) *Bacillus* species: 1, *B. subtilis;* 2, *B. subtilis "var niger";* 3, *B. circulans;* 4, *B. polymyxa.* (c) Coryneform organisms: 1, *Corynebacterium* sp.; 2, *Corynebacterium xerosis;* 3, *C. equi;* 4, *Kurthia zopfii;* 5, *Arthrobacter* sp. (d) Gram-negative organisms: 1, *Escherichia coli* K12; 2, *Enterobacter cloacae;* 3, *"Serratia indica";* 4, *Proteus vulgaris;* 5, *Klebsiella* sp.; 6, *Pseudomonas aeruginosa.* (e) Gram-positive cocci: 1, *Streptococcus faecalis;* 2, *"Staphylococcus albus";* 3, *S. aureus;* 4, *S. saprophyticus;* 5, *Micrococcus luteus;* 6, *M. roseus.* (f) Miscellaneous bacteria: 1, *Azotobacter vinelandii;* 2, *"Achromobacterium lacticum";* 3, *Vibrio* sp.; 4, *Aeromonas* sp.; 5, *Chromobacterium violaceum.* (From Wimpenny, J. W. T. and Waters, P., *J. Gen. Microbiol.*, 130, 2921, 1984. With permission.) B. Photograph of two-dimensional plate of *Serratia marcescens* NCTC 1377 in gradients of salt vs. pH and grown for 48 hr at 30°C.

*Experiments with Phototrophic Bacteria*

The two-dimensional gel-plate method was later developed for use with phototrophic bacteria by myself in collaboration with H. Gest.[18] A number of nonsulfur purple bacteria were examined when grown on a malate/mineral-salt/vitamin medium in pH/NaCl gradients. All plates were equilibrated at room temperature in an anaerobic hood before inoculation. They were then incubated in anaerobic Gas-Pak jars and illuminated with saturating light from tungsten filament bulbs. Plates were incubated for up to 72 hr at 28°C.

Profiles for representative species (Figure 9) show a range of salt sensitivities from about 1% w/v NaCl for *Rhodobacter capsulata* and *Rhodopseudomonas blastica* to around 10% w/v NaCl for *Rhodopseudomonas marina/agilis* and *R. sphaeroides.* All strains examined were capable of tolerating pH values more alkaline than pH 8.5; however, there was a range of sensitivity to low pH values from pH 5.1 to pH 6.8. *R. sphaeroides*, a common and ubiquitous organism, has been the subject of close systematic scrutiny[19] and nine strains were compared using two-dimensional plates (Figure 10). It is clear from the photographs that all the profiles have similar characteristic forms. On the other hand, there is a range of differences between strains as far as tolerance to NaCl and pH is concerned. These differences

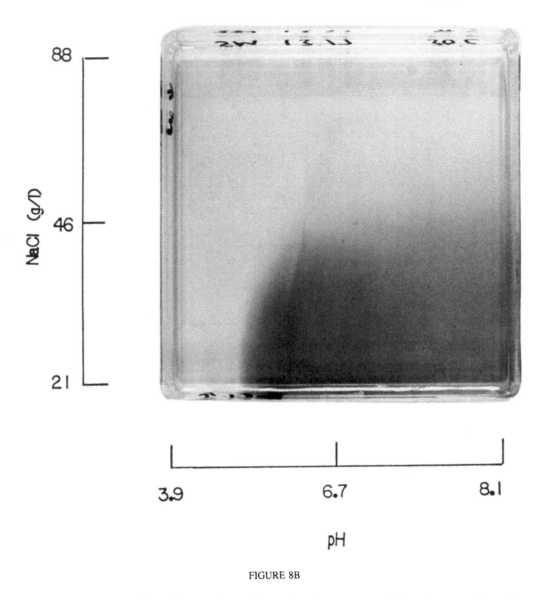

FIGURE 8B

were repeatable and consistent and provide another character on which to base a systematic investigation of this group.

*Two Dimensional Gradient Plates in Food Preservation Studies*

Lactobacilli are often the cause of spoilage of ambient stable preserved foods, and preservation options to reduce their growth are limited. There is also growing consumer pressure to reduce the amount of preservatives and additives in foods as well as a preference for milder, less salty products. This has led Martin Jones[23] to search for synergistic combinations of microbial inhibitors, and the two-dimensional gradient plate has proved to be a useful tool. Figure 11 shows the effect of salt and a sequestrant with or without the addition of a preservative on two species of lactobacilli. In this case, there is some synergy between salt and the sequestrant, but the effect of the preservative is apparently additive. Greater resistance is shown by *Lactobacillus plantarum* which is a common isolate from spoiled food.

**General Observations**

The technique just described could be even more valuable if further dimensions could be

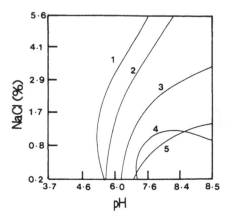

FIGURE 9.   Representative phototrophic species growing on pH/NaCl two-dimensional gradient plates. 1. *Rhodopseudomonas sphaeroides* 81.1; 2. *Rhodopseudomonas marina/agilis;* 3. *Rhodospirillum rubrum;* 4. *Rhodopseudomonas blastica;* 5. *Rhodobacter capsulata* EY1.

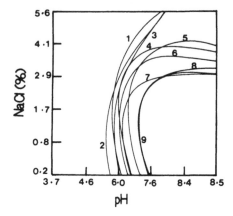

FIGURE 10.   Growth domains on pH/NaCl two-dimensional gradient plates for nine strains of *Rhodopseudomonas sphaeroides.* Strains numbers are as follows: 1, 2.4.7; 2, 81.4; 3, 81.1; 4, 81.2; 5, 81.3; 6, 2.4.9; 7, 2.4.4; 8, 2.4.3; 9, 2.4.1.

incorporated. This would be difficult using continuously varying solute concentrations in agar, since it would need a solid agar cube in which gradients could only be developed using diffusion techniques. On the other hand, it is not hard to devise "stepped" third or higher-order dimensions. For example, if a number of two-dimensional plates were incubated at different temperatures, the growth profiles for each plate could be stacked on top of one another to generate a solid object which represents the habitat domain for these three factors. Similarly, two-dimensional plates could be constructed in which all layers contained a different uniform concentration of a solute. Both approaches could be taken together so that temperature and solute concentration, as well as the initial two dimensions established as wedges in the plates, would allow four factors to be varied simultaneously. Such four-dimensional information could not be represented graphically, but could be analyzed using computational techniques. Four-dimensional experiments of this type would naturally need a considerable number of plates. If seven temperatures and seven solute concentrations were used, forty-nine two-dimensional plates would need to be poured. In practical terms, this

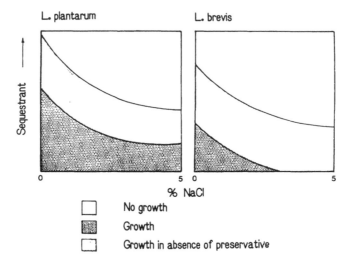

FIGURE 11.    Growth of *Lactobacillus plantarum* and *L. brevis* on salt vs. sequestrant two-dimensional gradient plates, with or without an added preservative. Courtesy of M. V. Jones.

is about as many as can conveniently be handled in a single experiment. It ought to be stressed that if a single two-dimensional plate has at least the resolution of a 10 × 10 array of individual tubes, such an experiment using 49 plates corresponds to 4900 separate tube assays!

## Experiments with Luminous Bacteria

Two-dimensional plates incubated over a range of times and temperatures have been used in Cardiff, Wales to investigate growth and bioluminescence of three species of luminous bacteria.[20] Luminosity is an activity-character and not to be strictly compared with growth. Thus, the position of luminosity on the region of growth was itself a time-dependent character which will probably depend not directly on the physical parameters of pH and NaCl concentration, but on the rate of metabolism and the depletion of energy source in the medium below the growing bacteria. Having said this, it is also clear that luminosity was a subset of the growth zone and that there were areas at lower NaCl concentrations and at pH values away from the more favorable acidic values where luminosity was never observed. Figure 12 shows one such plate where the culture is photographed by its own light. Figure 13 is a diagrammatic representation of *Vibrio fischeri* showing growth and luminosity patterns as a function of pH, NaCl concentration, time, and temperature.

## Three-Dimensional Maps of Heterotrophic Bacteria

Three-dimensional maps have been prepared for a number of strains of two species of bacteria. These are *Serratia marcescens* and *Micrococcus luteus*. Both organisms were maintained on nutrient agar slopes and grown in pH/NaCl gradients in the nutrient-broth-based medium described earlier.[17] Incubation temperatures ranged from 5 to 35°C.

Eight strains of *S. marcescens* and two of *M. luteus* were obtained from British stock culture collections. Photographs of the growth zone on each plate are shown for representative strains in Figure 14. It has proved possible to present some of the three-dimensional data in graphical form using a surface profile-drawing computer program. Figure 15 illustrates such profiles for the species of serratia. From all these results, the two micrococci are easy to distinguish, in particular by their sensitivity to NaCl. While the micrococci show similar temperature responses these are quite different from those of the serratia strains. The latter

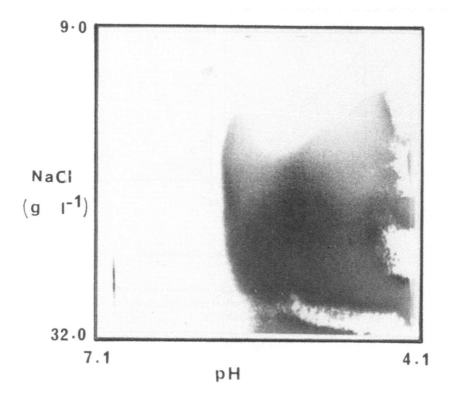

FIGURE 12.    Growth of *Vibrio fischeri* MJ-1 on pH/NaCl two-dimensional gradient plates. Photograph taken using luminosity of the growth zone alone.

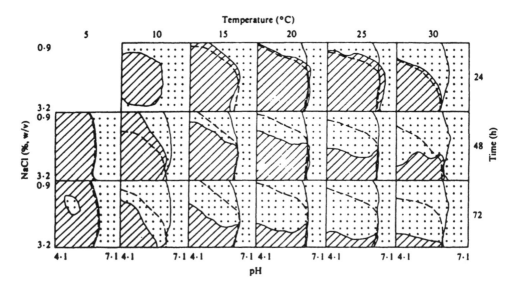

FIGURE 13.    Growth and luminosity profiles for *Vibrio fischeri* MJ-1 growing on pH/NaCl two-dimensional gradient plates at a number of different times and temperatures. Areas of perceptible growth indicated by solid lines; limits of heavy growth ( — — —); hatched areas indicate bioluminescence. The dots indicate 10 mm grid intersections. (From Waters, P. and Lloyd, D., *J. Gen. Microbiol.*, 131, 2865, 1985. With permission.)

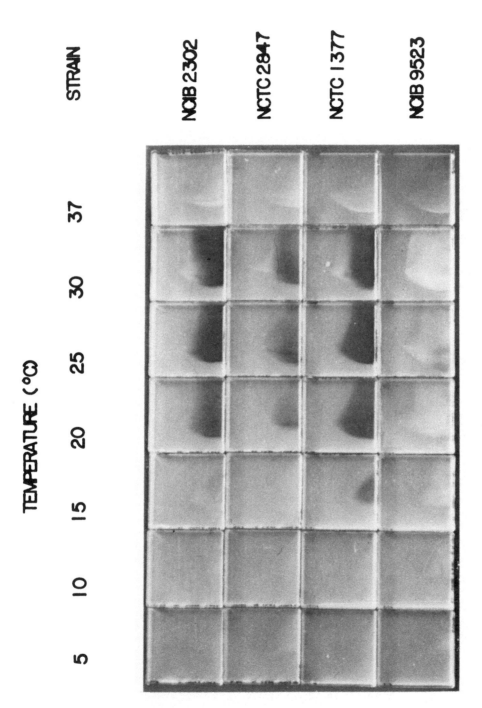

FIGURE 14. Growth of eight strains of *Serratia marcescens* and two strains of *Micrococcus luteus* on pH/NaCl two-dimensional gradient plates at seven different temperatures. Strain numbers are indicated on the photographs. This figure continues on the next two pages.

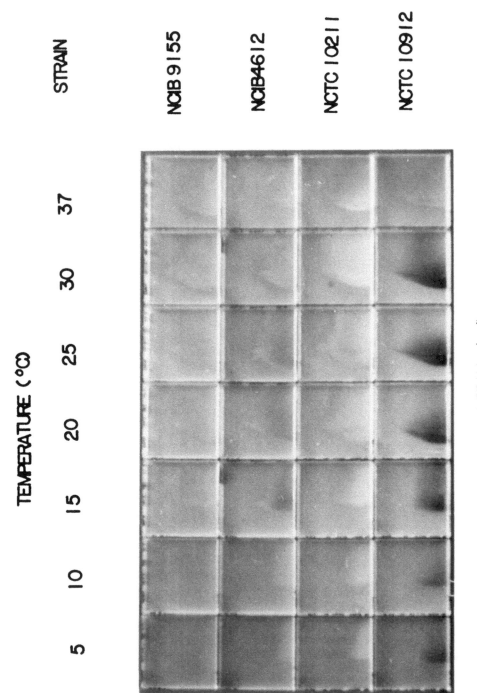

FIGURE 14 (continued).

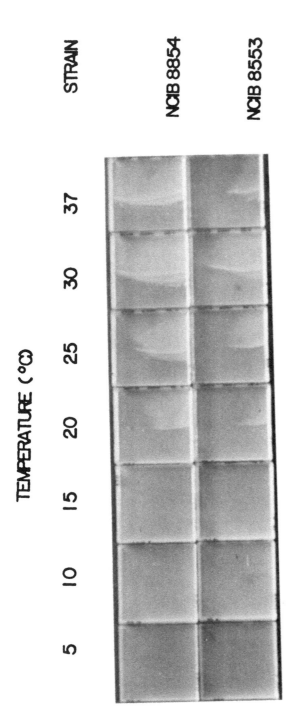

FIGURE 14 (continued).

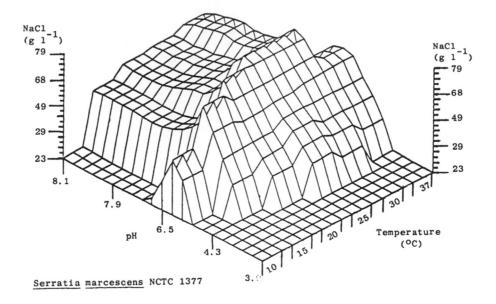

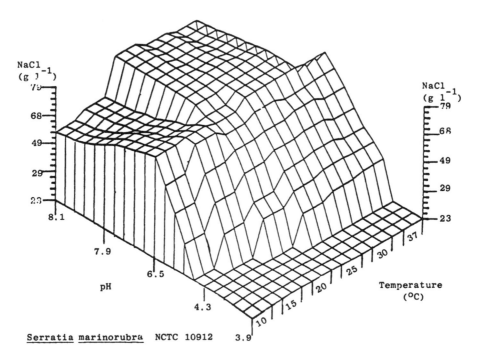

FIGURE 15.    Three-dimensional growth profile maps for two strains of *Serratia marcescens* growing on pH/NaCl two-dimensional gradient plates at seven different temperatures.

also show the extent of pigment production. It is well known that this phenomenon is temperature sensitive. In each of the strains examined here, pigment production ceases at temperatures between 30 and 35°C. Some of the strains are weak pigment producers, and here pigment production is also affected by pH and salinity. Even in well-pigmented strains, pigment production is sensitive to high salt concentration, especially at higher temperatures. What is very clear from the growth zones is the very significant difference between strains. Such data have been confirmed in replicate experiments, suggesting that these multidimensional response systems have a clear taxonomic value.

THE EFFECT OF VARYING 4 PHYSICO-CHEMICAL VARIABLES PH, NaCl,
TEMPERATURE and NaNO₃ ON THE GROWTH OF <u>SERRATIA MARCESCENS</u>
NCIB 9523

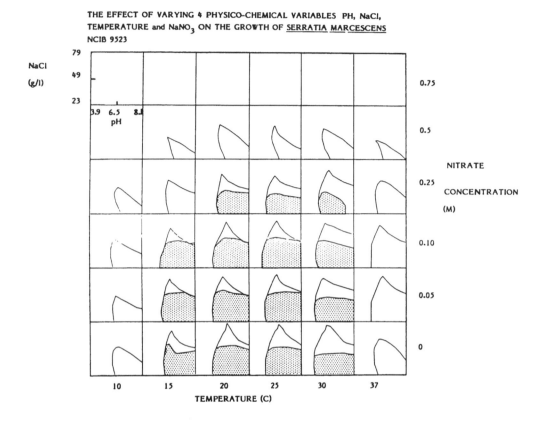

FIGURE 16. Results of a four-dimensional experiment using *Serratia marcescens* strain NCTC 10912, where pH, NaCl concentration, nitrate concentration, and temperature were all varied. 36 plates were poured. Six sets of six plates each contained nitrate at the concentrations indicated in the figure. Six sets, each containing the six different nitrate concentrations, were incubated at the six different temperatures. Results were recorded after 48 hr.

## Four Dimensions and Beyond

A representative strain of *S. marcescens* was grown on two-dimensional salt/pH gradient plates containing a range of nitrate concentrations up to 0.75 *M*. Six identical sets of these plates were incubated at different temperatures. The six sets of three-dimensional growth profiles are shown in Figure 16. Nitrate concentration inhibits both growth of this organism, and to a lesser extent, its ability to synthesize prodigiosin. Results clearly show the effect of temperature on pigment production. What cannot be seen from the drawings are the small but significant changes in pigment color as a function both of pH and nitrate concentration.

## FINAL COMMENTS

Is it possible to discern applications in any branch of microbiology for the sort of multidimensional systems described in this chapter? We believe that the answer is clearly "yes" and that applications are legion. In applied microbiology, there are certain areas where an understanding of factors that prevent microbial growth are of paramount importance. One of the fashionable phrases in food microbiology is "predictive microbiology". It is essential to determine appropriate conditions in food preservation, where harmful organisms grow very slowly if at all. At the same time, it is clear that the flavor of the product must not be impaired. It is best, therefore, to consider a multifactorial approach, where a combination of agents is needed to inhibit microbial proliferation. Multidimensional systems can constitute an almost perfect answer here.

Such combinatorial methods should have a part to play in detecting synergism or antagonism in combined drug therapy or in choosing agents needed to inhibit growth in industrial systems, fuel lines, engineering lubricants, oil storage tanks, and so on.

There seem to be potential benefits in selecting the correct conditions for gene transfer, since such experiments can take place over a wide range of conditions simultaneously, and it may become obvious what conditions are most suitable from a single experiment. Selection of organisms resistant to specific agents could be improved if a wide range of environmental variables is tested.

One of the most challenging areas is in ecology, however. It is suggested that multidimensional systems offer the best possible tool for selecting particular species from a complex ecosystem, since details of substrate concentration, pH, ionic strength, salinity, water activity, presence of trace minerals, and so on can be manipulated in a systematic fashion ensuring that at some position, on one plate or other, conditions will be right for the growth of a particular organism.

Preliminary experiments have indicated that multidimensional systems will prove a useful tool to examine competition between particular species. Much more work is needed in this area, in particular to quantify surface growth where two or more organisms are involved.

It must also be obvious that such techniques ought to be applied to an assessment of the properties of organisms isolated from different habitats. How different are the species found in a marine environment compared with those isolated from fresh water, especially as regards NaCl tolerance, for example? One approach is to plate a sample from each habitat onto two-dimensional plates and select as representative species from growth on the plate. Each organism should then be mapped on the same plates in pure culture and the habitat characteristics of the individuals from each population can then be compared. Preliminary experiments[24] suggest that this approach is valid.

On a more abstract plane, the ability to generate experimental systems that can give three or more dimensions of cultural information in a single experiment can contribute to an understanding of some aspects of the niche that an organism can occupy. The term niche has been variously defined however, one of the most useful definitions is due to Hutchinson,[21,22] who first pointed out that a niche could be imagined as an n-dimensional hyperspace, one dimension for each of the environmental variables which affected growth of the organism. Such a hyperspace could be potential or actual, depending on the presence of other organisms and the prevailing physico-chemistry of the niche space. It is likely that further development of the techniques described here may give an experimental framework to such definitions.

# REFERENCES

1. **Beijerinck, M.,** Auxanography, a method useful in biological research, involving diffusion in gelatine, *Arch. Neerl. Sci. Exactes Nat. Haarlem,* 23, 367, 1889.
2. **Baas-Becking, L. G. M. and Ferguson Wood, E. J.,** Biological processes in the estuarine environment. I. Ecology of the sulphur cycle, *Proc. K. Ned. Akad. Wet. Ser. B.,* 58, 160, 1955.
3. **Baas-Becking, L. G. M. and Ferguson Wood, E. J.,** Biological processes in the estuarine environment. VIII. Iron bacteria as gradient organisms, *Proc. K. Ned. Akad. Wet. Ser. B.,* 59, 398, 1956.
4. **Szybalski, W.,** Gradient plates for the study of microbial resistance to antibiotics, *Bacteriol. Proc.,* 36, 1952.
5. **Szybalski, W. and Bryson, V.,** Genetic studies on microbial cross resistance to toxic agents. I. Cross resistance of *Escherichia coli* to fifteen antibiotics, *J. Bacteriol.,* 64, 489, 1952.
6. **Bryson, V. and Szybalski, W.,** Microbial selection, *Science,* 116, 45, 1952.
7. **Sacks, L. E.,** A pH gradient agar plate, *Nature,* 178, 269, 1956.

8. **Davenport, R. R.,** An outline guide to media and methods for studying yeasts and yeast-like organisms, in *Biology and Activities of Yeasts,* Skinner, F. A., Skinner, S. M., and Davenport, R. R., Eds., Academic Press, London, 1980, 262.

9. **Weinberg, E. D.,** Double-gradient agar plates, *Science,* 125, 196, 1957.

10. **Halldal, P.,** Pigment formation and growth in blue green algae in crossed gradients of light intensity and temperature, *Physiol. Plantarum,* 11, 401, 1958.

11. **Halldal, P. and French, C. S.,** Algal growth in crossed gradients of light intensity and temperature, *Plant Physiol.,* 33, 249, 1958.

12. **Van Baalen, C. and Edwards, P.,** Light-temperature gradient plate, in *Handbook of Phycological Methods,* Stern, J. R., Ed., Cambridge University Press, 1973, 267.

13. **Caldwell, D. E. and Hirsch, P.,** Growth of microorganisms in two-dimensional steady-state diffusion gradients, *Can. J. Microbiol.,* 19, 53, 1973.

14. **Caldwell, D. E., Lai, S. H., and Tiedje, J. M.,** A two-dimensional steady-state diffusion gradient for ecological studies, *Bull. Ecol. Res. Comm. (Stockholm),* 17, 151, 1973.

15. **Wimpenny, J. W. T., Lovitt, R. W., and Coombs, J. P.,** Laboratory models systems for the investigation of spatially and temporally organised microbial ecosystems, *Symp. Soc. Gen. Microbiol.,* 34, 67, 1983.

16. **Wimpenny, J. W. T., Coombs, J. P. and Lovitt, R. W.,** Growth and interactions of microorganisms in spatially heterogeneous ecosystems, in *Current Perspectives in Microbial Ecology,* Klug, M. J. and Reddy, C. A., Eds., American Society for Microbiology, Washington, D.C., 1984, 291.

17. **Wimpenny, J. W. T. and Waters, P.,** Growth of microorganisms in gel stabilized two-dimensional diffusion gradient systems, *J. Gen. Microbiol.,* 130, 2921, 1984.

18. **Wimpenny, J. W. T., Gest, H., and Favinger, J.,** The use of two dimensional gradient plates in determining the responses of non-sulphur purple bacteria to pH and NaCl concentration, *Fed. Eur. Microbiol. Soc. Microbiol. Lett.,* 37, 367, 1986.

19. **Pellerin, N. B. and Gest, H.,** Diagnostic features of the photosynthetic bacterium, *Rhodopseudomonas sphaeroides, Current Microbiol.,* 9, 339, 1983.

20. **Waters, P. and Lloyd, D.,** Salt, pH and temperature dependencies of growth and bioluminescence of three species of luminous bacteria analysed on gradient plates, *J. Gen. Microbiol.,* 131, 2865, 1985.

21. **Hutchinson, G. E.,** *A Treatise on Limnology,* Vol. I, New York, John Wiley & Sons, 1957.

22. **Hutchinson, G. E.,** *The Ecological Theater and the Evolutionary Play,* New Haven, Conn., Yale University Press, 1965, 26.

23. **Jones, M.,** Personal communication.

24. **Wimpenny, J. W. T. and Reynolds, J.,** Unpublished observations.

# INDEX

9 781138 558366